Technology of
Bottled Water

TECHNOLOGY OF BOTTLED WATER

Second Edition

Edited by

DOROTHY SENIOR
Group Technical Adviser
Highland Spring Ltd
Blackford, Perthshire
UK

and

NICHOLAS DEGE
Supply Chain Manager
Nestlé Waters North America
Calistoga, California
USA

Blackwell
Publishing

© 2005 by Blackwell Publishing Ltd

Editorial Offices:
Blackwell Publishing Ltd, 9600 Garsington Road, Oxford OX4 2DQ, UK
  Tel: +44 (0)1865 776868
Blackwell Publishing Professional, 2121 State Avenue, Ames, Iowa 50014-8300, USA
  Tel: +1 515 292 0140
Blackwell Publishing Asia Pty Ltd, 550 Swanston Street, Carlton, Victoria 3053, Australia
  Tel: +61 (0)3 8359 1011

The right of the Author to be identified as the Author of this Work has been asserted in accordance with the Copyright, Designs and Patents Act 1988.

First edition published 1998 by Sheffield Academic Press
Second edition published 2005 by Blackwell Publishing Ltd

Library of Congress Cataloging-in-Publication Data
Technology of bottled water / edited by Dorothy A.G. Senior and Nick J. Dege.–2nd ed.
    p. cm.
  Includes bibliographical references and index.
  ISBN 1-4051-2038-X (hardback: acid-free paper)
  1. Bottling. 2. Bottled water. I. Senior, Dorothy A. G. II. Dege, Nick J.

TP659.T43 2004
663'.61–dc22
                                                                                              2004048433
ISBN 1-4051-2038-X

British Library Cataloguing-in-Publication Data:
A catalogue record for this title is available from the British Library

Set in 9.5/12pt Times New Roman
by Kolam Information Services Pvt. Ltd, Pondicherry, India
Printed and bound in Great Britain
by MPG Ltd, Bodmin, Cornwall

The publisher's policy is to use permanent paper from mills that operate a sustainable forestry policy, and which has been manufactured from pulp processed using acid-free and elementary chlorine-free practices. Furthermore, the publisher ensures that the text paper and cover board used have met acceptable environmental accreditation standards.

For further information on Blackwell Publishing, visit our website:
www.blackwellpublishing.com

Contents

vi CONTENTS

4 Hydrogeology of bottled waters 93

MIKE STREETLY, ROD MITCHELL
and MELANIE WALTERS

5 Water treatments 132
JEAN-LOUIS CROVILLE and JEAN CANTET

6 Bottling water – maintaining safety and integrity through the process 166
DOROTHY SENIOR

7 Filling equipment   189

FRED G. VICKERS and JOHN MEDLING

8 Cleaning and disinfection in the bottled water industry   227

WINNIE LOUIE and DAVID REUSCHLEIN

12 Microbiology of natural mineral waters 325
HENRI LECLERC and MILTON S. DA COSTA

13 Microbiology of treated bottled water 388
STEPHEN C. EDBERG

Index 403

Contributors

Mr Michael Barnett

Hydropure Group, Hydropure House, Alington Rd, St Neots, Cambridgeshire PE19 6WD, UK

Mr Jean Cantet

Nestlé Waters MT Process Technology Centre, 1020 Avenue George Clémenceau, BP101-F-88804 Vittel Cedex, France

Mr Jean-Louis Croville

Nestlé Waters MT Process Technology Centre, 1020 Avenue George Clémenceau, BP101-F-88804 Vittel Cedex, France

Professor Milton da Costa

Departamento de Bioquimica, Universidade de Coimbra, 3000 Coimbra, Portugal

Mr Nicholas J Dege

Nestlé Waters North America, 865 Silverado Trail, Calistoga, California 94515, USA

Professor Stephen Edberg

Department of Laboratory Medicine and Internal Medicine, Yale University School of Medicine, 333 Cedar Street, PO Box 3333, New Haven, CT 06510, USA

Mr Duncan Finlayson

Zenith International Projccts Ltd, 7 Kingsmead Square, Bath BA1 2AB, UK

Professor Henri Leclerc

Laboratoire de Microbiologie, Faculty of Medicine, University of Lille, 1 Place de Verdun, 59045 Lille Cedex, France

Ms Winnie Louie

Nestlé Waters North America Inc., 4330 20th Street, Zephyrhills, FL 3354, USA

Mr John Medling

Krones UK Ltd, Westregen House, Great Bank Road, Wingates Industrial Park, Westhoughton, Bolton BL5 3XB, UK

Mr Rod Mitchell   ENTEC UK Ltd, 160-162 Abbey Foregate, Shrewsbury, Shropshire SY2 6AL, UK

Mr David Reuschlein   Ecolab, 7907 Oakview Drive, Lenexa, Kansas 66215, USA

Mrs Dorothy Senior   Shinglenook, 41 Glenorchil View, Auchterarder, Perthshire PH3 1LU, UK

Mr Mike Streetly   ENTEC UK Ltd, 160-162 Abbey Foregate, Shrewsbury, Shropshire SY2 6AL, UK

Mr Bob Tanner   NSF International, Avenue Grand Champ 148, 1150 Brussels, Belgium

Mr Fred Vickers   1 Mayfield, Harwood, Bolton BL2 3LP, UK

Dr Melanie Walters   ENTEC UK Ltd, 160-162 Abbey Foregate, Shrewsbury, Shropshire SY2 6AL, UK

Preface to the Second Edition

The first edition of this book was published in 1998 with the aim of providing much needed guidance to regulators, beverage and packaging technologists, microbiologists and specialists in hygiene and food safety. It was originally written from a global perspective, to shed some light on the complexities of this burgeoning industry, prompted by the realization that there was much confusion in the minds of consumers and regulators alike regarding the quality, safety and identity of bottled waters.

Bottled water is, more than ever, a commodity of significance worldwide. While in the established markets of Europe and even more so in the United States, bottled water has continued to consolidate its position, in the emerging markets of Asia and Australasia growth in recent years has been phenomenal. This has resulted partly from the way in which the larger companies have extended their operations into the newer markets, but along with this has been the appearance of countless smaller bottlers, encouraged by the success of others and the relatively low cost of entry.

Against this background, it is more important than ever that the new entrants to the industry are well informed. The process of bottling water might, at first sight, seem to the uninitiated to be simple and risk-free – especially when compared with those more complex processes required for producing soft drinks, for example. In practice, however, because water is so sensitive to chemical, physical and microbiological contamination, it is one of the more difficult products to deliver to a consistently high standard.

The original edition was written primarily to provide practical advice on the technical challenges faced by anyone considering bottling water. Thus there were chapters covering hydrogeology, water treatments, filling technology and the methods and materials for bottling water. A further chapter dealt with the technology of bottled water coolers – already used extensively in the US, but also being used increasingly to dispense water in the emerging 'home and office' market in the Far East. Supporting information was provided in chapters discussing the development of the world market, the categories of bottled water and the legal requirements pertaining to them. Finally, chapters were included on the essential aspects of quality management, third party auditing and the microbiology of both natural mineral water and treated bottled water.

For the second edition, the original authors were invited to update their chapters to reflect the changes in the industry in recent years. There have been some changes in authorship: John Medling has revised the important chapter on filling equipment originally written by Fred Vickers, and, as the original author of the chapter on water treatments, Michael Wayman, has moved away from the industry, an entirely new chapter on this subject has been written by Jean Louis Croville and Jean Cantet.

Finally, we recognized that the original edition would have benefited by the inclusion of an additional chapter on the important subject of cleaning and disinfection, so essential to success in our industry. We are therefore pleased to have been able to address this in the second edition, with a new chapter by Winnie Louie and David Reuschlein.

The editors are, as always, indebted to the authors. They are all busy people and, in their own fields, eminent specialists from around the world. The combination of their contributions to this book has resulted in a work that can be a continuing point of reference to those within the bottled water industry. It remains, as far as we know, the only volume to address in one place the diverse technical aspects of bottling water, and, as the market continues to grow and improved technology supports greater efficiency and safety, we are convinced that publication of a second edition is appropriate to reflect further developments in this dynamic business.

The Editors would like to acknowledge the support and advice received from Joe Beeston, Chief Executive, Highland Spring Ltd and Thom Kleiss, Director of Quality Assurance, Nestlé Waters North America.

Dorothy Senior
Nicholas Dege

1 Introduction

Dorothy Senior and Nicholas J. Dege

1.1 Rationale for this book

In 1998, when this book was first published, the principal objective was to present in one volume some practical and technical advice for those involved in (and those considering entry to) the bottled water industry. A secondary objective was to provide guidance on the legal and technical aspects to those requiring it (regulators, technical managers, packaging technologists, micro-biologists). A powerful motivating factor for this was the general lack of practical information on this still young industry, and there was therefore an urgent need to redress the balance. Since that time, the market for bottled water has continued in its dramatic growth, and, as more and more people become involved in the industry, the need for a source of sound guidance has multiplied.

Human beings have a fundamental requirement for water, needing 1.8–2.0 litres/day to maintain good health under normal circumstances. Ancient man focused his life on access to water from springs, wells and rivers. As populations grew, civilisation and technology developed and there was an increase in use of water for domestic and industrial purposes. Delivery systems were developed, but it became necessary to treat water supplies effectively to ensure that they were safe for drinking by the individual and, even more important-antly, to prevent the spread of diseases that could be carried by water to the general population. However, even with the benefits of modern water supply systems, some types of chemical treatment and the composition of pipes can cause organoleptic changes to municipal water, giving it an unpleasant taste. There are also concerns in many parts of the world about potential pollution of municipal supplies, and for these reasons consumption of bottled waters has been increasing dramatically.

For very good reasons, there is an increasing awareness of environmental issues, and reputable companies include an environmental programme in their corporate agenda. Although the drivers towards this action can be legislation and consumer groups, it also makes good business sense to have sound environ-mental practice. Packaging materials are essential to the bottled water industry to ensure that product reaches the consumer in the best possible condition. Packaging costs account typically for about one-third of a company's turnover, so every effort is made to minimise them wherever possible.

Although in Western Europe returnable glass bottles are still used, the majority of bottles for packaging water are single-trip containers. The latter are, without doubt, the best option for maintaining product water integrity and, in the majority of cases, also the best environmental option.

For example, great savings can be made by transporting water in plastic containers as opposed to glass, and this is especially true in the case of the expanding home and office business, in which returnable polycarbonate bottles are used. As the industry continues to develop, much has been achieved in recent years to minimise the impact on the environment by improved manufacturing methods, rationalised distribution and reduction in packaging materials, for example by the lightweighting of containers.

1.2   The second edition

At the time of publication of this second edition, the principal task for the industry remains the same, namely addressing the technical challenges of finding, protecting and abstracting good supplies of water, followed by filling and distribution to the ultimate consumer of a packaged product that meets all quality, safety and legal requirements.

The remarkable expansion in the industry seems relentless, and by way of introduction to the rest of the book, Chapter 2 examines the development of the maturing bottled water market, taking into account historical and regional influences, as well as the major producers. Changes in packaging formats, reflecting lifestyle changes, and pressure from consumer groups are also shown to influence trends. The consumer may understandably also be surprised by the increasing choice of bottled water now available.

Different regions of the world continue to have a wide range of requirements and specifications for bottled water. This is examined in Chapter 3, primarily from the European and North American perspective, though other continents are also covered. A part of this chapter also discusses the work being done by the WHO/FAO Codex Alimentarius Commission in its endeavours to produce worldwide standards for bottled waters. Whatever the category of water, the label must be studied carefully to assist in making the right choice, since clear and colourless soft drinks (which may look like water, but which may also include sugars and other sweeteners, flavourings, acidity regulators and preservatives) may be positioned beside bottled waters. In addition to declaring the category of water, the label may also provide information on any treatment that the water has undergone and in many cases gives a typical analysis for a selected number of elements, thus enabling consumers with particular dietary needs or preferences to make an informed choice.

The activities of man can be responsible for pollution of water, for example through agriculture, industry, road and rail construction, and special awareness and control is needed to protect vulnerable groundwater from undesirable changes in quality where there is an intention to bottle it without treatment. Chapter 4 describes the evaluation of groundwater sources, discussing varying geological influences. Development and protection of boreholes, and management of catchment zones and water yields are also covered.

Water may come from various sources or supplies, so, in many instances, treatments are used either for safety or legal reasons or to change the compositional quality of the water. Chapter 5 looks at the many options and possible reasons for choosing one or a combination of several treatment processes, although the choice available is always subject to local regulations.

The susceptibility of water to change, chemically, microbiologically or organoleptically, brings challenge to the bottling process. The inherent properties of water, the raw and packaging materials available and the equipment used all have profound implications for the safety and quality of the finished product. In Chapter 6, practical advice is given concerning the factors to be considered in order to protect the integrity of water throughout the process.

Central to the whole operation, of course, is the activity of filling bottles, and with increasing capacity comes increased complexity. Chapter 7 is therefore devoted to filling equipment, describing in detail the technology behind this most critical stage of the process.

In view of the need for rigorous standards in good manufacturing practices, and more specifically for hygiene when dealing with bottled water, Chapter 8 introduced for the second edition deals with cleaning and disinfection, describing why and when this is needed. Cleaning-in-place schedules, employee training and safe use of chemicals are also discussed.

Chapter 9 on quality management alludes to the important role of risk assessment and the use of hazard analysis as the foundation of a management system. Areas of process control, in which operators undertake monitoring of quality parameters, as well as the more technical work performed in the quality assurance laboratory, are described.

More and more bottled water is used in the home and office, and in densely populated areas, where the quality of municipal water is more affected chemically and organoleptically, it is also used for making beverages such as tea and coffee and in cooking and food preparation. Many offices and public places now frequently provide water dispensers, commonly referred to as water coolers. These can incorporate facilities for chilled and hot water and in some cases, sparkling water. Water coolers are an important and growing part of the bottled water market. Although in some respects there are similar considerations to those for the retail sector, there are some additional priorities associated with their distribution, as well as the design of the dispensing equipment, all of which are discussed in Chapter 10.

Although companies producing bottled waters may adopt best practices for the industry, it is often the independent audit of the bottling process and systems used which give it credibility with customers and regulatory authorities. Chapter 11 sets out to detail the philosophy behind, and the steps taken during, the process of third-party auditing.

All bottled waters must be safe to drink and are required to be free from any pathogenic (disease-causing) micro-organisms. Some, such as natural mineral waters and spring waters, are required to be free from pathogens without treatment, and compliance with this requirement is monitored by testing for the absence of indicator organisms, as specified by applicable legislation. On the other hand, some bottled waters, especially those originating from surface or municipal supplies, may be treated to kill any harmful bacteria and make them safe to drink; indicator organisms are again the means of monitoring this.

In the case of groundwater, there is also a natural population of indigenous harmless bacteria. In some markets, these naturally present bacteria are simply monitored to ensure that the normal condition of the water is not compromised; in others, it is a requirement that they remain within specified limits, both in the source and at the time of bottling. Thereafter, even though in still (non-carbonated) water, the number of these organisms grows logarithmically within days of bottling and can remain high for many months, these benign bacteria are not detrimental to the keeping quality of the water or to the well-being of the healthy consumer.

The difference in microbiological status between municipal or mains water and bottled waters, is often used in alarmist articles in the media, where the two products are compared. Such a comparison is perfectly understandable and justifiable, but the assumption that the same qualitative standards apply to both products is not. All waters for consumption must be safe to drink. Municipal water achieves and maintains this status through chemical treatments and the presence of residual chlorine disinfection at the point of use. In the case of bottled water, such chemical residues are not only undesirable, as they impart an unpleasant taste and odour, but are also prohibited by legislation, as they contravene the 'standard of identity' of the product. The fact that bottled waters are usually governed by legislation different from that which applies to municipal water demonstrates recognition by governments that these products are different. It is therefore no accident that both Chapters 12 and 13 discuss the subject of microbiology, one dealing with water bottled without treatment and the other for which treatment is used.

With respect to carbonated water, the addition of carbon dioxide lowers the pH, and this has an inhibitory effect on the growth of micro-organisms; consequently, numbers of bacteria in sparkling water are usually very low.

Finally, although not covered in detail in the main body of the book, it is worth giving some consideration to the way in which the finished product is used by the consumer. Throughout distribution, warehousing and retailing, bottled waters are held at ambient temperatures, although care is taken that this does not reach freezing point. The expansion of water in freezing can cause bursting of bottles, and this is especially so for sparkling products. For this reason, the consumer is advised never to put bottled water into a freezer, even for a short time, to cool it quickly, as bursting could lead to personal injury. Similarly, sparkling products should always be stored with the bottles upright, in the manner of champagne. Lying bottles on their sides increases the area of head-space, and this can make the bottles more liable to risk of explosion. The consumer may continue to store bottled water in its unopened state at ambient temperature at home for the duration of its prescribed shelf-life. Once opened, however, it is advisable to reseal the bottle and place it in a refrigerator, using the remaining contents within a few days.

Bottled water may be served in various ways. It can be used as it comes or as a mixer with alcoholic drinks and fruit cordials. There are numerous options for developing cocktails incorporating still or sparkling water. For the purist there is nothing like a good quality water on its own, served either at ambient tempera-ture or gently chilled. Where ice is used at home or in a hotel, restaurant or bar, it is preferable to prepare the ice from the same bottled water. In restaurants, ice should be offered, rather than assuming that it is wanted. Some people have a preference for water served with a slice of lemon or lime. In the public domain it is good to be offered this choice, but it is all too often assumed that it will be preferred. When a bottle of water is requested in a hotel or restaurant, it should, like wine, be opened at the table in the presence of the consumer. Whatever the preference of the individual for style of consumption, bottled water will provide much needed nutriment and refreshment, and add to the pleasure and enjoyment of life.

2 Market development of bottled waters

Duncan Finlayson

2.1 Introduction

When asked to revise this chapter for the second edition, I was struck immediately how the perspective of bottled water has changed between 1997 and now (2003). Then, we were looking forward to the emergence of Asia and especially China. Now, Asia is poised to become the biggest regional market ahead of Western Europe. The development of the soft drinks giants The Coca-Cola Company and PepsiCo into major bottled water players, which was only foreseen, is now a fact. In 1997, we were tracking water cooler developments in Europe but we did not predict the explosive expansion which took place or the dramatic consolidation this provoked in 2002 and 2003.

During 1994–2002, the world bottled water market has grown from 58 to 144 billion litres. The Asian/Australasian market has multiplied by more than five. Even geriatric Western Europe has managed a 37% increase, whilst dynamic North America has added 141%.

These statistics are impressive, but behind them lie a number of factors which contribute to growth differently in different segments of the market. Moreover, parts of the world have diverse traditions of water consumption, which means that they respond to market pressures in distinct ways. Overlaying them all are modern global factors such as health and well-being. Important questions in a market analysis include: What is the recent history? Where is the market today? Which are the important market drivers? What will the future be? Providing answers requires knowledge of these factors and traditions.

This chapter starts by giving a historical perspective, which touches also on the relationship between bottled waters and other soft drinks. Section 2.3 discusses product and market attributes to make the following sections intelligible. These definitions are dealt with more fully in Chapters 3 and 10. Section 2.4 introduces the big four global bottled water companies. A brief global review that puts bottled water into the context of other beverages is then followed by three sections looking at different markets by way of example – the USA, Europe and China. The final section looks at future trends.

2.2 The historical background

You could say that bottled water once had an Old World and a New World. The Old World is Europe, West and East, extending into Russia. The New World has its older markets – in the USA – and the very youngest such as China. The New World does not have the same traditions, and is driven by modern concerns. Today the Old World, although still exerting a strong influence on the industry and home to Nestlé and Danone (the two largest global bottled water companies), is giving way to the New.

Let us first consider the Old World. Every pupil knows that water is essential for life. Throughout history we have taken in water to survive, but have added to this use, at every opportunity, its role as the base of something more convivial such as wine and beer. Nevertheless, it is possible to discern an early trend for drinking water on its own, of two kinds. The first is a highly mineralised water, prized for its health-giving attributes and possibly for its medicinal properties. This water would often be naturally carbonated from an effervescent spring, and might well be hot on emergence. The second is a cool, fresh water, drunk for its purity and cleansing properties. Here these are named *Mineral Waters* and *Spring Waters* respectively – terms that should not be confused with modern legal definitions such as Natural Mineral Water and Spring Water, which are referred to later.

The doyenne of the Old World would be a place like Vichy in France. The history of Vichy reads as a history of Europe: first exploited by the Gallo-Romans during the first two centuries AD, the town and spa became part of the Bourbon estates of Louis II. The cures became famous during the Renaissance, but were really first developed for leisure during the 2nd Republic when the Parc des Sources was created by order of Napoleon. A second great period of construction followed during the Belle Époque (1890–1930). The two bottled waters of Vichy are Vichy St-Yorre and Vichy Celestins, both highly mineralised. Commercialisation on a major scale started in the 1860s. Both are now in the stable of Groupe Castel, the number three bottled water company in France.

We now turn to the UK, where, as in many markets, the development of bottled water consumption has been closely linked with that of soft drinks. The following brief review is adapted from a publication in the Shire series by Colin Emmins (1991).

The Romans developed various spas including Bath (Aquae Sulis) and Buxton (Aquae Arnemetiae), but more for bathing than drinking. By the eighteenth century, spa resorts were once again flourishing. The properties of various mineral waters became well known including those of the Epsom Spa (from which Epsom salts were extracted). Even the Bath water was drunk in the elegant Pump Room. Spring waters had been bottled from Tudor times, and by the year 1700, flasks of spring water were being taken from Hampstead Wells for sale in Fleet Street.

The technique of carbonation (adding carbon dioxide gas to water) was discovered by Dr Joseph Priestly in the late 1760s, a technique which turned out to be the spur to the creation of commercially manufactured soft drinks. Interestingly, much of the development was in artificial or manufactured mineral waters. Soda waters as well as artificial Seltzer, Spa and Pyrmont waters were on sale by 1800. Schweppes was set up in 1792. By the early 1800s, carbonated spring waters were being offered for sale. However, the tone was still overwhelmingly medicinal and the market still small.

During the nineteenth century, the market changed to one of much wider consumption, developing as much if not more in lemonade, ginger beer and other flavour-based soft drinks than in soda and Seltzer waters. This trend towards an increasing share for non–bottled water soft drink consumption accelerated in the UK because of the giant leaps made in the safety and palatability of the public water supply: progress not reflected on the Continent. Mineral waters, such as Apollinaris from Germany, were still fashionable mixers in the 1890s, but did not have a mass market. By 1902, arguments about naturalness had arisen. Apollinaris was taken to court for claiming that it was a 'natural mineral water' (*Davenport* v. *Apollinaris Co Ltd*), when the composition in the bottle did not exactly match that in the spring (a case rejected, see Chapter 3 for more details on modern natural mineral waters).

Nevertheless, the trends in the twentieth century, especially in the UK as opposed to Continental Europe, remained inexorably of rising consumption of nonwater soft drinks. Fruit squashes as dilutables were introduced just before World War I. Coca-Cola was introduced in the 1930s, remaining a modest item until the arrival of American troops for World War II, when consumption became firmly established.

Thus the UK did not go the way of the Old World, because of the good quality of tap water. By 1980, a distinctive European pattern of consumption had been established, with the UK as an atypical Anglo-Saxon outlier, strongly influenced by the USA (Table 2.1).

And what of the New World? Here bottled water has developed as a safe, reliable, consistent, refreshing and convenient alternative to tap water. As consumers have become more health conscious, bottled water seems a better choice

Table 2.1 European patterns of consumption for 1980

Country	Annual bottled water sales (million litres)	Annual bottled water sales (litres/person)	Annual soft drinks sales (million litres)	Bottled water as percentage of soft drinks
UK	30	0.5	4840	0.6
Spain	800	21	3050	26
Germany	2550	41	8450	30
Italy	2350	42	4100	57
France	3125	68	5715	55

Source: Zenith International © Zenith International 2003.

than soft drinks. In some parts of the world where tap water is not universally available, or may be unsafe, bottled water is not a luxury.

2.3 Market segmentation

An analysis of trends in the bottled water market requires market segmentation. In addition, bottled water is itself a segment of the overall soft drinks market. Segmentation is needed because different product/market combinations react in different ways in different countries. Various approaches have been used. Here the main segments, with some element of hierarchy, have been taken as follows:

For bottled water

packaged	vs.	water coolers
high mineralisation	vs.	low mineralisation
still	vs.	carbonated
Natual Mineral Water	vs.	Spring Water
purified/re mineralised	vs.	Natural Mineral Water and Spring Water
brands	vs.	own label
international brands	vs.	others
global brands	vs.	international brands
mainstream	vs.	flavoured and lightly sparkling
glass	vs.	plastic
PET	vs.	other plastics
nonreturnable (NR)	vs.	returnable (R)
supermarkets	vs.	other outlets

Some of the complexity behind this selection is indicated in Table 2.2. Each of these attributes is worth considering more fully. However, as many are inter-related, the following segmentation is necessarily simplified.

- *Product type.* Packaged water is sold in containers of not usually more than 5–10 litres capacity, directly for consumption. Water coolers are refrigerating units which dispense water from a large bottle into a cup. The water cooler market worldwide has developed from the US market, where poly-carbonate returnable bottles, originally of 3, 5 or 6 US gallon capacity (11.4, 18.9 or 22.7 litres), are supplied to rented cooler units. Outside the USA, the 22 litre container hardly exists. The US market is established both in offices and the home. Transfer to the European and other markets is mainly post 1985, and in Europe is so far confined to offices. Bulk water, such as in tankers, is progressively being taken over by the 19 litre container, in some cases through specialist water shops known as water stores or water

Table 2.2 Elements contributing to market segmentation

Attribute	Categories	Subcategories
Product type	Packaged water Water cooler Bulk water	
Water type	Still Carbonated (sparkling)	Naturally carbonated Lightly carbonated
Mineralisation	Range high to low	
Flavourings	Natural low calorie Sweet/artificial	Becomes a soft drink 'clear flavoured drink'
Functionality	Many, also known as water plus	Becomes a soft drink
Legal status	Natural Mineral Water Spring Water Other waters	Purified mains water Treated well water Remineralised water
Container	Glass Plastic	R NR PET (R & NR) PET multilayer PVC (now rarely seen) Polycarbonate (R) Polyethylene
Positioning	Premium Lifestyle Mainstream Budget Staple	
Branding	Global International National Regional Retailer own label	
Distribution	Supermarkets HORECA[a] Offices Independents Door-to-door Others	

[a] Hotels, restaurants and catering.
Source: Zenith International © Zenith International 2003.

stations. Mexico, for example, is the second largest bottled water market in the world, with relatively few coolers but widespread sales in 19 litre bottles. Water is drawn from the bottle through valves or ceramic pots. In poorer markets, PET is often used for the large bottles, although shrinkage and a lower washing temperature make it less than ideal.

- *Water type.* The still/carbonated split dates back to the original mineral water/spring water differentiation. However, artificial carbonation, widely available from the nineteenth century, 'muddied the waters' as it removed the correspondence between naturally carbonated waters which were mostly highly mineralised and still waters which were generally low in mineral content. The water that bridged the two, and proved to be the outstanding success of the twentieth century, was Perrier. Nevertheless, post 1990, the trend has been towards still waters for home consumption.

- *Mineralisation.* It is a truism in the European context that, as you travel eastward, the palate for water becomes stronger. Generally speaking, US consumers are interested in the absence of microbes, minerals, and certainly contaminants. A typical water would have a dry residue of 200 mg/litre. (Dry residue is the solids left after heating to 180°. It is a measure of dissolved minerals, which is usually less than total dissolved solids. It is used because it is directly measurable in a single test.) In France, highly mineralised waters such as Contre and Vichy Celestins have long had a special role. These are in the range 1700–3000 mg/litre. Even mainstream Vittel (Grande Source) has a dry residue of 850 mg/litre. However, Evian, Volvic and Valvert from neighbouring Belgium have been the success of the 1990s, and these are all low mineralisation. The German market has long been based on relatively highly mineralised waters (although in the new millennium this is finally changing. Germany has been a relatively successful market for Bonaqua, a re-mineralised water from Coca-Cola). In the former Soviet Union, waters were categorised into medicinal/mineral, mineral and fresh. The major brand was Borjomi, from Georgia, a sodium bicarbonate water with dry residue in the range 5000–6500 mg/litre.

- *Flavourings.* Perrier with a twist (of orange, lemon, lime or berry) was introduced to Europe in the 1980s, following its successful launch in America. These are natural flavourings with negligible calorie content, and though legally classified as soft drinks, can be legitimately presented as modified waters. More recently, clear flavoured drinks have entered the market, of which Clearly Canadian is the prime example. Some of these are not waters at all, having calorie contents as high as regular colas.

- *Functionality.* Functional offerings are relatively new but increasingly numerous. They are one of the fastest developing categories of soft drink at the time of writing, but still quite small in terms of volume. Some of these are on a water platform, otherwise known as near waters. Examples include calcium fortified water, sports waters and calming waters.

- *Legal status.* Obviously legal status depends on which part of the world a water originates from and (less and less) is marketed in. Full details are given in Chapter 3. The categories in Table 2.2 correspond to the 2003 European regulations. Particularly interesting is the market for remineralised waters, because this is being developed by the soft drinks majors (Coca-Cola and PepsiCo) in South America, Eastern Europe, the CIS and other newer markets.
- *Container.* Glass has always been the choice for premium products. In some countries such as Germany and Austria, a tradition exists for returnable glass of a standard format. This has been promoted as environmentally friendly, and even used as an excuse for excluding nonreturnable products. However, the jury is out on whether the life cycle cost of returnable glass really is lower, because of the weight of the glass container, the energy required for washing and the chemicals used. Returnable PET is well established. In the nonreturnable field, PVC has all but disappeared in favour of PET. Difficult applications, such as the new Perrier bottle, require multilayer technologies. Polyethylenes are used for larger containers of budget products, and polycarbonates have captured the water cooler bottle market. PET is used in large numbers for the lower-priced bulk market in essentially the same format as that used for water coolers (and does also appear on coolers). Bottled waters are also marketed in cans for the airline trade and vending machines, and cups, again for airlines and in fast food outlets.
- *Positioning.* Generally speaking, there is a relationship between price and total market size. Premium products in high-quality, individually designed glass bottles aim at profitable, small-volume niche markets. The mainstream is aimed at the wider market. Lifestyle/convenience positioning has been adopted by Coca-Cola and PepsiCo with great success in the USA. Budget brands are a feature of supermarkets in western countries and also of 19 litre distribution elsewhere. Staple products are a direct tap water replacement.
- *Branding.* Branded products cost more for consumers, a differential they are willing to pay provided they perceive added value, either in product quality or product values or both. Even own-label products have become bound up with the branding of the supermarkets themselves. In the third millennium, branding has a new dimension, which we categorise as international versus global. It used to be said that water is a local business because it is tied to a source. Then the big international brands were created – Perrier, Vittel, Evian, San Pellegrino, Volvic. Now Nestlé has signalled the first global brand Pure Life, a staple water produced by reverse osmosis or steam distillation and remineralisation at locations all round the world.
- *Distribution.* Distribution patterns have been quite different across countries, although the trend is towards supermarkets being the dominant

outlet for bulk consumption. Nevertheless, in Western markets, HORECA is disproportionately important to market value because of the loading on premium brands and smaller sizes. In Asia, for example, direct delivery of the 19 litre package can be very important, whether delivered by truck or two by two on the back of a bicycle.

2.4 Global giants and local leaders

In bringing this chapter up to date, the starting point has to be the development of the big four global bottled water companies: Nestlé Waters, Danone, Coca-Cola and PepsiCo. Only in the last few years has it been possible to piece together a picture, as the world has become smaller and more information is available on the Internet. Now we can see that the share of these four has reached 30% and is still rising (Fig. 2.1). In the case of Coca-Cola and PepsiCo, their share includes water brands owned by bottler partners or franchisees.

However, underlying this is a more startling picture of the ambitions of The Coca-Cola Company. The traditional foundation to both Coca-Cola and PepsiCo has been a relationship between brand owner (the Company) and bottler, which allowed the company to concentrate on marketing and the bottler on production and distribution. The Company maintained control and income by providing the proprietary syrup formulation on which the product is based. Bottled waters were an incidental activity of the bottler, with the brand not owned by the Company at all. Two trends have overtaken this. The first is the consolidation of bottling into regional anchor bottlers in which the Company has a substantial or majority stake – deals that in places have included the transfer of water brands to the Company. The second is the launch and

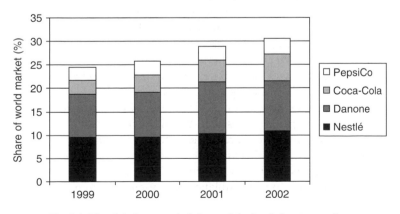

Fig. 2.1 The global companies' share of the bottled water market.
(*Source*: Zenith International © Zenith International 2003.)

expansion of the Company's own brands: e.g. in the case of Coca-Cola, Dasani and Bonaqua; for PepsiCo, Aquafina and Aqua Minerale. The result is a marked switch into own water brands as opposed to bottler-owned products (Fig. 2.2). In 2003, the proportion of own-brand water for Coca-Cola took another giant leap when Panamco (Mexico) switched from Risco to the Coca-Cola brand Ciel, following a multimillion-dollar transaction.

There is no doubt that Coca-Cola has a strategy of expanding its mainstream business into still drinks in general and bottled water in particular. How successful it will be against the bottled water specialists is an open question. In the USA, Nestlé Waters North America is more than holding its ground (see Section 2.6).

A small cadre of international brands sells round the world on a large scale: Perrier, Vittel, San Pellegrino, Evian, Volvic. Other than these, bottled water is a local business, because it is not economic to transport product over long distances. In the Old World (see Section 2.2), this meant also local brands, because the brand is tied to the source. In the New World such restraint is missing. A link still exists for natural waters, although it is tenuous. For example, the two giant natural spring water brands in America – Arrowhead and Poland Spring – both come from multiple sources. Poland Spring sources are all in Maine, but Arrowhead now has sources in Canada and California. No such link exists for manufactured water products. Coca-Cola and PepsiCo waters are typically treated and then remineralised, and can therefore be produced, in principle, at any soft drinks plant (although quality problems can arise if water is not bottled on a dedicated line).

So the Europeans can only develop regional brands, whilst Coca-Cola and PepsiCo are free to create global brands – but is that the picture? Not really – for several reasons. First, it is the Europeans who hold all the international brands

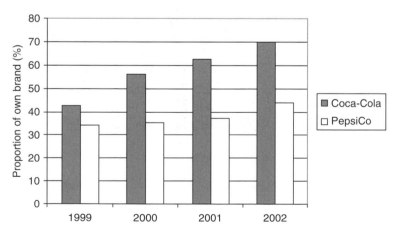

Fig. 2.2 Share of Coca-Cola and PepsiCo water as Company brand.
(*Source*: Zenith International © Zenith International 2003.)

mentioned above, which can be thought of as global although they can never be produced around the world. Second, Nestlé is determined to exploit its own brand, Nestlé Pure Life and Nestlé Aquarel. Pure Life is the first truly global brand. Aquarel perhaps has only regional ambitions in Europe. Third, Coca-Cola has several brands in different countries for its treated water, and recently had been buying natural water brands in Europe!

What is clear is that Danone and Nestlé Waters have changed from being French companies at heart with a natural water ethos to global entities on multiple platforms – a transition certainly encouraged if not dictated by the activities of Coca-Cola and PepsiCo.

2.5 Global review

Globally, hot drinks dominate world consumption of beverages, with an estimated 495 billion litres to be drunk in 2003 (globaldrinks.com). When examining trends for bottled water, it is useful to create a context which includes other beverages. The normal classification of the global sectors is as follows: hot drinks, milk drinks, soft drinks and alcohol. Bottled water is a segment of soft drinks. Figure 2.3 compares bottled water with the rest of the soft drinks segment

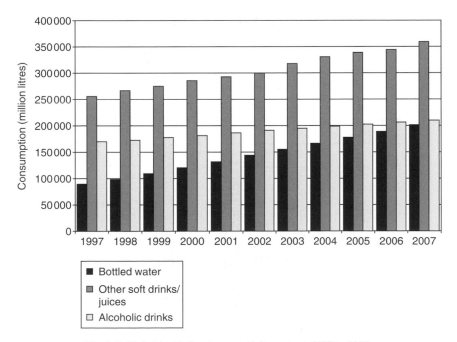

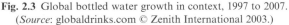

Fig. 2.3 Global bottled water growth in context, 1997 to 2007.
(*Source*: globaldrinks.com © Zenith International 2003.)

and with alcoholic drinks over the historic period from 1997 to 2002 and projected from 2003 to 2007.

By the end of this period, bottled water will be poised to overtake alcohol, and be well over 50% of the volume of other soft drinks. What factors are driving this high growth rate compared to other beverages? The mix will vary according to the part of the world, but the following influences probably come into play everywhere to some degree.

Wealth	Increasing prosperity allows the Chinese to buy bottled water as a staple, and the Americans to buy water on the go for its image and lifestyle choice.
Health	Greater affluence has combined with more attention to healthy living and bottled water has the attraction of no calories, additives or alcohol. In the USA especially, obesity has become a major issue which will drive both consumers and producers away from sugar-based soft drinks – the latter because of questions of liability.
Lifestyles	Bottled water's image has also been developed to fit contemporary lifestyles, with emphasis on nature and purity.
Marketing	Effective marketing and packaging have enhanced this appeal and so enabled premium prices for certain brands.
Quality	Even in the West there are many incidents of tap water pollution, which receive extensive media coverage, each giving bottled water a powerful boost. In many countries, the public water supply is unsafe or of variable quality.
Availability	Some countries are unable to extend the public water supply to their populations. In others, water infrastructure development cannot keep up with the pace at which cities are expanding.
Occasion	Bottled water is the only all-day beverage.
Taste	People are becoming more sensitive to discolouring or off-tastes in tap water.
Habit	Once the change has been made, people are unlikely to move back unless their confidence in bottled water is shattered or their economic circumstances decline – the new habit gains its own momentum.

In fact the fastest growing markets are those in Asia and the Pacific Rim which come from a low base of consumption, have experienced rapid economic growth and where the public supply infrastructure is either poor or struggling to keep pace (Fig. 2.4). Moreover, it is striking how low that base really is in terms of per capita consumption (Fig. 2.5).

To enlarge this perspective, the sections that follow pick three markets to look at in more detail: the USA, Europe and China. First, however, some

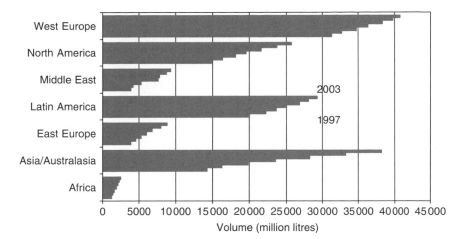

Fig. 2.4 Regional bottled water growth, 1997 to 2003.
(*Source*: globaldrinks.com © Zenith International 2003.)

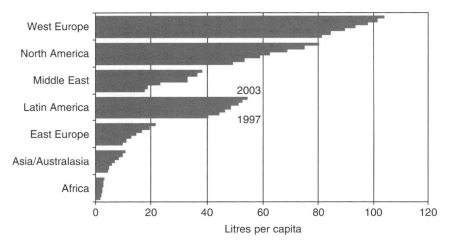

Fig. 2.5 Regional bottled water per capita growth, 1997 to 2003.
(*Source*: globaldrinks.com © Zenith International 2003.)

remarks about opposing forces to balance the picture of inexorable and natural development. As the industry grows and particularly as the major companies become more visible, bottled water will become a target for vested interests who seek to hinder its development or feel they can attack the industry to promote their own agenda. In some of the developed countries the water utilities are one such vested interest who see the public's confidence in bottled water as undermining their status and image. An example is the UK, where bottled water is regularly attacked in the press, particularly around the time that the Drinking Water Inspectorate releases its annual report on the status of the tap water supply.

Bottled water developments based on natural waters can provoke vociferous opposition. Such waters best come from unspoilt environments, so there can be a tension between preserving the landscape and constructing a factory. On the other hand, bottling plants often provide employment in rural areas where opportunities are scarce, and depopulation is threatening the environment in a different way. They also have low environmental impact and use a renewable resource. Communities have to resolve these legitimate issues through their planning and permitting procedures. Whatever the merits of a particular situation, the industry's visibility and the large size of the major companies encourage both campaigners and lawyers.

The use of water resources can also be the cause of dispute. In a locality this may well be a legitimate concern, again dealt with through the permitting system. However, claims of impact on a regional or national scale can hardly be justified, because even highly developed bottled water markets represent only a fraction of 1% of total water consumption. In 2002, the water utilities in Italy delivered nearly 50 billion litres/day to a population of 57.5 million, who also consumed just over 170 litres per capita of bottled water, i.e. 0.05% of the country's water use.

2.6 USA

At over 24 billion litres in 2002, the USA has the largest bottled water market in the world, ahead of Mexico (14 billion litres), China (13 billion litres) and Italy (10 billion litres). Besides being the largest, it is in many ways also the most developed. How so, cry the Europeans, whose consumption per capita is more than 20% higher, and whose brands are the best known on the planet? Because the USA has a much more integrated market than Europe, because the water cooler market is mature with a high penetration into homes and because the Americans have gone further in the adoption of modern technology. It is also a key battleground between the two global water companies Nestlé and Danone on the one hand, and the soft drinks giants Coca-Cola and PepsiCo on the other. Figure 2.6 gives the trend in the market since 1999, showing solid growth at over 12% for 2000, 7% for 2001 and nearly 12% again for 2002.

The growth is by no means evenly spread. 'Greater than 10 litres', which is almost exclusively Home and Office Delivery (HOD), is growing rather slowly at 4–5%. And the separate line for 'still to 1 litre' shows that this segment is providing nearly all of the growth in packaged water. Thus, the US market is very focused on lifestyle and convenience, i.e. on the small packs which can be bought at outlets everywhere and consumed on the go.

This condition has been influenced partly by the entry of Aquafina (PepsiCo) into national distribution in 1997 and Dasani (Coca-Cola) in 1999. Both

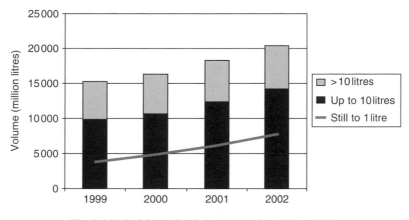

Fig. 2.6 United States bottled water market, 1999 to 2002.
(*Source*: Zenith International © Zenith International 2003.)

corporations have applied their marketing talents to exploit this convenience orientation and an already existing distribution infrastructure. So how did their entry affect the other bottled water companies? Figure 2.7 indicates that the smaller companies suffered first – losing market shares to the big four. Second, Danone has been squeezed. In June 2002, the company announced a partnership with Coca-Cola for the production, marketing and distribution of Groupe Danone's retail bottled water throughout the USA. Danone contributed the assets of its US businesses, licences for the Dannon and Sparkletts brands and ownership of several value brands. Coca-Cola contributed cash and runs the new company Danone Waters of North America with a 51% share. In 2002, Danone and Coca-Cola also signed a master distribution agreement under which Coca-Cola

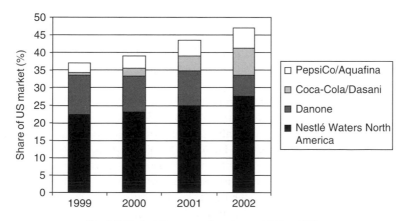

Fig. 2.7 United States market shares, 1999 to 2002.
(*Source*: Zenith International © Zenith International 2003.)

manages marketing execution, sales and distribution for the Evian brand in North America. Figure 2.7 allows for the shift in activity from Danone to Coca-Cola.

The third trend in Fig. 2.7 is the steady gain in market share by Nestlé Waters North America. Formerly Perrier Group of America, the Connecticut-based company has been the leading bottled water company in the USA since the late 1980s with a stable of important sources, including the giant brands Poland Spring in Maine and Arrowhead in California and Canada. Most of the company's products are natural spring water. Interestingly, none of its brands is truly national and the water brand is the primary platform. The name of Nestlé Waters is present in publicity but very much at a secondary level. The company's success has been attributed in part to its technological development, especially in the HOD segment where customers can manage their accounts online.

2.7   Europe into the new millennium

In Table 2.1, European patterns of consumption as they existed in 1980 are illustrated for sample countries. It is interesting to see how these have changed in the 15 years up to 1995 and then on to 2002, a comparison which is made in Table 2.3.

The German figures are distorted by unification. The 1980 figure is for West Germany, the 1995 for unified Germany. The per person consumption dropped from 86 to 72 litres in 1990, the year of unification, because of the low consumption of the former East Germany.

Table 2.3 presents a remarkable picture of growth which has continued from 1995 into the new millennium. From 1980 to 1995, Italy experienced a compound rate of 8.5%; France 5.2%; the UK 24%. From 1980 to 2002, the growth has been 6.8%, 4.8% and 20% respectively, a startling record over such a long period. Per capita consumption has now exceeded 170 litres/year in Italy, but is still less than 30 in the UK. Nevertheless, the UK consumption exceeds the 1980 figure for Spain and is approaching that of Germany and Italy in 1980!

Table 2.3 Changes in European consumption between 1980 and 2002

Country	Annual bottled water sales (million litres)			Annual bottled water sales (litres per person)		
	1980	1995	2002	1980	1995	2002
UK	30	800	1770	0.5	13	29
Spain	800	3100	4801	21	79	121
Germany	2550	7935	9289	41	97	111
Italy	2350	7950	9917	42	139	171
France	3125	6670	8813	68	115	147

Source: Zenith International © Zenith International 2003.

How did the industry fare in 2002, given the pressures of global economic slowdown? There has been a discernible impact on most European water markets, although its effect has been more acute in some countries than others. Despite uncertainty within the world economy, a downturn in tourism because of fears of terrorist attacks and unremarkable weather, European bottled water consumption managed to grow by a volume of 153 million litres, corresponding to an increase of 4% between 2001 and 2002 to reach 39 billion litres. However, rates of growth have shown decline in recent years; in 1999 hot weather helped to advance consumption by 6.2%, but in 2001 average market growth was down to 5.3%. Denmark joined Ireland and the UK as the only countries registering double digit growth in 2002. Lower-than-average growth was recorded in the following five countries: Belgium, France, Italy, Norway and Sweden.

Italy, the largest European market and the most saturated, showed the lowest growth at 1.6%; Ireland showed the highest growth at 18%. Germany's economic position has stood out as having deteriorated markedly recently, yet both bottled water and cooler growth have remained encouraging. Norway and Sweden displayed low levels of growth with the degenerating national economy being cited as a significant factor behind the ebbing market. None of the 16 European countries recorded a decline in 2002, but this pattern looked set to change in 2003. The German market was expected to experience a fall in sales of over 3%, faced with both an economic downturn and trading difficulties surrounding a new controversial law brought out in January 2003. This law imposes an obligatory deposit on nonreturnable plastic bottles and cans for bottled water, soft drinks and beer. In fact a spectacular summer across most of Europe more than counteracted the German situation as well as giving a seasonal lift to most countries' sales for the year as a whole.

And how has the market been changing?

- Sparkling water is holding its volume but all the growth is in still water (Fig. 2.8).
- There are still some countries that hold onto their sparkling tradition (Fig. 2.9).
- Bottled water coolers, although still only a small part of the market, have shown dramatic growth (Fig. 2.10).
- PVC packaging has disappeared in favour of PET. Glass has reduced substantially. Polycarbonate packaging has reached nearly 3% of the market because of water coolers (Fig. 2.11).

It is the water cooler market that has seen a dramatic consolidation coming to a peak in early 2003. As a result, Nestlé Waters and Danone have changed from small or non-existent positions to the number one and two spots respectively (Fig. 2.12).

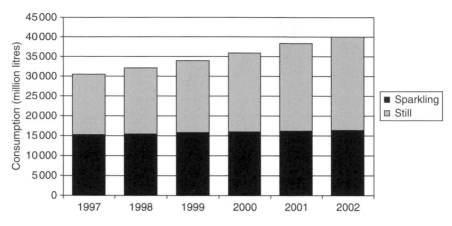

Fig. 2.8 West European still/sparkling split, 1997 to 2002.
(*Source*: Zenith International © Zenith International 2003.)

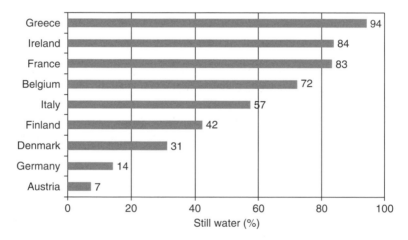

Fig. 2.9 West European percentage of still water, 2002.
(*Source*: Zenith International © Zenith International 2003.)

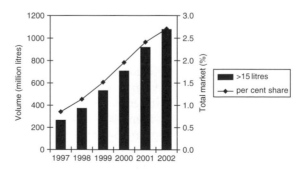

Fig. 2.10 Growth in the West European water cooler market, 1997 to 2002.
(*Source*: Zenith International © Zenith International 2003.)

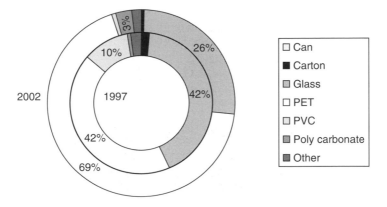

Fig. 2.11 West European packaging split, 1997 and 2002.
(*Source*: Zenith International © Zenith International 2003.)

The events (by announcement date) were as follows:

- December 2001, Ionics sells the US and European Aqua Cool business to Nestlé Waters – 80 000 coolers in Europe for €247m. (exchange rate on the day).
- October 2002, Ondeo (the water division of Suez) sells Chateau d'eau to Danone for an undisclosed sum, totalling 130 000 coolers in Europe.
- November 2002, Danone buys Sparkling Spring of Canada, including a European portfolio of 55 000 coolers.
- March 2003, Nestlé acquires Powwow from Hutchison Wampoa for €560m., thus gaining 230 000 coolers.
- Danone announces a joint venture (JV) with Eden Springs, to put Eden's 155 000 West European coolers with Danone's 210 000, the JV to be run by Eden.

2.8   China

China is vast and information difficult to come by. Despite these difficulties, all indicators and industry sources point towards a bottled water market in China that has really developed its potential in recent years. Chinese consumers have embraced the bottled water concept. Some consumers have taken to using bottled water for cooking as well as regular consumption. However, per capita consumption, and therefore penetration, is still low at 8.7 litres in 2002, and consumption is polarised towards the cities.

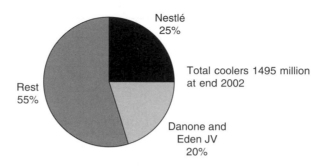

Fig. 2.12 European water cooler consolidation by August 2003.
(*Source*: Zenith International © Zenith International 2003.)

The market was pioneered in the Guangdong region in 1986 at a time when the area was spearheading growth for the country as a whole – assisted by close political and economic links with Hong Kong. In recent years, Beijing and Shanghai in particular have become increasingly important for growth across the cooler and retail pack size segments. Beijing, Guangdong, Hebei, Jiangsu, Shanghai and Zhejiang are the highest yielding soft drinks areas, and consequentially pivotal for bottled water activity. China's policy of reform and the opening up of markets greatly benefited bottled water. As a result of corporate restructuring, asset reorganisation and a strong brand strategy, the market saw an influx of major bottled water players in the late 1990s. Coupled with consolidation amongst existing local players, this changed the shape of China's bottled water industry.

In 1999 Nestlé began construction of a new production facility in Shanghai in the first stage of a US$50m. three-year investment in China. Nestlé's arch-rival Groupe Danone first entered the Chinese market in 1996 when it became a shareholder in market leader Wahaha. Danone also acquired number four, Health, in 1998. In March 2000, Danone added market number two, Robust, and in December 2000, Danone agreed to acquire 50% of another important bottled water cooler player, Shanghai Aquarius. The major difficulty currently being faced by the larger bottled water groups is the downward pricing spiral that has drawn in most active bottled water players. Figure 2.13 shows the characteristics of the market and demonstrates clearly that the large water cooler size bottle is the format that dominates, and is also rapidly growing.

In urban areas, households have embraced water cooler bottles simply because there is often no infrastructure in place for reliable municipal water. This is particularly so in outlying areas of rapidly growing cities. Here water for cooler delivery has assisted the rapid commercial expansion of cities such as Shanghai, and to a lesser extent Beijing.

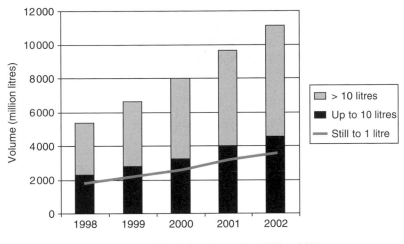

Fig. 2.13 China bottled water market, 1998 to 2002.
(*Source*: Zenith International © Zenith International 2003.)

There are a large number of bottled water producers, probably more than 2000, with new entrants joining all the time and companies falling out because of the competitive pressures on sales pricing. The degree of regional diversity within China means that the typical producer focuses sales and distribution activity on one region or locality. Informal and formal trade barriers between different regions also contribute. There is very little national distribution, with the larger 'national brands' usually only present in the major outposts of bottled water activity. Indeed, the rural market remains very much untapped at present. Launching a bottled water product does not therefore mean immediate access to 1.3 billion potential consumers. All imported water comes under the scrutiny of the authorities and must be distributed and marketed by a domestic Chinese agent, usually via equity or contractual JVs.

Purified water represents approximately 70% of the market, with mineral water the remainder. Sparkling water has been on the market only since 1997. Wahaha purified water from Danone is the country's biggest brand with about 13% share. Robust takes overall second place, followed by Nong Fu San Quan from Yang Sheng Tang. However, it must be noted that the significance of brand share varies significantly from region to region.

The HOD arena has enabled local operators to make notable volume gains, supplying 5 gallon bottles to the expanding number of offices and homes in China's major cities. In Shanghai, important HOD players include Zheng Guanghe, Yanzhong and Angel. In Beijing, important HOD companies include

Yanjing, Wahaha and Angel. Other HOD operators of note include Lebaishi
and Tian Tian Tong Jing, particularly in Wuhan, C'est Bon, Idea, Health (non-
Danone) and Zhong Liang Shan.

There are, of course, a staggering number of other players. Some of the more
established, not previously mentioned, include Cool, Coca-Cola's Sensation and
Heaven & Earth brand offerings, Jiulong, Kesai, Laoshan, Pure, Shizhong,
Sparkling, Waterman, Watson's and XiShiBao.

2.9   Trends for the future

The prospect for the future continues to be one of exciting and dynamic change
on a global scale. Whereas in the 1990s we talked of two bottled water groups,
now there are four: Nestlé Waters, Danone, The Coca-Cola Company and
PepsiCo. Their grip on the world market will probably increase in terms of
market share. However, bottled water has always been a product for which a
local producer can serve a local market, or where entrepreneurial spirits can
carve a niche with a new idea. So, diversity of companies and brands seems to be
assured.

At the end of the 1990s, competition was between the different product
values of the mature Western European and North American markets, according
to the history and players in each region. Thus, Coca-Cola and PepsiCo already
had bottled water brands that were based on a concept of manufacturing and
commercialisation not unlike the colas and other essence-based drinks: the brand
is the asset, not the source; manufacturing is based on applying a supplied
essence or composition to a demineralised water, which can be from any source;
production is franchised or carried out by a separate bottling company. This
approach was, and remains at odds with, the European concept of using untreated natural water, which has to be free from all traces of pollution, bottled at
source. Such attributes are obtained by using special sources that have particular
levels of protection from environmental pressures, as endowed by their hydrogeology. Naturally protected sources come at a cost in development and in
distribution from fewer locations which tend to be in remoter areas. The question is: Will the public be prepared to pay for this?

Since 2000, these genetic distinctions have almost gone. The major players are
acting globally in every segment, and are seeking to leverage their company brand
onto bottled water products. Product attributes are tuned to the marketplace.
Nestlé, the champion of natural waters, is market leader in Pakistan with Pure
Life, a water produced by reverse osmosis. In July 2003, Coca-Cola Hellenic
Bottling Company, a European anchor bottler for The Coca-Cola Company,
acquired Römerquelle, a flagship Austrian company with brands on a natural
water platform including enhanced water.

The future probably does hold more conflict between bottled water and vested interests, which the industry will have to resist through high-quality standards and good governance. In May 2001, the World Wide Fund for Nature declared that, in the light of a new independent study, the conservation organisation was urging people to drink tap water – which is often as good as bottled water – for the benefit of the environment and their wallets. The study was actually a World Wildlife Fund(WWF)-commissioned review, which in itself was uncontroversial. It seems that the WWF thinks bottled water is easing the pressure on governments to provide potable water supplies. In July 2003, a government official in Sharjah, UAE, declared that bottled water producers were exporting a valuable resource from the region. Of course, governments have their responsibilities independent of commerce, and the proportion of water used in the UAE for bottled water is tiny – fractions of a per cent of the country's supply. Nevertheless, bottled water is visible and so will be a target.

A final dimension not touched on previously in any detail is the extension of bottled water by adding flavour or functionality, but essentially on a pure water platform. This will provide opportunities for all but especially for small, entrepreneurial concerns. Whatever the developments, it should be an exciting time for those in the industry – small, medium and large.

References

Emmins, C. (1991) *Soft Drinks – their Origins and History*, Shire Publications, UK.

Zenith International Ltd (2003) *Zenith Report on Asia/Middle East Bottled Water Market – June 2003*, Zenith International, Bath, UK.

Zenith International Ltd (2003) *Zenith Report on West Europe Bottled Water Market – September 2003*, Zenith International, Bath, UK.

Zenith International Ltd (2003) *Zenith Report on West Europe Water Cooler Market – July 2003*, Zenith International, Bath, UK.

3   Categories of bottled water

Nicholas J. Dege

3.1   Introduction

As the bottled water market continues to expand, more and more consumers are faced with the choice of which waters to buy. The way they choose will be influenced by availability, habit and, to an increasing extent, their knowledge of the product. However, the range of names applied to bottled waters of various origins, some having undergone treatments and others untreated, can continue to surprise.

What, for example, is the difference between a Natural Mineral Water (NMW) sold within the European Union (EU) and a mineral water sold in the USA? What, if any, are the similarities between the principles applied to the exploitation of NMW and Spring Water (SW), and how do these compare qualitatively with the tap water available from municipal suppliers within the same community? Furthermore, the distinction between, and relative merits of, NMW, mineral water, SW, 'natural' water, table water, flavoured water, purified/demineralised waters and so on can be difficult to determine.

This chapter is an attempt to clarify these differences and the reasons why bottled water, which on the face of it, is a universal product, should be marketed under such a wide variety of names. It is not possible here to carry out a complete country-by-country study of water worldwide, so the focus is on principles, with examples where appropriate. The bulk of the work concentrates on European waters, as it was in Europe that the market for bottled water first developed. By way of comparison, the approach in the USA is also examined, as is the position in principle of other important markets, which have tended to follow one or the other of two distinct philosophies.

Some waters, especially those originating as groundwater from a protected source, may require little or no treatment to guarantee fitness for consumption. Such supplies will always be valued, whether the means of abstraction is through a naturally flowing spring, a sunken well or a borehole. In Europe, many of these sources underwent development, not only to provide for visitors but also to supply bottled waters for distribution further afield. To guarantee the consistency of quality, the emphasis lay not in treatment, but in protection of the original unpolluted state of the water and prevention of pollution, even up to the point of consumption at a place and time far removed from the original source.

In other areas, particularly where the supply is principally from the surface, the wholesomeness of water can only be guaranteed by treatment to modify its chemical and microbiological condition. The extent of such treatment will more or less vary according to the nature of the supply, the technology available and the quality required. Hence in Europe and in other parts of the world, systems have developed to ensure that all waters for public supply, regardless of origin, comply with given standards at the point of consumption. Although specific standards for different parameters have varied, technological development in the nineteenth and twentieth centuries have enabled operators and regulators to compile a comprehensive list of standards (a parametric list) to cover microbiological condition, major, minor and trace elements and indicators of pollution. The content and extent of the list has also varied in accordance with local requirements and the acquisition of knowledge, while the pertinent regulatory framework has also differed markedly. This has meant that in some areas, the quality of water was and is heavily controlled and dictated by the regulators and enforcement agencies through national regulations, and in other countries fewer quality standards have existed. To some extent and particularly in markets that are heavily regulated, this philosophy of control is equally applicable both to water supplied through the tap and to bottled water.

Thus there have evolved two distinct regimes for ensuring water quality. On the one hand, naturally wholesome water from a protected source is supplied without treatment other than precautionary measures during abstraction and bottling to ensure that it is not polluted chemically or microbiologically. On the other, water from a range of sources and often of dubious quality is taken and treated to comply with a list of chemical and microbiological standards at the point of consumption.

This is to a large degree illustrated by the publication in 2001 of two standards by the Codex Alimentarius Commission (of which more later). These standards were written in recognition of the growing significance of bottled water worldwide, both as a commodity and as an alternative (frequently necessary) to tap water. Of these, the second – the *Code of Hygienic Practice for Bottled/Packaged Drinking Waters* (other than *Natural Mineral Waters*), contains an introduction that explicitly recognises the reasons why such a standard is required, namely the rapid increase in the international trade in bottled water. This has been driven not only by the growing traditional bottled water market but also by the need for reliable supplies of potable water in geographical areas of risk and, with improved transport capacity, the increasing ease with which such supplies can now be made available. Even in those areas where the public supply is normally of good quality, the role played by bottled water has been increasingly significant in times when, due to natural or man-made disaster, public supplies have been disrupted.

3.2 Europe

Although many of the European waters have been exploited since classical times, the present European bottled water market stems from a tradition originating in the late Middle Ages. During this time the great 'spa' towns developed, patronised by visitors from far and wide, who were attracted by the various therapeutic claims made about the different waters.

There is no doubt that one of the key reasons for their popularity was the dubious quality of the surface water supplies available at the time, so that any source of good, clean water would have been particularly valued. In addition of course, the mineral content of many of the waters was regarded as a positive aid to health. In the absence of modern medicine, different waters became recognised for their own particular health benefits: cures for kidney, urinary and digestive disorders were claimed of highly mineralised waters, while total hydrotherapy treatments, including immersion combined with dosed drinking of the water, were used to alleviate arthritis, rheumatism and respiratory problems.

By the eighteenth and nineteenth centuries, many of the 'spa' towns, including the original Spa in Belgium, Baden-Baden in Germany, Vittel and Vichy in France and Malvern and Buxton in England, had grown prosperous and fashionable through the influx of wealthy visitors in search of good health. With the development of improved water supplies and with it the advances in modern medicine, many of the great spas declined, most notably in Britain. In other parts of Continental Europe, however, particularly in France, Belgium, Germany and Italy, the custom of drinking waters for their health benefits was so deep-rooted that the practice of bottling water for distribution both in the country of origin and abroad continued to grow, eventually resulting in a huge bottled water market. It is not surprising, therefore, that the methodology for exploitation and the criteria for 'recognition' of waters should be rooted in the European, and particularly the French, tradition.

In 1856 the French government drew up a law decreeing that a water source could be declared *d'interêt public*; this was the first procedure by which the health benefits of waters could be officially recognised by the authorities and was equally applicable to the water at source and as bottled. Many further developments took place, but these original principles played a key part in the evolution of the EEC Directives governing NMW and (subsequently) SWs.

The dual evolution outlined in Section 3.1 is well illustrated in Continental Europe, where the two philosophies have developed side by side, undergoing many changes and adaptations, but culminating in 1980 with the publication of two EEC Directives:

(1) Council Directive of 15 July 1980 relating to the exploitation and marketing of NMWs (80/777/EEC) (known as the Natural Mineral Water Directive), which has been subsequently amended by two further Council Directives, 96/70/EC and 2003/40/EC.

(2) Council Directive of 5 July 1980 relating to the quality of water intended for human consumption (80/778/EEC) (known as the Drinking Water

Directive), replaced on 3 November 1998 by an updated Directive (98/83/EC) with the same title.

The Natural Mineral Waters Directive 80/777/EEC, published in 1980, consists of 17 Articles, covering definitions, conditions for exploitation, permitted treatments, microbiological standards and methods. The text details the requirements for packaging and labelling and explains the process for incorporation of the Directive into national regulations. The Directive also has three annexes, which contain the detailed requirements for recognition, exploitation and marketing. In addition, Annex III provides a list of 'indications', which may appear on the label, together with the criteria for such indications. This Directive served the community for many years, but developments in scientific methods, the need to clarify the criteria for certain treatments and, most importantly, the need to bring into the same legislation a prescription specifically for SWs, led to the publication in 1996 of an amendment (Directive 96/70/EC). In May 2003, a further Commission Directive (2003/40/EC) was adopted, 'establishing the list, concentration limits and labelling requirements for the constituents of natural mineral waters, and the conditions for using ozone-enriched air for the treatment of natural mineral waters and spring waters'. Therefore, the result is that for the time being at least, NMWs are regulated through three Directives: 80/777/EEC, 96/70/EC and 2003/40/EC, and SWs are regulated through 80/777/EEC, 96/70/EC and (as we will see later) the Drinking Water Directive 98/83/EC.

3.2.1 Natural Mineral Waters
Annex I, Section I of 80/777/EEC defines NMW as follows.

1. 'Natural mineral water' means microbiologically wholesome water, within the meaning of Article 5 [which deals with microbiological standards], originating in an underground water table or deposit and emerging from a spring tapped at one or more natural or bore exits.

   Natural mineral water can be clearly distinguished from ordinary drinking water:

   (a) by its nature, which is characterized by its mineral content, trace elements or other constituents and, where appropriate, by certain effects;
   (b) by its original state, both characteristics having been preserved intact because of the underground origin of such water, which has been protected from all risk of pollution.

2. These characteristics, which may give natural mineral water properties favourable to health, must have been assessed:

   (a) from the following points of view:

1. geological and hydrological,
2. physical, chemical and physico-chemical,
3. microbiological,
4. if necessary, pharmacological, physiological and clinical;

(b) according to the criteria listed in Section II; [see below].

(c) according to scientific methods approved by the responsible authority.

SECTION II – REQUIREMENTS AND CRITERIA FOR APPLYING THE DEFINITION [OF NMWS]

1.1   Requirements for geological and hydrogeological surveys
There must be a requirement to supply the following particulars:

1.1.1 the exact site of the catchment with indication of its altitude, on a map with a scale of not more than 1 : 1000;
1.1.2 a detailed geological report on the origin and nature of the terrain;
1.1.3 the stratigraphy of the hydrological layer;
1.1.4 a description of the catchment operations;
1.1.5 the demarcation of the area or details of other measures protecting the spring against pollution.

1.2   Requirements for physical, chemical and physico-chemical surveys
These surveys shall establish:

1.2.1   the rate of flow of the spring;
1.2.2   the temperature of the water at the source and the ambient temperature;
1.2.3   the relationship between the nature of the terrain and the nature of the type of minerals in the water;
1.2.4   the dry residues at 180°C and 260°C;
1.2.5   the electrical conductivity or resistivity, with the measurement temperature having to be specified;
1.2.6   the hydrogen ion concentration (pH);
1.2.7   the anions and cations;
1.2.8   the non-ionized elements;
1.2.9   the trace elements;
1.2.10 the radio-actinological properties at source;
1.2.11 where appropriate, the relative isotope levels of the constituent elements of water, oxygen (^{16}16O–^{18}O) and hydrogen (protium, deuterium, tritium);
1.2.12 the toxicity of certain constituent elements of water, taking account of the limits laid down for each of them.

1.3 Criteria for microbiological analyses at source

These analyses must include:
1.3.1 demonstration of the absence of parasites and pathogenic micro-organisms;
1.3.2 quantitative determination of the revivable colony count indicative of faecal contamination:

(a) absence of *Escherichia coli* and other coliforms in 250 ml at 37°C and 44.5°C;
(b) absence of faecal streptococci in 250 ml;
(c) absence of sporulated sulphite-reducing anaerobes in 50 ml;
(d) absence of *Pseudomonas aeruginosa* in 250 ml.

1.3.3 determination of the revivable total colony count per ml of water:

(i) at 20°C to 22°C in 72 hours on agar-agar or onagar-gelatine mixture,
(ii) at 37°C in 24 hours on agar-agar.

1.4 Requirements for clinical and pharmacological analyses

1.4.1 The analyses, which must be carried out in accordance with scientifically recognized methods, should be suited to the particular characteristics of the natural mineral water and its effects on the human organism, such as diuresis, gastric and intestinal functions, compensation for mineral deficiencies.

1.4.2 The establishment of the consistency and concordance of a substantial number of clinical observations may, if appropriate, take the place of the analyses referred to in 1.4.1. Clinical analyses may, in appropriate cases, take the place of the analyses referred to in 1.4.1 provided that the consistency and concordance of a substantial number of observations enable the same results to be obtained.

Thus, the fundamental principles are that the water should be taken from a single protected source, be naturally wholesome and of consistent mineral character, and must be bottled in its unaltered state.

3.2.1.1 *Recognition of Natural Mineral Waters*

The notion of source recognition is also embodied in the Directive: in France this is carried out jointly by the Ministries of Health and of Mines; in the UK by the local Trading Standards and Environmental Health Authorities. To enable this recognition, particulars must be established in the following areas:

- Source Protection (hydrogeological description/abstraction method)
- Physical and chemical characteristics of the water (proof of stability)
- Microbiological analyses (to confirm that the water is wholesome in its natural state)
- Freedom from pollution (absence of toxic substances)
- Chemical and Pharmacological analyses (where required)

Source protection. The Directive demands that the hydrogeology of the source be established and appropriate measures implemented to prevent pollution. As a minimum this means that:

- the catchment area must be identified, the level of risk to the aquifer evaluated, and appropriate controls put in place to eliminate the risk.
- the aquifer and subterranean route for the water must be established.
- the source or point of emergence must be protected against pollution, such as surface water or chemical contaminants.

In practice, different sources require different kinds of management because conditions such as size and characteristics of catchment, type of aquifer and travel time vary significantly.

Physical and chemical characteristics. The chemistry of the water will depend on several factors, including the types of soil and rock through which it passes, the depth of the aquifer and the travel time. The composition of waters can vary significantly, from the very heavily mineralised ones beloved in Germany, to the lighter ones generally preferred in the UK. Some contain naturally occurring carbon dioxide, which originates in rocks with relatively recent volcanic activity, or perhaps due to alteration of limestone.

Regardless of the particular chemistry, it is a key requirement as stated in Annex I, Section I that:

> 3. The composition, temperature and other essential characteristics of natural mineral water must remain stable within the limits of natural fluctuation; in particular, they must not be affected by possible variations in the rate of flow.

In most cases, the recognition process for a 'new' NMW requires a minimum of two years' evidence of stability; once established, however, the consistency must be demonstrated on an ongoing basis. As a minimum, this requires a regular (at least annual) analysis against a scheduled list within the Directive. In practice, it is recommended that newly established sources are subject to a much more rigorous regime to ensure that minor seasonal variations are understood, and that more radical changes indicative of major alterations are recognised at the earliest opportunity.

In the case of 'back-up' boreholes sunk in order to increase the yield from a NMW aquifer, the process can be shorter (a matter of months), but contiguity with existing boreholes must be demonstrated. Continuous monitoring at the source of temperature, pH, conductivity, flow rate and water level provides an 'early warning' system of any potential changes in the source. In addition, regular full chemical analyses provide comprehensive evidence of continuing consistency.

Microbiological analyses. Paragraphs 1 and 2 of Article 5 and Section II of Annex I give details of the microbiological requirements for NMW; key to the recognition procedure and also essential during marketing is the need to demonstrate the following:

(1) The water is free from specified parasites and pathogenic micro-organisms (see 1.3 under Section II above).
(2) The number of naturally occurring benign bacteria, though present, is at low levels; this is done by determination of the revivable total colony count – often referred to as the total viable colony count (TVC) per millilitre of water after incubation on agar media, thus:

> At source, these values should not normally exceed 20 per millilitre at 20–22°C in 72 hours and 5 per millilitre at 37°C in 24 hours respectively, on the understanding that they are to be considered as guide figures and not as maximum permitted concentrations.

It is important to note that the revivable total colony count – often referred to as the total viable colony count (TVC) – will develop and change with time, because sterilisation or disinfection of the water is prohibited. In recognition of this, the Directive specifies limits on counts, but only within 12 hours of bottling, the water being maintained at 3–4°C during this time. Thereafter, according to paragraph 3 of Article 5:

- the revivable total colony count of a natural mineral water may only be that resulting from the normal increase in the bacteria content which it had at source,
- the natural mineral water may not contain any organoleptic defects.

Consequently, counts in still waters, which will be very low at source and at the time of bottling, can rise exponentially throughout the shelf-life of the product to 10^5 and even 10^6 organisms per ml within days of bottling before ultimately declining several months later.

The constituents of the natural flora, sometimes known as autochthonous (indigenous) bacteria, are important, because their presence in the source provides evidence that chemical pollution is *absent*. In addition, it is commonly accepted that the original natural flora stabilises the water microbiologically and renders it less susceptible to the growth of undesirable organisms.

Freedom from pollution. Although it is an inherent requirement that freedom from pollution should be established, the Directive specifies only microbiological standards; no standards for indicators of chemical pollution are included. It has therefore been left to Member States to compile appropriate standards for such substances and to determine whether a water complies. In practice, most states have adopted a list of volatile organic compounds (VOCs), polycyclic aromatic hydrocarbons (PAHs) and pesticides, to act as indicator parameters. However, with the continuing development of analytical methods, enabling detection of constituents at lower and lower trace levels, the Commission is presently considering a definition of 'freedom of pollution', and the European trade association is also developing a common position paper.

Limits on certain constituents. In the original Directive, the only requirement was that NMWs should be free from pollution and not contain toxic substances, but without specifying what these substances were or what the acceptable limits should be. It was therefore left to Member States to establish this list of parameters. However, modifying Directive 96/70/EC (Article 11) introduced for the first time the proposal to apply limits for the levels of some naturally occurring constituents, and 2003/40/EC specifies this list in Annex I.

ANNEX I

Constituents naturally present in natural mineral waters and maximum limits, which, if exceeded, may pose a risk to public health.

Constituents	Maximum limits (mg/l)
Antimony	0.005
Arsenic	0.010 (as total)
Barium	1.000
Boron	For the record*
Cadmium	0.003
Chromium	0.050
Copper	1.000
Cyanide	0.070
Fluorides	5.000
Lead	0.010
Manganese	0.500
Mercury	0.001
Nickel	0.020
Nitrates	50.000
Nitrites	0.100
Selenium	0.010

* The maximum limit for boron will be fixed, where necessary, following an opinion of the European Food Safety Authority and on a proposal from the Commission by January 2001.

The deadline for compliance is 1 January 2006 (with the exception of fluoride and nickel, for which the deadline is extended until 1 January 2008). It is now required for Member States to include this requirement in national regulations.

Clinical and pharmacological analyses. According to paragraph 1 of Article 9:

2. (a) All indications attributing to a natural mineral water properties relating to the prevention, treatment or cure of a human illness shall be prohibited.

   (b) However, the indications listed in Annex III [reproduced on page 42] to this Directive shall be authorised if they meet the relevant criteria laid down in that Annex or, in the absence thereof, criteria laid down in national provisions, and provided that they have been drawn up on the basis of physico-chemical analyses and, where necessary, pharmacological, physiological and clinical examinations carried out according to recognized scientific methods, in accordance with Section I, paragraph 2 of Annex I.

   (c) Member States may authorize the indications 'stimulates digestion', 'may facilitate the hepato-biliary functions' or similar indications. They may also authorize the inclusion of other indications, provided that the latter do not conflict with the principles stated in (a) and are compatible with those stated in (b).

Thus, while the Directive rules out claims that may be of a curative or medicinal nature, it does implicitly suggest that characteristics beneficial to health *may* be demonstrated by appropriate scientific methods; the requirements for these analyses are outlined in paragraph 1.4 of Annex I, Section II.

This aspect of the Directive gave rise to some differences between Member States, in that some, particularly France, Italy, Spain and to some extent Germany translated this section as a *requirement* when drawing up national regulations – i.e. they incorporated the word 'must' in the second paragraph of the definition in Annex I; hence: 'These characteristics, which *must* give natural mineral water properties favourable to health . . .' This was clearly at odds with the English language version and with the desires of the UK, Ireland and Holland.

This matter was examined by the European Court of Justice in 1997, the Advocate General's final opinion being that the original discussions leading to the publication of the Directive in 1980 were such that it was not intended to make 'properties favourable to health' a compulsory requirement.

Where such physiological and medical studies are performed, however, the objective has traditionally been to establish by trials the effects of mineral intake on health and the particular benefits to be derived from specific waters. Such trials have been performed since the origin of spa treatments in the eighteenth century and continue, with the aid of modern medical methods. There is good and increasing evidence that waters rich in calcium, magnesium and sulphates can have measurable benefits and in Continental Europe, consumers purchase waters with these particular characteristics in mind.

In addition to the requirements for clinical and pharmacological analyses, Section II of Annex I prescribes the requirements for geological, hydrogeological, physical, chemical and physicochemical surveys and the criteria for microbiological analyses at source.

Derecognition of NMWs. If, for some reason, following the recognition process, a NMW is found to have fundamentally changed its character, its status as a NMW may be jeopardised, since stability and safety are prerequisites for recognition. In order to provide guidance on this, modifying Directive 96/70/EC includes a new Article 10a, which states:

> 1. Where a Member State has detailed grounds for considering that a natural mineral water does not comply with the provisions laid down in this Directive, or endangers public health, ... that Member State may temporarily restrict or suspend trade in that product within its territory. It shall immediately inform the Commission and other Member States thereof and give reasons for its decision.

There have been some instances in which waters that had originally been recognised as NMWs were subsequently found to have become polluted following extreme climatic changes or local contamination of the aquifer, and have been forced to cease operations pending remedial work. In more extreme cases, the NMW registration has been removed altogether. Additionally, the exploiter of the source has the right voluntarily to request derecognition of a source; this has been done in several instances by bottlers wishing to exploit a source for multiple purposes, including other non-NMWs.

3.2.1.2 Exploitation of Natural Mineral Waters

Having gained recognition for an NMW, it is imperative that it is abstracted and packaged in a way that does not pose a threat to its quality or change its natural characteristics. Maintenance of the absence of pollution can only be assured during exploitation by adherence to several strict principles.

In paragraph 2 of Annex II, the Directive makes a general prescription for the methods of abstraction, suitability of materials used for product contact and the actions to be taken for an NMW source proved to be polluted:

> 2. Equipment for exploiting the water must be so installed as to avoid any possibility of contamination and to preserve the properties, corresponding to those ascribed to it, which the water possesses at source.
>
> To this end, in particular:
>
> (a) the spring or outlet must be protected against the risks of pollution;
> (b) the catchment, pipes and reservoirs must be of materials suitable for water and so built as to prevent any chemical, physico-chemical or microbiological alteration of the water;
> (c) the conditions of exploitation, particularly the washing and bottling plant, must meet hygiene requirements. In particular, the containers must be so treated or manufactured as to avoid adverse effects on the microbiological and chemical characteristics of the natural mineral water;
> (d) the transport of natural mineral water in containers other than those authorized for distribution to the ultimate consumer is prohibited.

However, point (d) need not be applied to mineral waters exploited and marketed in the territory of a Member State if, in that Member State at the time of notification of this Directive, transport of the natural mineral water in tanks from the spring to the bottling plant was authorized.

3. Where it is found during exploitation that the natural mineral water is polluted and no longer presents the microbiological characteristics laid down in Article 5, the person exploiting the spring must forthwith suspend all operations, particularly the bottling process, until the cause of pollution is eradicated and the water complies with the provisions of Article 5.

4. The responsible authority in the country of origin shall carry out periodic checks to see whether:

   (a) the natural mineral water in respect of which exploitation of the spring has been authorized complies with Section I of Annex I;
   (b) the provisions of paragraphs 2 and 3 are being applied by the person exploiting the spring.

Point 2.(d) is a specific reference to the tankering of waters for bottling. The final comment on 15 July 1980, that this restriction need not be applied to waters already being tankered, was written to prevent a constraint on trade for those waters; however, it is not applicable to NMW whose exploitation commenced after that date. Hence, in those Member States where this allowance was incorporated into national regulations, there are still some NMWs that legitimately can be tankered and others that cannot.

The principle behind these general requirements is one of good manufacturing practice, bearing in mind the prohibition on any treatment calculated to improve or alter the quality of the water, as specified in Article 4.

Article 4

1. Natural mineral water, in its state at source, may not be the subject of any treatment other than:

   (a) the separation of its unstable elements, such as iron and sulphur compounds, by filtration or decanting, possibly preceded by oxygenation, in so far as this treatment does not alter the composition of the water as regards the essential constituents which give it its properties;
   (b) the separation of iron, manganese and sulphur compounds and arsenic from certain natural mineral waters by treatment with ozone-enriched air in so far as such treatment does not alter the composition of the water as regards the essential constituents which give it its properties, and provided that:

      • the treatment complies with the conditions of use to be laid down in accordance with the procedure laid down in Article 12 and following consultation of the Scientific Committee for Food . . . [whose role is to advise on matters likely to have an effect on public health]

- the treatment is notified to, and specifically controlled by, the competent authorities

(c) the separation of undesirable constituents other than those specified in (a) or (b), in so far as such treatment does not alter the composition of the water as regards the essential constituents which give it its properties, and provided that:

- the treatment complies with the conditions of use ... [as in (b) above]

(d) the total or partial elimination of free carbon dioxide by exclusively physical methods.

2. Natural mineral water, in its state at source, may not be the subject of any addition other than the introduction or the reintroduction of carbon dioxide under the conditions laid down in Annex I, Section III.

3. In particular, any disinfection treatment by whatever means and, subject to paragraph 2, the addition of bacteriostatic elements or any other treatment likely to change the viable colony count of the natural mineral water shall be prohibited.

Thus, a water can be subject to physical or mechanical filtration to remove 'unstable' elements, such as iron, manganese or sulphur compounds, which might otherwise precipitate in the bottle. The Directive also allows pre-oxygenation and the use of ozone-enriched air to reduce the solubility of such compounds (also including arsenic) to aid removal. This use of ozone-enriched air was introduced into the legislation with the amending Directive 96/70/EC. The method had been employed previously in some Member States, but had always been the cause of much discussion, as ozone is known to have bactericidal properties and hence its use seems to contravene the prohibition of disinfection treatments in paragraph 3 of Article 4. However, ozonation is permitted only if the exploiter can demonstrate the need for it and only when its use is approved by the enforcing authorities. In addition, details of its use must appear on the label (see Section 3.2.1.3).

With regard to the removal of the other undesirable constituents, as specified in 2003/40/EC, any methods other than the use of ozone have not yet been identified in the Directives, but, as in the case of ozone-enriched air, the use of any proposed methods must be justified and approved at the European Community level, and again, details must appear on the label.

3.2.1.3 Labelling of Natural Mineral Waters
There are six key requirements for the labelling:

(1) The *Commercial Designation* must include the name of the source or the place of exploitation in letters at least one and a half times larger than any other part of the commercial designation.

(2) The *Description* must be one of Natural Mineral Water, or as defined in Annex I, Section III, in the case of *effervescent (carbonated)* natural mineral waters:

SECTION III – SUPPLEMENTARY QUALIFICATIONS RELATING TO EFFERVESCENT NATURAL MINERAL WATERS

At source or after bottling, effervescent natural mineral waters give off carbon dioxide spontaneously and in a clearly visible manner under normal conditions of temperature and pressure. They fall into three categories to which the following descriptions respectively shall apply:

(a) 'naturally carbonated natural mineral water' means water whose content of carbon dioxide from the spring after decanting, if any, and bottling is the same as at source, taking into account where appropriate the reintroduction of a quantity of carbon dioxide from the same water table or deposit equivalent to that released in the course of those operations and subject to the usual technical tolerances;

(b) 'natural mineral water fortified with gas from the spring' means water whose content of carbon dioxide from the water table or deposit after decanting, if any, and bottling is greater than that established at source;

(c) 'carbonated natural mineral water' means water to which has been added carbon dioxide of an origin other than the water table or deposit from which the water comes.

Part (a) simply means that the water contains only naturally produced carbon dioxide from the original source and at the original 'natural' concentration. Part (b) means that the carbon dioxide originates at the source, but may have been tapped off and recombined with the water at a higher concentration than that naturally occurring. Part (c) refers to a still water to which artificially manufactured carbon dioxide has been added.

(3) *Composition* must be indicated by means of a list of the major components in accordance with the officially recognised analysis, expressing their concentration in mg/l. This list will normally include the major anions and cations, together with the 'dry residue' – sometimes also referred to as the total dissolved solids (TDS).

(4) *Information on any treatments* using ozone-enriched air for removal of iron, manganese, sulphur or arsenic, or any other treatments designed to remove undesirable constituents.

(5) It is not permitted to sell an NMW which is not marked or labelled in accordance with the regulations. No water other than an NMW can be labelled in such a way as to cause confusion with an NMW.

(6) No therapeutic claims attributing properties relating to the prevention, treatment or cure of human disease can be made.

However, Annex III does permit the use of 'indications' on the label, provided that they are based on '... physico-chemical analyses and, where necessary, pharmacological, physiological and clinical examinations carried out according to recognized scientific methods in accordance with Section I, paragraph 2 of Annex I (reproduced in 3.2.1 above).

ANNEX III

INDICATIONS AND CRITERIA LAID DOWN IN ARTICLE 9

Indications	Criteria
Low mineral content	Mineral salt content, calculated as a fixed residue, not greater than 500 mg/l
Very low mineral content	Mineral salt content, calculated as a fixed residue, not greater than 50 mg/l
Rich in mineral salts	Mineral salt content, calculated as a fixed residue, greater than 1500 mg/l
Contains bicarbonate	Bicarbonate content greater than 600 mg/l
Contains sulphate	Sulphate content greater than 200 mg/l
Contains chloride	Chloride content greater than 200 mg/l
Contains calcium	Calcium content greater than 150 mg/l
Contains magnesium	Magnesium content greater than 50 mg/l
Contains fluoride	Fluoride content greater than 1 mg/l
Contains iron	Bivalent iron content greater than 1 mg/l
Acidic	Free carbon dioxide content greater than 250 mg/l
Contains sodium	Sodium content greater than 200 mg/l
Suitable for the preparation of infant food	—
Suitable for a low-sodium diet	Sodium content less than 20 mg/l
May be laxative	—
May be diuretic	—

Directive 80/777/EC was written in 1980 to provide general guidance to Member States for incorporation in national regulations, to cover a market sector that had hitherto been regulated (if at all) in a sporadic and uncoordinated manner. Following the amendments in 1996 and 2003, there is now improved clarity of the intent and a consistency in the regulation of NMWs within the EU. There are of course some unresolved issues, most notably that concerning the allowable treatments for the removal of naturally occurring undesirable elements. The fact that there are three Directives also presents some difficulties for Member States implementing the changes into their own national regulations, but there are at present no plans to combine the Directives.

3.2.2 Spring Water

For many years, SWs (and other bottled non-NMWs) were not subject to their own regulation, had no official definition and were thus required simply to meet food safety standards; in practice, this meant that they had only to comply with parametric standards laid down in Council Directive 80/778/EEC, relating to the quality of water intended for human consumption (referred to as the Drinking Water Directive). This was written primarily to prescribe the standards for tap waters, but in some European states, it was for many years the only legislation applicable to bottled waters other than NMW. Since that time, however, two key changes have occurred.

(1) The term 'Spring' water was for the first time incorporated in Article 4 of the NMW Directive, through the modifying Directive 96/70/EC. Thus, many of the strictures that previously only applied to NMWs were also applied to SWs.

(2) Directive 80/778/EEC on the quality of water intended for human consumption has itself been revoked and replaced with a new Directive 98/83/EC that contains a modified list of microbiological and chemical standards, of which the latter remain fully applicable to SWs.

In summary, the main requirements for SW are:

(1) A single name for a single source.

(2) Unlike an NMW, no 'source recognition' procedure is required. This is one of the most important differences between NMW and SW.

(3) Unlike an NMW, SW is not required to have a stable composition.

(4) Naturally wholesome; no treatments permitted other than those for an NMW. However, according to paragraph 4b of Article 9, in the absence of Community provisions on the treatment of SWs, Member States may maintain national provisions for treatments. In practice, this means that in some countries, other treatments, such as ion exchange for removal of nitrates, are still permitted.

(5) Must meet the conditions of exploitation laid down in paragraphs 2 and 3 of Annex II of the NMW Directive, (see above); in particular, source protection, suitable materials, methods and hygienic practices so as to prevent any alteration of the water.

(6) Must be bottled at source (although see notes following Table 3.2).

(7) SW may not be labelled as having any health benefits.

(8) As a minimum, must meet the microbiological standards applicable to NMWs and the chemical standards for drinking waters as specified in the three Annexes to Directive 98/83/EC. Annex I details the parameters to be monitored, divided between (A) microbiological parameters, including those specific to bottled waters; (B) chemical parameters, based on health risks; and (C) indicator parameters, which are not necessarily health-related but give warning of a deteriorating condition. These lists include additional notes explaining the bases for the values and any developments pending. A simplified version of the tables in Directive

98/83/EC is reproduced in Table 3.1. Annexes II and III deal with monitoring and analytical methods respectively.

3.2.3   Other bottled waters in Europe

For other waters, packaged as Drinking Water under various descriptions such as table water, there are no requirements for source recognition, and it cannot be labelled in any way that allows confusion with NMWs or SWs. A fixed composition is not mandatory and they can be taken from any source and transported by tanker prior to bottling. Any treatment is permissible that ensures compliance with Directive 98/83/EC on the quality of water intended for human consumption. Labelling of such waters is regulated at national level – no harmonisation between Member States currently exists.

3.2.4   Implementation of the Directives in Europe

The previous sections examined the overall European position, with specific reference to the Directives that are binding in principle for all Member States, but which are left to the individual States to implement within their own regulations. Note that a 'Directive' is an instruction to Member States to incorporate the agreed legislation into their own national regulations, in order to harmonise legislation within the community. However, there are different types of Directives. Sometimes they just define the goal the Members States have to reach, and Member States remain free to decide how to reach the targets fixed by the EU. In other cases they give very prescriptive instructions and it is virtually impossible for Member States to deviate from the European document. Directive 80/777/EC and its modifying Directives belong to the latter category, meaning that the freedom of Member States is limited and that they are expected to respect the letter and spirit of all articles.

Table 3.1 Simplified list of parameters from Annex I of Directive 98/83/EC relating to the Quality of water indended for human consumption (official Journal No L 330/32)

Part A: Microbiological parameters

Parameter	Parametric value
Escherichia coli (E. coli)	0/100 ml
Enterococci	0/100 ml
The following applies to water offered for sale in bottles or containers	
Escherichia coli (E. coli)	0/250 ml
Enterococci	0/250 ml
Pseudomonas aeruginosa	0/250 ml
Colony count 22°C	100/ml
Colony count 37°C	20/ml

Part B: Chemical parameters

Parameter	Parametric value (μg/l)
Acrylamide	0.1
Antimony	5
Arsenic	10
Benzene	1
Benzo(a)pyrene	0.01
Boron	1
Bromate	10
Cadmium	5
Chromium	50
Copper	2
Cyanide	50
1,2-dichloroethane	3
Epichlorhydrin	0.1
Fluoride	1.5
Lead	10
Mercury	1
Nickel	20
Nitrate	50*
Nitrite	0.5*
Pesticides	0.1
Pesticides - Total	0.5
Polycyclic aromatic hydrocarbons	0.1
Selenium	10
Tetrachlorethene and Trichlorethene	10
Trihalomethanes - Total	100
Vinyl chloride	0.5

* mg/l

Part C: Indicator parameters

Parameter	Parametric value
Aluminium	200 μg/l
Ammonium	0.5 mg/l
Chloride	250 mg/l
Clostridium perfringens (including spores)	0/100 ml
Colour	Acceptable to consumers and no abnormal change
Conductivity at 20°C	2500 μS cm^{-1}
Hydrogen ion concentration	≥ 6.5 and ≤ 9.5 pH units
Iron	200 μg/l
Manganese	50 μg/l
Odour	Acceptable to consumers and no abnormal change
Oxidizability	5 mg/lO$_2$
Sulphate	250 mg/l
Taste	Acceptable to consumers and no abnormal change
Colony count at 22°C	No abnormal change
Coliforms	0/100 ml
Total organic carbon (TOC)	No abnormal change
Turbidity	Acceptable to consumers and no abnormal change

RADIOACTIVITY

Parameter	Parametric value
Tritium	100 Bq/l
Total indicative dose	0.10 mSv/year

Sometimes, however, Member States do not respect harmonisation directives (and the European Commission is at liberty to launch an infringement procedure against such a country), but such infringements are usually applicable only to those that do not export to other countries. Additionally, the translation and interpretation process sometimes leads to minor variation between the different versions, influenced also by historical practice, cultural differences and the state of the market. Table 3.2 summarises the key requirements for NMW, SW and drinking water, and the text that follows highlights some differences in application within Europe.

Table 3.2 Comparing NMWs, SWs and other drinking waters in the European Union

	NMWs	SWs	Other drinking waters, including tap water
Must demonstrate source protection	Yes	No	No
Source recognition	Yes	No	No
Must be bottled at source	Yes	Yes	No
Source must be specified on the label	Yes	No	No
One brand-named water comes only from a single source	Yes	Yes	No
Must have a constant composition	Yes	No	No
Must specify mineral content	Yes	No[a]	No
Must be wholesome in its untreated state	Yes	Yes	No
Treatment permitted for removal of unstable constituents	Yes	Yes	Yes
Treatment permitted for removal of undesirable substances	Yes	Yes	Yes
Treatment permitted for removal of pathogenic micro-organisms	No	No	Yes
Safe to drink	Yes	Yes	Yes

[a] Although, as there is no prohibition for SWs, a 'typical' composition is sometimes included.
Source: Based on a table supplied by the Natural Mineral Water Information Service, UK.

A full list of the national regulations implementing the NMW, SW and drinking water Directives in the Member States appears in Table 3.3.

Table 3.3 List of regulations implementing European Directives in EU Member States

Council Directive 98/83/EC of 3 November 1998 on the quality of water intended for human consumption

Belgium	Arrêté du 24 janvier 2002 du Gouvernement de la Région de Bruxelles-Capitale relatif à la qualité de l'eau distribuée par réseau ref: MB du 21/02/2002, page 6600 Arrêté royal du 14 janvier 2002 relatif à la qualité des eaux destinées à la consommation humaine qui sont conditionnées ou qui sont utilisées dans les établissements alimentaires pour la fabrication et/ou la mise dans le commerce de denrées alimentaires ref: MB du 19/03/2002, page 11443–11458 Decreet van 24.05.2002 (Vlaams Parlement) betreffende water bestemd voor menselijke aanwending ref: MB du 23.07.2002 p.32838 (C- 2002/35862) Besluit van de Vlaamse regering van 13.12.2002 houdende reglementering inzake de kwaliteit en levering van water, bestemd voor menselijke consumptie ref: MB du 28/01/2003, Ed 2 p. 2907 Décret du 12/12/2002 relatif à la qualité de l'eau destiné à la consommation humaine. ref: MB du 14/01/2003 p. 1135 (C - 2003/27001) (A/2003/04937 du 22/05/2003)

Denmark	Bekendtgorelse om vandkvalitet og tilsyn med vandforsyningsanlaeg ref: Vandkvalitetdavidakt, 21/09/2001 (entrée en vigueur 15/10/2001
Germany	Verordnung zur Novellierung der Trinkwasserverordnung vom 21. Mai 2001 ref: BGBl. Teil I Nr. 24 vom 28/05/2001 page 959
Greece	Décision ministérielle commune Y2/2600/2001 ref: FEK B n° 892 du 11/07/2001, page 10865
	Instrument légal Y4a/9019/2001 ref: FEK B n° 1082 du 14/08/2001, page 14987
Spain	Real Decreto 1074/2002, de 18 de octubre, por el que se regula el proceso de elaboración, circulación y comercio de aguas de bebida envasadas. ref: BOE n° 259 de 29/10/2002, p. 37.934 – 37.949. (SG(02) A/10990 du 08/11/02). Real Decreto 140/2003, de 7 de febrero, por el que se establecen los criterios sanitarios de la calidad del agua de consumo humano. ref: BOE n° 45 de 21/02/2003 p. 7228 – 7245. (SG(03) A/2242 du 27/02/2003). Corrección de erratas del Real Decreto 140/2003 por el que se establecen los criterios sanitarios de la calidad del agua de consumo humano. ref: BOE n° 54 de 04/03/2003 p. 8469. (SG(03) A/4426 du 06/05/2003).
France	Code de la Santé Publique, annexes au décret n°2003-461 du 21 Mai 2003 et n°2003-462 du 21 Mai 2003. Décret n° 2001-1220 du 20 décembre 2001 relatif aux eaux destinées à la consommation humaine, à l'exclusion des eaux minérales naturelles ref: JORF du 22 décembre 2001, page 20381 -2001 Art.4-I, Art. 50 et 51
Ireland	European Communities (Drinking Water) Regulations, 2000 ref: S.I. n° 439 of 2000
Italy	DECRETO LEGISLATIVO 2 Febbraio 2001, n.31 Attuazione della Direttiva 98/83/CE relativa alla qualità delle acque destinate al consumo umano Supplemento ordinario alla "Gazzetta Ufficiale" n° 52 del 3 Marzo 2001 - Serie Generale DECRETO LEGISLATIVO 2 Febbraio 2002, n° 27 Modifiche ed Integrazioni al D. Lgs. 2 febbraio 2001, n°31, recante attuazione della Direttiva 98/83/CE relativa alla qualità delle acque destinate al consumo umano Pubblicato nella Gazzetta Ufficiale 9 Marzo 2002, n. 58
Luxembourg	Règlement grand-ducal du 7 octobre 2002 relatif à la qualité des eaux destinées à la consommation humaine ref: Mémorial n° 115 du 11/10/2002 p. 2816
Netherlands	Besluit van 31 juli 1998, houdende regels voor bronwater, natuurlijk mineraalwater en andere verpakte waters (Warenwetbesluit Verpakte waters)
	Besluit van 23 september 1999, houdende wijziging van het Warenwetbesluit Bereiding en behandeling van levensmiddelen en van het Warenwetbesluit Verpakte Waters – ref: Staatsblad nr 429 van 1999 van 29/06/1999 – 02. Regeling materialen en chemicaliën leidingwatervoorziening – ref: Staatscourant n° 241 du 13/12/2002 p. 25 (SG(2003)A/02109 du 25/02/2003) – –

Austria

Verordnung des Bundesministers für sociale Sicherheit und Generationen über die Qualität von Wasser für den menschlichen Gebrauch (Trinkwasserverordnung-TWV) ref: BGBl. Teil II Nr. 304/2001, 21/08/2001 page 1805
Verordnung der Landesregierung über eine Änderung der Bautechnikverordnung ref: LGBI n° 64/2001 (Vorarlberg) du 27/12/2001, Seite 223

Portugal

Decreto-Lei n° 243/2001, de 5 de Setembro ref: Diário da República I série A n° 206 du 05/09/2001 page 5754

Finland

Laki terveydensuojelulain muuttamisesta (441/2000) (19/05/2000) ref: SSK n° 441/2000
Sosiaali- ja terveysministeriön asetus talousveden laatuvaatimuksista ja valvontatutkimuksista (461/2000) (19/05/2000) ref: SSK n° 461/2000
Terveydensuojelulaki (763/1994) 20 § (19/08/1994) ref: SSK n° 763/1994
Alands landskapsstyrelses beslut om ändring av Alands landskapsstyrelses beslut angaende tillämpning i landskapet Aland av vissa riksförfattningar om hushallsvatten ref: Alands Författningssamling 2001, nr 3
Landskapslag om hälsovarden ref: Alands Författningssamling 1967, page 119, nr 36
Landskapslag om ändring av landskapslagen om hälsovarden ref: Alands Författningssamling 1997, nr 99, page 269
Sosiaali- ja terveysministeriön asetus talousveden valvontatutkimuksia tekevistä laboratorioista ref: Suomen Säädöskoelma n° 173/2001
Ålands landskapsstyrelses beslut om ändring av Ålands lanskapsstyrelses beslut angående tillämpning i landskapet Åland av vissa riksförfattningar om hushållsvatten (18/01/2001) ref: ÅFS n° 3/2001

Sweden

Statens livsmedelsverks föreskrifter om dricksvatten ref: SLVFS 2001:30 du 13/12/2001

United Kingdom

Water, England and Wales 2000 N° 3184, The Water Supply (Water Quality) Regulations 2000 ref: S.I. n° 3184 of 2000
Water Supply : The Water Supply (Water Quality) (Scotland) Regulations 2001 ref: S.I. n° 207 of 2001, 04/06/2001
Bill for an ordinance to amend the Public Health Ordinance in order to provide for the transposition into the law of Gibraltar Council Directive 98/83/EC of 3 November 1998 on the quality of water intended for human consumption ref: Third supplement to the Gibraltar Gazette, N° 3219 of 19/04/2001
Public Health Ordinance (Amendment) Ordinance n° 19 of 2001 (07/06/2001) ref: First supplement to the Gibraltar Gazette n° 3,227 date 07/06/2001 n° de page 117
Public Health (Potable water) rules 1994 (Amendment)Rules 2001 - Legal notice n° 61 of 2001 (07/06/2001) ref: Second supplement to the Gibraltar Gazette n° 3227 of 07/06/2001, page 201
The Water supply (water quality) (Amendment) Regulation 2001 - Water, England and Wales (05/09/2001) ref: S.I. n° 2885 of 2001
The Water Supply (Water Quality) Regulations 2001- Wales ref: S.I. n° 3911 (W.323) of 2001 (in force 1/01/2002)
The Water Supply (Water Quality) Regulations (Northern Ireland) 2002 ref: S.R. n° 331 in force 28/11/2002 (SG(2003)A/00074 du 09/01/2003)

Council Directive 80/777/EEC of 15 July 1980 on the approximation of the laws of the Member States relating to the exploitation and marketing of natural mineral waters

Belgium	arrêté royal du 8 février 1999 concernant les eaux minérales naturelles et les eaux de source.
Denmark	Miljøministeriets bekendtgørelse nr. 463, Ministerialtidende af 05/09/1984
Germany	Verordnung über natürliches Mineralwasser, Quellwasser and Tafelwasser (Mineral- und Tafelwasser-Verordnung) vom 01/08/1984 ref: Bundesgesetzblatt Teil I vom 02/08/1984 Seite 1036 i.d.f v. 06. August 2002 (BGBI I Nr 57 vom 14. August 2002, S.3082, 3099)
Greece	Décision ministérielle n° 433/83 ref: FEK n° 163 du 09/11/1983
Spain	Real Decreto n° 2119/81 de 24/06/1981 por el que se aprueba la Reglamentación Técnico-Sanitaria para la elaboración, circulación y comercio de aguas de bebida envasadas ref: BOE 226 de 21/09/1981. Real Decreto n° 1164/91 de 22/07/1991, por el que se aprueba la Reglamentación Técnico-Sanitaria para la elaboración, circulación y comercio de aguas de bebida envasadas ref: BOE n° 178 de 26/07/1991 Página 24818 (Marginal 19201) Real Decreto n° 781/1998 de 30/03/1998 que modifica el Real Decreto 1164/91 de 22/07/1991 que aprueba la Reglamentación Técnico-Sanitaria para la elaboración, circulación y comercio de aguas de bebida envasadas.
France	Code de la Santé Publique, annexes au décret n°2003-461 du 21 Mai 2003 et n°2003-462 du 21 Mai 2003.
Ireland	European Communities (Natural Mineral Water) Regulations, 1986 ref: S.I. n° 11 of 1986
Italy	Decreto ministeriale del 01/03/1982, nuove norme per le etichette delle acque minerali ref: GURI - Serie generale - del 10/02/1983 n. 40 pag. 1099 Decreto legislativo del 25/01/1992 n. 105, attuazione della direttiva n. 80/777/CEE relativa alla utilizzazione e alla commercializzazione delle acque minerali naturali ref: Supplemento ordinario n. 31 alla GURI - Serie generale - del 17/02/1992 n. 39
	Art. 8 e 9 della legge del 1° marzo 2002, n° 39 ref: Supplemento ordinario n° 54/L alla GURI - Serie generale - n° 72 del 26/03/2002, page 8
Luxembourg	Règlement grand-ducal du 08/10/1983 concernant l'exploitation et la mise dans le commerce des eaux minérales naturelles, Mémorial Grand-Ducal A Numéro 88 du 27/10/1983 Page 2001
Netherlands	Besluit van 31 juli 1998, houdende regels voor bronwater, natuurlijk mineraalwater en andere verpakte waters (Warenwetbesluit Verpakte waters)
Austria	Verordnung des Bundesministers für Gesundheit, Sport und Konsumentenschutz über natürliche Mineralwässer (Mineralwasserverordnung), Bundesgesetzblatt für die Republik Österreich, Nr. 552/1994

Gesetz vom 26/06/1997 mit dem das Heilvorkommen- und Kurortegesetz geändert wird, Landesgesetzblatt für Kärnten, Nr. 104/1997 herausgegeben am 28/10/1997 03. Verordnung der Bundesministerin für Frauenangelegenheiten und Verbraucherschutz über natürliche Mineralwässer und Quellwässer. ref: BGBl, 09/09/1999, nr 309. SG(1999)A/13707

Portugal   Decreto-Lei n. 283/91 de 09/08/1991. Transpõe para a ordem jurídica interna a Directiva n. 80/777/CEE, do Conselho, de 15 de Julho, relativa à exploração e à comercialização de águas naturais, Diário da República I Série A n. 182 de 09/08/1991 Página 4012

Decreto Regulamentar n. 18/92 de 13/08/1992. Aprova o regulamento sobre águas minerais naturais, Diário da República I Série B n. 186 de 13/08/1992 Página 3941

Portaria n. 703/96 de 06/12/1996. Define as regras técnicas relativas às respectivas denominações, definições, acondicionamento e rotulagem das bebidas refrigerantes, Diário da República I Série B n. 282 de 06/12/1996 Página 4387

Decreto Regulamentar n. 8/97 de 18/04/1997. Altera algunas disposições do Decreto Regulamentar n. 18/92, de 13 de Agosto, em matéria de rotulagem das águas minerais, Diário da República I Série B n. 91 de 18/04/1997 Página 1746

Finland   Kauppa- ja teollisuusministeriön päätös eräiden elintarvikkeita koskevien Euroopan yhteisöjen direktiivien täytäntöönpanosta (1312/93) 20/12/1993

Landskapslag om tillämpning i landskapet Åland av riksförfattningar om livsmedel (43/77) 23/05/1977

Landskapsförordning om tillämpning i landskapet Åland av riksförfattningar om livsmedel (63/95) 27/07/1995, ändring (41/96) 14/05/1996

Sweden   Statens livsmedelsverks kungörelse med föreskrifter och allmänna råd om naturligt mineralvatten, Statens livsmedelsverks författningssamling (SLVFS) 1993:29

Livsmedelslag, Svensk författningssamling (SFS) 1971:511

United Kingdom   The Natural Mineral Water, Spring Water and Bottled Drinking Water Regulations 1999 Statutory Instrument 1999 No. 1540

The Natural Mineral Water, Spring Water and Bottled Drinking Water (Amendment) Regulations 2003 -Statutory Instrument 2003 No. 666 for England/ Statutory Rule 2003 No. 182 for Northern Ireland / Scottish Statutory Instrument 2003 No. 139 for Scotland

---

Directive 96/70/EC of the European Parliament and of the Council of 28 October 1996 amending Council Directive 80/777/EEC on the approximation of the laws of the Member States relating to the exploitation and marketing of natural mineral waters

---

Belgium   arrêté royal du 8 février 1999 concernant les eaux minérales naturelles et les eaux de source.

Denmark	Bekendtgørelse nr. 67 af 30/01/1998 om naturligt mineralvand og kildevand
Germany	Verordnung über natürliches Mineralwasser, Quellwasser and Tafelwasser (Mineral- und Tafelwasser-Verordnung) vom 01/08/1984 ref: Bundesgesetzblatt Teil I vom 02/08/1984 Seite 1036 i.d.f v. 06. August 2002 (BGBI I Nr 57 vom 14. August 2002, S.3082, 3099)
Greece	Décision ministérielle numéro Y2/OIK.329 du 02/02/1998, FEK B numéro 114 du 12/02/1998 Page 1128
Spain	Real Decreto número 781/98, de 30 de abril, por el que se modifica el Real Decreto 1164/91, de 22 de julio, por el que se aprueba la reglamentación técnico-sanitaria para la elaboración, circulación y comercio de aguas de bebidas envasadas, Boletín Oficial del Estado número 121 de 21/05/1998 Página 16808 (Marginal 11790)
France	Code de la Santé Publique, annexes au décret n°2003-461 du 21 Mai 2003 et n°2003-462 du 21 Mai 2003. Décret n° 2001-1220 du 20 décembre 2001 relatif aux eaux destinées à la consommation humaine, à l'exclusion des eaux minérales naturelles ref: JORF du 22 décembre 2001, page 20381 -2001 Art.4-I, Art. 50 et 51
Ireland	European Communities (Natural Mineral Waters) (Amendment) Regulations, 1998, Statutory Instruments number 461 of 1998
Italy	Decreto legislativo del 25/01/1992 n. 105, attuazione della direttiva n. 80/777/CEE relativa alla utilizzazione e alla commercializzazione delle acque minerali naturali, Supplemento ordinario n. 31 alla Gazzetta Ufficiale - Serie generale - del 17/02/1992 n. 39
Luxembourg	Règlement grand-ducal du 24/05/1998 concernant l'exploitation et la mise dans le commerce des eaux minérales naturelles – ref : Mémorial A Page 6
Netherlands	Besluit van 31 juli 1998, houdende regels voor bronwater, natuurlijk mineraalwater en andere verpakte waters (Warenwetbesluit Verpakte waters)
Austria	Verordnung der Bundesministerin für Frauenangelegenheiten und Verbraucherschutz über natürliche Mineralwässer und Quellwässer
Portugal	Portaria n. 156/98 de 06/06/1998, Diário da República I Série A n. 131 de 06/06/1998 Página 2593
Finland	Kauppa- ja teollisuusministeriön päätös eräiden elintarvikkeita koskevien Euroopan yhteisöjen direktiivien täytäntöönpanosta annetun kauppa- ja teollisuusministeriön päätöksen muuttamisesta (935/97) 15/10/1997, Suomen säädöskokoelma 22/10/1997 Landskapslag om tillämpning i landskapet Åland av riksförfattningar om livsmedel (43/77) 23/05/1977, Ålands författningssamling Landskapsförordning om tillämpning i landskapet Åland av riksförfattningar om livsmedel (63/95) 27/07/1995, ändringar (36/97) 27/03/1997, (50/97) 05/06/1997, (67/97) 21/08/1997, (71/97) 18/09/1997 och (89/97) 30/10/1997

Sweden	Statens livsmedelsverks kungörelse om ändring i kungörelsen (SLV FS 1993:29) med föreskrifter och allmänna råd om naturligt mineralvatten, Statens livsmedelsverks författningssamling (SLVFS) 1997:35
United Kingdom	The Natural Mineral Water, Spring Water and Bottled Drinking Water Regulations 1999 Statutory Instrument 1999 No. 1540
	The Natural Mineral Water, Spring Water and Bottled Drinking Water (Amendment) Regulations 2003 -Statutory Instrument 2003 No. 666 for England/ Statutory Rule 2003 No. 182 for Northern Ireland / Scottish Statutory Instrument 2003 No. 139 for Scotland

With regard to implementation in the Member states, the following differences and requirements apply.

FRANCE

(1) *Eaux Minérales Naturelles* (NMW). In addition to general compliance with the Directives, it is also required that the compatibility of the container material with the water is confirmed by the Ministry of Health.
(2) *Eaux de Source* (SW or source waters). These must comply with the same basic requirements (naturally wholesome without treatment, bottled at source, etc.).
(3) *Eaux rendues potable par traitement* (waters made potable by treatment). Here, any treatment is permitted, provided that the final product complies with 98/83/EC. The treatment received must be indicated on the label.

GERMANY

(1) *NMW*. Regulated according to the NMW Directives.
(2) *Quellwasser* (SW). This is naturally wholesome without treatment other than those permitted for NMW. Its composition may vary. Limits for substances are based upon Directive 98/83/EC.
(3) *Tafelwasser* (table water). This is a manufactured water, made with NMW or drinking water and authorised *added* minerals.
(4) *Trinkwasser* (drinking water). This may be treated and must meet the quality requirements in Directive 80/778/EEC.

THE NETHERLANDS

(1) *NMW*.
(2) *SW*.

No restrictions on tankering, so bottling away from source is permitted.

UK

(1) *NMW*.
There is at present no formal mechanism for establishing or recognising pharmacological benefits.
The UK has adopted the derogation permitting the tankering of NMWs

that were being tankered for bottling prior to the date of the publication of the NMW 1985 regulations.

(2) *SWs and other bottled drinking waters.*

The terms 'natural water', 'natural spring water' and 'table water' are presently allowed as descriptions. However, these must not be confused with NMWs.

The derogation applicable to NMWs tankered prior to the publication of the 1985 regulations has also been applied to SWs tankered before the publication of 96/70/EC.

In the UK, an SW may at present also be bottled as a non-SW under any number of names.

3.2.5   Future developments in Europe

Directive 80/777/EEC contains in Annex III a reference to the indication 'Suitable for the preparation of infant food'. This has to some extent been implemented in national regulations in mainland Europe, but has yet to be permitted in the UK, where the Food Standards Agency has only recently begun the consultation process with other regulatory agencies and the industry on how such an indication could be permitted and enforced.

For some years now, there has been some growth in a category of drink frequently referred to as 'water', which is actually very often in effect a soft drink (e.g. 'XXX natural mineral water with a touch of fruit. . . . '), containing vitamins and sometimes additional minerals, but also all the classical additional ingredients of a soft drink: sugar, flavours, citric acid and so on.

Preparatory work is also under way on a new Directive governing limits on radioactivity in NMWs and SWs, though there is as yet no timetable for implementation.

3.3   North America

3.3.1   United States

From the middle of the nineteenth century, spas developed in many areas of the USA where good supplies of wholesome groundwater (often in the form of spectacular hot springs) were available. Unlike in mainland Europe, however, the spas declined during the first half of the twentieth century and, though some of the more famous brands continued to be bottled, the market became polarised between the rare exotic imported waters from Europe and the ubiquitous larger glass (later polycarbonate) bottles used to supply water coolers for domestic and business premises – the so-called 'home and office' market.

In the last quarter of the twentieth century, a move to a healthier lifestyle made bottled waters more popular, and the market, prompted by the success of imported waters, grew steadily, encouraging the exploitation of the many local or regional springs that dominate the market today. In parallel with the ongoing growth of SW, other bottled waters have also found a market. In the latter case, emphasis has generally been on the compliance of the water to parametric standards, rather than the European 'non-treated' approach. Consequently a wide variety of different bottled water types developed, with different methods and levels of treatment and to some extent different nomenclature.

The regulation of water for human consumption in the USA falls under the jurisdiction of two agencies:

(1) Water supplied through private and public treatment and distribution systems (tap water) is regulated by the Environmental Protection Agency (EPA), which is also responsible for establishing minimum standards of quality.
(2) Bottled water is regulated at the Federal level, but the individual states and the industry itself also have a role to play.

3.3.1.1 Federal Regulation

The US Food and Drug Administration (FDA), through the Federal Food, Drug and Cosmetic Act has overall responsibility for setting quality criteria, establishing labelling standards ('standards of identity') and specifying good manufacturing practices (GMPs) for all bottled waters (which are legally considered a food). Since the early 1970s, the FDA, in cooperation with the American Bottled Water Association (later to become the International Bottled Water Association, IBWA), has promulgated operating codes and quality standards for bottled waters. This culminated in a 'Final Rule on a Standard of Identity for Bottled Waters', published in the Federal Register, Vol. 60, No. 218 on 13 November 1995, which not only provided definitions for many different types of bottled waters, but also finalised standards of quality. The FDA's standards of identity are contained in Part 165.110(B) of Title 21 Code of Federal Regulations 2002 (CFR 21) and the requirements for processing and bottling are specified in Part 129. The FDA Final Rule also confirms the principle of pre-emption. This ensures that the FDA's rules take precedence in the case of interstate trade, in that any bottled water crossing state boundaries for sale must comply with Federal requirements as a minimum. On the other hand, in states where the standards of quality are higher than the Federal regulation, the state requirements also apply to imported waters.

3.3.1.2 State Regulation

The individual states have responsibility for sampling, analysis, source permitting and (where appropriate) certification of factory laboratories. They are also responsible for inspecting sources and bottling premises to monitor compliance. Additionally, anyone wishing to import bottled water from another state must obtain a permit to do so, based on analyses for compliance. States can also write regulations that are more rigorous than those of the FDA, and these in effect become applicable to any bottled water sold within that state. States are also at liberty to impose additional requirements on those wishing to bottle water within the state; for example, in Texas, it is a requirement that all water transported for bottling (whether by tanker or even through a pipeline directly to the bottling plant) must be chlorinated during distribution, and the chlorine removed prior to bottling.

3.3.1.3 Industry Regulation
At the Industry level, the IBWA has worked with the FDA in developing a Model Bottled Water Regulation (also known as the Model Code). This provides specific guidance to bottlers on legal requirements, GMPs, quality standards, monitoring procedures and labelling requirements. All members of the IBWA are expected to abide by the Code, and membership is contingent upon bottlers submitting to unannounced third-party auditing of sources and bottling premises.

The definitions which appear in CFR Part 165.110 (B) are summarised below.

SW (natural spring water)
- Water derived from an underground formation from which water flows naturally to the surface of the earth.
- ... collected only at the spring or through a borehole tapping the underground formation feeding the spring.
- Natural flow to the surface (even if tapped via a borehole).
- There must be a measurable hydraulic connection between any borehole and the natural outlet or spring, and the quality and composition of the water from the borehole must be the same as that from the spring.
- Water must continue to flow at the spring during exploitation.
- Location of the spring must be identified (though not necessarily on the label).
- Water must be naturally wholesome; at source and after bottling it must meet the compositional and quality standards laid down in CFR 21 (see below). Treatments are limited to those that control the hygienic condition of the water during distribution and packaging. Thus, microfiltration, UV treatment and ozonation are permitted, but other treatments that may alter the composition are not.
- Sources (springs, and if used, boreholes) must be approved and licensed by State regulators.

Natural spring water has long been the major category of bottled water in the USA. For many years, uncertainty about the quality of public water supplies and a rigorous control of the integrity of sources used for SW has ensured a high level of consumer loyalty, even in the face of competition from increasingly processed waters from other sources.

Artesian water
- Water from a well, tapping a confined aquifer in which the water level stands at some height above the top of the aquifer.
- ...may be collected with the assistance of external force to enhance the natural underground pressure. [The external force may not draw the level down below the level of the top of the aquifer, or change its natural state.]

Mineral water
- Water containing not less than 250 mg/litre total dissolved solids. [Consumer expectation is that minerals will be present.]
- ...from one or more boreholes or springs originating from a geologically and physically protected underground water source.
- Constant level and relative proportions of minerals and trace elements at the point of emergence, within natural fluctuations.
- Labelling must include the following: in the case of a water with a TDS of < 500 mg/litre – 'low mineral content'; in the case of a water with a TDS of > 1500 mg/litre – 'high mineral content'

Mineral waters are also exempt from the allowable levels for colour, odour, TDS, chlorine, iron, manganese, sulphate and zinc.

Sparkling bottled water
- ...water, that after treatment and possible replacement with carbon dioxide, contains the same amount of carbon dioxide that it had at emergence from source. [This definition is equivalent to that for a naturally carbonated NMW; however, waters containing artificial carbon dioxide, which were described previously as 'sparkling', can continue to be so described.]

Purified water
- Water that has been produced by distillation, de-ionization, reverse osmosis, or other suitable processes.[This meets the definition of the term 'purified water' in the most recent edition of the United States Pharmacopoeia, 23rd revision, 1 Jan, 1995. The terms 'demineralized', 'distilled' and 'reverse osmosis water' may also be used.]

This category of bottled water has grown significantly in recent years, resulting from the relative ease with which municipal waters can be taken and treated in any number of ways to achieve a consistent finished product. (In addition to those above, microfiltration, UV treatment and ozonation are also used.) Unlike the case of SW, the only criterion is that the water must be wholesome at the point of consumption; no 'source identity' (as in the case of spring water) is required. However, the treatment methods may also be detailed on the label, so the descriptions, 'purified drinking water' (sometimes with added minerals), 'reverse osmosis drinking water', etc. may be used.

Water for infant use
- Some waters, even ones not guaranteed to be sterile, are supplied for infant use. In this case, the label should read 'Not sterile. Use as directed by physician or by labelling directions for use of infant formula'.

3.3.1.4 Quality standards

With the exception of the particular requirements for mineral water, compositional and quality standards for bottled water in the USA are also unified through CFR 21. In all cases, the parameters in question are to be analysed using the methods described in *Standard Methods for the Examination of Water and Wastewater* (published jointly by the American Public Health Association, the American Water Works Association and the Water Pollution Control Federation) as follows.

Microbiological quality; according to Part 165.110 (b) (2):

(1) *Multiple-tube fermentation method.* Not more than one of the analytical units in the sample shall have a most probable number (MPN) of 2.2 or more coliform organisms per 100 ml and no analytical unit shall have an MPN of 9.2 or more coliform organisms per 100 ml; or

(2) *Membrane filter method.* Not more than one of the analytical units in the sample shall have 4.0 or more coliform organisms per 100 ml and the arithmetic mean of the coliform density of the sample shall not exceed one coliform per 100 ml.

Note that the only organism mentioned is coliform; this is regarded as the principal indicator organism for microbiological contamination, and the FDA has not deemed it necessary to specify others. On the other hand, it is required of bottlers through monitoring to demonstrate that their process and product is under microbiological control.

Physical quality; Part 165.110 (b) (3):

(i) The turbidity shall not exceed 5 units.
(ii) The color shall not exceed 15 units.
(iii) The odor shall not exceed threshold odor No.3.

Chemical quality; Part 165.110 (b) (4) (reproduced in simplified form in Table 3.4) provides allowable levels for (A) inorganic substances, (B) volatile organic compounds (VOCs), (C) pesticides and other synthetic organic chemicals (SOCs), and (D) certain chemicals for which EPA has established secondary maximum contaminant levels in its drinking water regulations.

There are also maximum levels at which fluoride can be added to bottled water, and an additional table specifies the allowable levels for residual disinfectants and disinfection by-products. In particular, the latter refers to those substances (organochlorines and related substances) potentially resulting from chlorination, and those formed as a consequence of ozonation, which, if not

Table 3.4 Code of Federal Regulations, Title 21 – Standard of Identity for Bottled Waters – simplified list of maximum allowable levels

'Bottled water shall...meet standards of chemical quality and shall not contain chemical substances in excess of the following concentrations.'

Substance	Concentration (mg/l)
Arsenic	0.05
Chloride[1]	250.0
Iron[1]	0.3
Manganese[1]	0.05
Phenols	0.001
Total dissolved solids[1]	500.0
Zinc[1]	5.0

[1] Mineral water is exempt from allowable level. The exemptions are aesthetically based allowable levels and do not relate to a health concern.
Source: From the Code of Federal regulations, Title 21; U.S. Government Printing Office: Washington, DC, 2002.

(ii) (A) Bottled water packaged in the USA to which no fluoride is added shall not contain fluoride in excess of the levels in Table 1 and these levels shall be based on the annual average of maximum daily air temperatures at the location where the bottled water is sold at retail.

TABLE I

Annual average of maximum daily air temperatures (°F)	Fluoride concentration (mg/l)
53.7 and below	2.4
53.8–58.3	2.2
58.4–63.8	2.0
63.9–70.6	1.8
70.7–79.2	1.6
79.3–90.5	1.4

(B) Imported bottled water to which no fluoride is added shall not contain fluoride in excess of 1.4 mg/l.
Source: From the Code of Federal regulations, Title 21; U.S. Government Printing Office: Washington, DC, 2002.

(C) Bottled water packaged in the USA to which fluoride is added shall not contain fluoride in excess of levels in Table 2.

TABLE 2

Annual average of maximum daily air temperatures (°F)	Fluoride concentration in milligrams per litre
53.7 and below	1.7
53.8–58.3	1.5
58.4–63.8	1.3
63.9–70.6	1.2
70.7–79.2	1.0
79.3–90.5	0.8

(D) Imported bottled water to which fluoride is added shall not contain fluoride in excess of 0.8 mg/l.

Source: From the Code of Federal regulations, Title 21; U.S. Government Printing Office: Washington, DC, 2002.

(A) The allowable levels for inorganic substances are as follows:

Contaminant	Concentration (mg/l or as specified)
Antimony	0.006
Barium	2.0
Beryllium	0.004
Cadmium	0.005
Chromium	0.1
Copper	1.0
Cyanide	0.2
Lead	0.005
Mercury	0.002
Nickel	0.1
Nitrate	10 (as nitrogen)
Nitrite	1 (as nitrogen)
Total nitrate and nitrite	10 (as nitrogen)
Selenium	0.05
Thallium	0.002

Source: From the Code of Federal regulations, Title 21; U.S. Government Printing Office: Washington, DC, 2002.

(B) The allowable levels for volatile organic chemicals (VOCs) are as follows

Contaminant	Concentration (mg/l)
Benzene	0.005
Carbon tetrachloride	0.005
o-Dichlorobenzene	0.6
p-Dichlorobenzene	0.075
1,2-Dichloroethane	0.005
1,1-Dichloroethylene	0.007
cis-1,2-Dichloroethylene	0.07

(Continued)

(*Continues*)

trans-1,2-Dichloroethylene	0.1
Dichloromethane	0.005
1,2-Dichloropropane	0.005
Ethylbenzene	0.7
Monochlorobenzene	0.1
Styrene	0.1
Tetrachloroethylene	0.005
Toluene	1.0
1,2,4-Trichlorobenzene	0.07
1,1,1-Trichloroethane	0.2
1,1,2-Trichloroethane	0.005
Trichloroethylene	0.005
Vinyl chloride	0.002
Xylenes	10

Source: From the Code of Federal regulations, Title 21; U.S. Government Printing Office: Washington, DC, 2002.

(C) The allowable levels for pesticides and other synthetic organic chemicals (SOCs) are as follows

Contaminant	Concentration (mg/l)
Alachlor	0.002
Atrazine	0.003
Benzo(a)pyrene	0.0002
Carbofuran	0.04
Chlordane	0.002
Dalapon	0.2
1,2-Dibromo-3-chloropropane	0.0002
2,4-D	0.07
Di-(2-ethylhexyl)adipate	0.4
Dinoseb	0.007
Diquat	0.02
Endothall	0.1
Endrin	0.002
Ethylene dibromide	0.00005
Glyphosate	0.7
Heptachlor	0.0004
Heptachlor epoxide	0.0002
Hexachlorobenzene	0.001
Hexachlorocyclopentadiene	0.05
Lindane	0.0002
Methoxychlor	0.04
Oxamyl	0.2
Pentachlorophenol	0.001
PCBs (as decachlorobiphenyl)	0.0005
Picloram	0.5
Simazine	0.004
2,3,7,8-TCDD (Dioxin)	3×10^{-8}
Toxaphene	0.003
2,4,5-TP (Silvex)	0.05

Source: From the Code of Federal regulations, Title 21; U.S. Government Printing Office: Washington, DC, 2002.

(D) The allowable levels for certain chemicals for which EPA has established secondary maximum contaminant levels in its drinking water regulations (40CFR part 143) are as follows

Contaminant	Concentration (mg/l)
Aluminum	0.2
Silver	0.1
Sulfate[1]	250

[1] Mineral water is exempt from allowable level. The exemptions are aesthetically based allowable levels and do not relate to a health concern.
Source: From the Code of Federal regulations, Title 21; U.S. Government Printing Office: Washington, DC, 2002.

(H) The allowable levels for residual disinfectants and disinfection byproducts are as follows

Substance	Concentration (mg/l)
Disinfection byproducts:	
Bromate	0.010
Chlorite	1.0
Haloaceticacids (five) (HAA5)	0.060
Total trihalomethanes (TTHM)	0.080
Residual disinfectants:	
Chloramine	4.0 (as Cl_2)
Chlorine	4.0 (as Cl_2)
Chlorine dioxide	0.8 (as ClO_2)

Source: From the Code of Federal regulations, Title 21; U.S. Government Printing Office: Washington, DC, 2002.

properly controlled, oxidises naturally occurring bromide to bromate in some types of water.

Mineralisation, though not regulated (with the exception of mineral waters), is often given consideration by the consumer when choosing which waters to buy. In general, the market preference in the USA is for relatively low mineralisation, unlike in Europe, where typical mineral levels range from 250–1500 mg/l (see Chapter 2). The more favoured brands in the USA are those in the range of 75–250 mg/l.

3.3.2 Canada

In Canada, bottled water is regulated under the Food and Drugs Act and the Food and Drug Regulations Part B, Division 12, governing 'Prepackaged Water and Ice'. This contains some basic definitions, but the active regulation of the industry occurs through the work of the Canadian Food Inspection Agency (CFIA), which is charged with inspection and monitoring of products and premises. Additionally, the Canadian Bottled Waters Association (CBWA) has drafted a Model Bottled Water Code somewhat similar in structure and content to that used by its sister organisation in the USA. This expands on definitions, quality, GMPs, monitoring and labelling requirements.

Bottled water falls into two main categories:

(1) Mineral water and Spring Water
(2) Bottled waters, not represented as mineral water or Spring Water (other waters)

According to the definitions in the Model Code, the following requirements must be met.

Spring Water:
- It must be collected from an underground source from which water may flow naturally to the surface, and can be collected either with a spring catchment at natural emergence or with the use of a borehole.
- The TDS content may not exceed 500 mg/l +/− 10%.

Mineral water:
- Must be collected from an underground source exactly as in the case of Spring Water.
- Must have a TDS content above 500 mg/l +/− 10%.

Both of the above may also be labelled 'natural', provided that they meet the following definition:

'Natural water' shall be water obtained from an underground or approved natural source or sources, and

- Must maintain the same general composition and characteristics at the time of bottling that it possessed at the point(s) of collection.
- May not be obtained from a public community water system.
- Can be treated with ozone or other 'acceptable and suitable' disinfection processes, provided that this does not result in a change in its general composition and characteristics.
- Can be treated to remove or reduce the concentration of dissolved gases or undissolved solids.
- Can be treated to remove or reduce the concentration of unstable and undesirable substances.
- At the point of emergence, it must, without treatment, be free of exogenous organisms or harmful substances.
- Must comply, at emergence and without treatment, with Maximum Allowable Concentrations as defined in the Guidelines for Canadian Drinking Water Quality. (These are shown in the Monitoring Matrix from the CBWA Model Code, which is reproduced in Table 3.5.)
- It must also have, at emergence and without treatment, a stable composition, taking into account natural fluctuations in the concentrations of the major minerals.

Table 3.5 Monitoring Matrix from the Canadian Bottled Water Code (comparing the CBWA Standards of Quality and Guidelines with FDA bottled water requirements and Canadian Drinking Water Standards)

Appendix A
Monitoring Matrix
CBWA Model Code Monitoring Requirements

Monitoring parameter group (*individual group analytes*)	Monitoring frequency	SOQs, MACs, IMACs, and Guidelines (all results in mg/l except as noted)		
Inorganic chemicals (IOCs)	Annually (product & source)	CBWA SOQ	FDA SOQ	CDWG MAC
Antimony[11]		0.006	0.006	0.006*[5]
Arsenic		0.025	0.05	0.025*
Barium		1.0	2.0	1.0
Beryllium[11]		0.004	0.004	
Bromate[1]		0.010	0.010	0.010*
Cadmium		0.005	0.005	0.005
Chlorine[1]		0.1	4.0	
Chloramine[1]		3.0	4.0	3.0
Chlorine dioxide[1]		0.8	0.8	
Chlorite[1]		1.0	1.0	
Chromium		0.05	0.1	0.05
Cyanide[11]		0.1	0.1	0.2
Fluoride		1.5	2/1.3[2]	1.5[6]
Lead		0.005	0.005	0.010[5]
Mercury		0.001	0.002	0.001
Nickel[11]		0.1	0.1	
Nitrate-N		10	10	45[7]
Nitrite-N		1	1	[7]
Total nitrate and nitrite		10	10	
Selenium		0.01	0.05	0.01
Thallium[11]		0.002	0.002	

Secondary inorganic parameters	Annually (product & source)	CBWA SOQ	FDA SOQ	CDWG MAC
Aluminium		0.2	0.2	[4]
Chloride		250	250	250**
Copper		1.0	1.0	1.0**
Iron		0.3	0.3	0.3**
Manganese		0.05	0.05	0.05**
Silver		0.025	0.1	
Sulfate		250	250	500**[8]
Total dissolved solids (TDS)		500	500	500**
Zinc		5.0	5.0	5.0**

(*Continued*)

(*Continues*)

Volatile organic chemicals (VOCs)	Annually (product & source)	CBWA SOQ	FDA SOQ	CDWG MAC
1,1,1-Trichloroethane		0.03	0.2	
1,1,2-Trichloroethane		0.003	0.005	
1,1-Dichloroethylene		0.002	0.007	0.014
1,2,4-Trichlorobenzene		0.009	0.07	
1,2-Dichloroethane		0.002	0.005	0.005*
1,2-Dichloropropane		0.005	0.005	
Benzene		0.001	0.005	0.005
Carbon tetrachloride		0.002	0.005	0.005
cis-1,2-Dichloroethylene		0.07	0.07	
trans-1,2-Dichloroethylene		0.1	0.1	
Ethylbenzene		0.0024	0.7	0.0024**
Methylene chloride (Dichloromethane)		0.003	0.005	0.05
Monochlorobenzene		0.05	0.1	0.08
o-Dichlorobenzene		0.20[9]	0.6	0.20[9]
p-Dichlorobenzene		0.005[9]	0.075	0.005[9]
Haloaceticacids (HAA5)[1]		0.06	0.06	
Styrene		0.1	0.1	
Tetrachloroethylene		0.001	0.005	0.03
Toluene		0.024	1.0	0.024**
Trichloroethylene		0.001	0.005	0.05
Vinyl chloride		0.002	0.002	0.002
Xylenes (total)		0.3	10.0	0.3**
Bromodichloromethane				
Chlorodibromomethane				
Chloroform				
Bromoform				
Total trihalomethanes (TTHM)[1]		0.01	0.08	

Semivolatile organic chemicals (SVOCs)	Annually (product & source)	CBWA SOQ	FDA SOQ	CDWG MAC
Benzo(a)pyrene		0.0002	0.0002	0.00001
Di(2-ethyhexyl)adipate		0.4	0.4	
Di(2-ethyhexyl)phthalate		0.006		
Hexachlorobenzene		0.001	0.001	
Hexachlorocyclopentadiene		0.05	0.05	
Total recoverable phenolics		0.001	0.001	

Synthetic organic chemicals (SOCs)	Annually (product & source)	CBWA SOQ	FDA SOQ	CDWG MAC
2,4,5-TP (Silvex)		0.01	0.05	
2,4-D (Dichlorophenoxy acetic acid)		0.07	0.07	0.1*

Alachlor		0.002	0.002	
Aldicarb		0.003	0.003	0.009
Aldicarb sulfone		0.003	0.003	
Aldicarb sulfoxide		0.004	0.004	
Atrazine		0.003	0.003	0.005*[6]
Carbofuran		0.04	0.04	0.09
Chlordane		0.002	0.002	
Dalapon		0.2	0.2	
Dibromochloropropane (DBCP)		0.0002	0.0002	
Dinoseb		0.007	0.007	0.01
Endrin		0.002	0.002	
Ethylene dibromide		0.00005	0.00005	
Heptachlor		0.0004	0.0004	
Heptachlor epoxide		0.0002	0.0002	
Lindane		0.0002	0.0002	
Methoxychlor		0.04	0.04	0.9
Oxamyl (vydate)		0.2	0.2	
Pentachlorophenol		0.001	0.001	0.06
Picloram		0.19	0.5	0.19*
Polychlorinated biphenyls (PCBs)		0.0005	0.0005	
Simazine		0.004	0.004	0.01*
Toxaphene		0.003	0.003	

9 Contaminant SOCs	Annually (product & source)	CBWA SOQ	FDA SOQ	CDWG MAC
Dioxin (2,3,7,8-Tetrachlorodibenzo-p-dioxin)		3×10^{-8}	3×10^{-8}	
Diquat		0.02	0.02	0.07
Endothall		0.1	0.1	
Glyphosate		0.28	0.7	0.28*

Additional regulated contaminants	Annually (product & source)	CBWA SOQ	FDA SOQ	CDWG MCL
Methyl tertiary butyl ether (MTBE)		0.07		

Water properties	Annually (product & source)	Guideline	Guideline	Guideline
Colour		5 Units	5 Units	15 TCU
Turbidity		0.5 NTU	0.5 NTU	1.0 NTU
pH		5–7/6.5–8.5	6.5–8.5	6.5–8.5
Odour		3 T.O.N.	3 T.O.N.	Inoffensive

(*Continued*)

(*Continues*)

Radiological contaminants		CBWA SOQ	FDA SOQ	CDWG MAC
Gross alpha } Gross beta }	Annually (product & source)	15 pCi/l	15 pCi/l	
		50 pCi/l	50 pCi/l	
Radium$_{226}$	When gross alpha exceeds 5 pCi/l	5 pCi/l	5 pCi/l	0.6 Bq/l
Radium$_{228}$	When gross alpha exceeds 5 pCi/l	5 pCi/l	5 pCi/l	0.5 Bq/l
Strontium$_{90}$	When gross beta exceeds 8 pCi/l	8 pCi/l	8 pCi/l	5 Bq/l
Tritium and other man-made nuclides	When gross beta exceeds 50 pCi/l			

Microbiological contaminants	Product: daily Source: weekly	CBWA SOQ	FDA SOQ	CDWG MAC
Total Coliform/E. coli Health Canada, Standard of Quality: any coliform bacteria, as determined by official method MFO-15, Microbiological Examination of Water in Sealed Containers (excluding Mineral and Spring Water) and of Prepackaged Ice, November 30, 1981; more than 100 total aerobic bacteria per millilitre, as determined by official method MFO-15, Microbiological Examination of Water in Sealed Containers (excluding Mineral and Spring Water) and of Prepackaged Ice, November 30, 1981.	Note: Confirmation and validation of all positive total coliform results required. See Appendix B of the CBWA model.	No *Escherichia coli* detectable in a 100 ml portion/ sample. No validated total coliform detectable in a 100 ml portion/ sample as substantiated by retesting.	*MPN:* < 9.2 organisms/ 100 ml *MF:* < 4 CFU/ml	*Total Coliform:* 0 organisms/ 100 ml but no sample should contain more than 10 CFU. 1 out of 10 with coliform OK but not consecutive samples. Over 10 CFU resample. *E. Coli:* zero or thermo-tolerant coliforms.

All results in mg/l except as noted.

Glossary of acronyms

FDA: US Food and Drug Administration
CDWG: Canadian Drinking Water Guidelines

SOQ: standard of quality
MCL: maximum contaminant limit
MAC: maximum acceptable concentrations
IMAC: interim maximum acceptable concentration
AO: aesthetic objective
mg/l: milligrams per litre
μg/l: micrograms per litre
ppm: parts per million
ppb: parts per billion
Example: $0.08\,mg/l = 0.08\,ppm = 80\,μg/l = 80\,ppb$
TCU: true colour units
NTU: nephelometric turbidity unit
pCi/l: picocuries per litre
Bq/l: Becquerel per litre. A Becquerel (Bq) is the unit of activity of a radioactive substance, or the rate at which transformations occur in the substance. One Becquerel is equal to one transformation per second and is approximately equal to 27 picocuries (pCi).
MPN: most probable number
MF: membrane filtration
CFU: colony forming unit

Footnotes

* Indicates the SOQ is an IMAC.
** Indicates the SOQ is an AO.
[1] Included in D/DBP rule. See D/DBP monitoring requirements on page 18, Appendix A for details.
[2] FDA SOQ for fluoride: 2.0 mg/l if naturally occurring, 1.3 mg/l if added.
[3] 2° IOCs are guidelines that are classified by the USEPA as Secondary Drinking Water Contaminants, i.e. aesthetic, not health-related, and non-enforceable.
[4] A health-based guideline for aluminum in drinking water has not been established. However, water treatment plants using aluminum-based coagulants should optimize their operations to reduce residual aluminum levels in treated water to the lowest extent possible as a precautionary measure. *Operational guidance values* of less than 100 μg/l total aluminum for conventional treatment plants and less than 200 μg/l total aluminum for other types of treatment systems are recommended. Any attempt to minimize aluminum residuals must not compromise the effectiveness of disinfection processes or interfere with the removal of disinfection by-product precursors.
[5] Because first-drawn water may contain higher concentrations of metals than are found in running water after flushing, faucets should be thoroughly flushed before water is taken for consumption or analysis.
[6] It is recommended, however, that the concentration of fluoride be adjusted to 0.8–1.0 mg/l, which is the optimum range for the control of dental caries.
[7] Equivalent to 10 mg/l as nitrate–nitrogen. Where nitrate and nitrite are determined separately, levels of nitrite should not exceed 3.2 mg/l.
[8] There may be a laxative effect in some individuals when sulphate levels exceed 500 mg/l.

[9] In cases where total dichlorobenzenes are measured and concentrations exceed the most stringent value (0.005 mg/L), the concentrations of the individual isomers should be established.
[10] These are the other five analytes not included in the heading '9 Contaminant SOCs' as defined by the US FDA Contaminant List.
[11] One of the 9 Contaminant SOCs referred to in Appendix A – Monitoring Parameter Group.

Source: Reproduced by permission of the Canadian Bottled Water Association.

In the case of both SWs and mineral waters, source approval based upon hydrogeological surveys and watershed surveillance are required, together with evaluation of source construction and auditing of production and packaging facilities.

In addition, the water may be labelled:

- 'Carbonated' or 'Sparkling' when the original carbonation level has been supplemented to make the water effervescent.
- 'Naturally Carbonated' or 'Naturally Sparkling' when the water contains at the time of bottling the same level of carbon dioxide as that which naturally occurs at emergence.

Bottled waters other than mineral waters or SWs may come from any approved source, including a municipal supply. They can be treated to make them fit for consumption (must also in their packaged state comply with the Guidelines for Canadian Drinking Water Quality), but any significant modifying treatments must be detailed on the label, thus:

- Demineralized drinking water – water with TDS content not exceeding 10 mg/l.
- Drinking water – water with TDS content equal to or more than 10 mg/l but not above 500 mg/l +/− 10%.
- Mineralized drinking water – water with TDS content above 500 mg/l +/− 10%.

They can also be labelled with the following qualifiers where applicable:

- 'Carbonated' or 'Sparkling' when carbon dioxide has been added to make the water effervescent.
- In the case when the water has been drawn from a public water system and the general composition has not been significantly modified, a statement must appear conspicuously on the principal display panel stating 'Drawn directly from public water supply without significant processing.'

Packaged waters from several sources may also be blended:

- Blended water, in which composition enters two or more categories of water, must be labelled accordingly: thus, Demineralized Drinking Water, Drinking Water or Mineralized Drinking Water, as appropriate, based on the TDS content of the resulting blended water.
- Blended water in which composition is drawn from one category cannot be labelled as a different category.

In addition to the requirements for source water and finished product monitoring, the Model Code also provides guidance on the management, cleaning and monitoring of bulk water tankers (tankering is permitted for all categories) and lays down requirements for cleaning and disinfection of water processing and filling equipment.

3.4 Codex Alimentarius

There is as yet no universal standard for any category of bottled waters, and even the meaning of some of the words used to describe different waters can differ from country to country as can the inferences drawn. For example, in Europe, as we have seen, an NMW is not required to contain any particular level of minerals, whereas in other parts of the world, a specific minimum level of minerals is a prerequisite.

However, the first attempts have now been made to establish some universal standards, the most significant of which are happening through the Codex Alimentarius Commission. Codex Alimentarius in Latin means 'food code' and the Codex Alimentarius Commission, which was set up in 1962 through collaboration between the Food and Agriculture Organization (FAO) of the United Nations and the World Health Organization (WHO), has as its aim 'to guide and promote the elaboration and establishment of definitions and requirements for food, to assist in their harmonization and, in doing so, to facilitate international trade'.

Codex has a membership exceeding 140 countries and has produced numerous sets of standards and guidelines aimed at ensuring the wholesomeness of food. Codex standards, which may be regional or national, but which can also be international, have no legal force as such, and usually have the status of codes of practice when adopted. However, in developing countries where no food laws currently exist, Codex standards are increasingly seen as quasi-legal and do form the basis of legislation as it develops.

The original Codex Standard for Natural Mineral Waters was written in the early 1980s as a European Regional Standard. It mirrored the principles laid down in Directive 80/777/EEC, in that it required exploitation of water from a protected source, bottled without treatment. During the 1990s, conscious of the need to provide standards not only for NMWs but also for all bottled waters, the Codex commission published two pairs of standards in parallel, one for NMWs and one for bottled/packaged waters other than NMWs.

3.4.1 Codex and Natural Mineral Waters

The standard for NMWs (but not including non NMWs) (Codex Stan 108 Rev 1) was adopted in mid-1997 by the Codex Commission as a worldwide standard. Like the European directive, the Codex standard contains a definition, a recognition requirement and specifications for treatment and handling. In another Codex document – International Code of Hygiene Practice for the Collecting, Processing and Marketing of Natural Mineral Waters (CAC/RCP 33-1985) – hygiene rules are outlined, packaging and labelling requirements are covered, and reference is also made to the methods of analysis and sampling appropriate to NMW (published in other Codex documents). Principal differences from the Directive are as follows:

- Product recognition, but no requirement for geological/hydrogeological surveys to permit *source* recognition.
- The removal of unstable compounds containing iron, manganese, sulphur or arsenic is permitted using 'decantation, and/or filtration, if necessary, accelerated by previous aeration.' The specific use of ozone is not mentioned, although it is explicitly stated that any treatment must not alter the natural composition of the water.
- Unlike in the Directive, which requires details of the *treatment*, if any such treatments are applied, the *results* of that treatment must be declared on the label – hence 'arsenic removed' etc.
- Must be packaged at source (no tankering).
- The product is to be labelled as one of the following:

  Natural Mineral Water
  Naturally carbonated Natural Mineral Water
  Decarbonated Natural Mineral Water
  Natural Mineral Water fortified with carbon dioxide from the source
  Carbonated Natural Mineral Water

- No claims concerning medicinal (preventative, alleviative or curative) effect can be made on the label, but 'claims of other beneficial effects related to the health of the consumer shall not be made unless true and not misleading'. However, the method of establishing such benefits is not touched upon.

- The label must bear the indications 'contains fluoride' in the case of a water containing more than 1 mg/l of fluoride, or 'this product is not suitable for infants and children under the age of seven years' in the case of a water containing more than 2 mg/l of fluoride.

Like the NMW Directive, the Codex contains health-related limits for some substances. These are similar to those in 2003/40/EC, and Sections 3 and 4, concerning composition and quality factors and microbiological standards, are shown in Tables 3.6 and 3.7.

Table 3.6 Section 3 of Codex Standard 108 for Natural Mineral Waters

3.2 Health-related limits for certain substances

Natural Mineral Water in its packaged state shall contain not more than the following amounts of the substances indicated hereunder.

Parameter number	Substance	Amount (mg/l)
3.2.1	Antimony	0.005
3.2.2	Arsenic	0.01, calculated as total As
3.2.3	Barium	0.7
3.2.4	Borate	5.0, calculated as B
3.2.5	Cadmium	0.003
3.2.6	Chromium	0.05, calculated as total Cr
3.2.7	Copper	1.0
3.2.8	Cyanide	0.07
3.2.9	Fluoride	[See labelling indications above]
3.2.10	Lead	0.01
3.2.11	Manganese	0.5
3.2.12	Mercury	0.001
3.2.13	Nickel	0.02
3.2.14	Nitrate	50, calculated as nitrate
3.2.15	Nitrite	0.02, calculated as nitrite[2]
3.2.16	Selenium	0.01

The following substances shall be below the limit of quantification[3] when tested, in accordance with the methods prescribed in Section 6 [ISO methods]:

3.2.17 Surface active agents[4]
3.2.18 Pesticides and PCBs[4]
3.2.19 Mineral oil[4]
3.2.20 Polynuclear aromatic hydrocarbons[4]

[2] Set as a quality limit (except for infants).
[3] As stated in the relevant ISO methods.
[4] Temporarily endorsed pending elaboration of appropriate method(s) of analysis.

Source: Reproduced by permission of the Food and Agriculture Organization of the United Nations.

Table 3.7 Section 4 of Codex Standard 108 for Natural Mineral Waters – Microbiological Requirements for Natural Mineral Waters

During marketing, natural mineral water
   (a) shall be of such a quality that it will not present a risk to the health of the consumer (absence of pathogenic micro-organisms); and
   (b) furthermore, it shall be in conformity with the following microbiological quality specifications.

First examination			Decision
E. coli or thermotolerant coliforms	1 × 250 ml	}	Must not be detectable in any sample
Total coliform bacteria	1 × 250 ml	} if ≥ 1 or ≤ 2	Second examination is carried out[1]
Faecal streptococci	1 × 250 ml	}	
Pseudomonas aeruginosa	1 × 250 ml	} if > 2	Rejected
Sulphite-reducing anaerobes	1 × 50 ml	}	

[1] According to a statistical sampling plan – not reproduced here.
Source: Reproduced by permission of the Food and Agriculture Organization of the United Nations.

3.4.2   Codex and Non-Natural Mineral Waters

For waters other than NMWs, Codex has drafted a second pair of standards: The general Standard for Bottled/Packaged Drinking Waters (other than Natural Mineral Waters) Codex Stan 227-2001, and Code of Hygienic Practice for Bottled/Packaged Drinking Waters (other than Natural Mineral Waters) CAC/RCP 48-2001.

Under the General Standard, a distinction is drawn between 'Waters Defined by Origin' (whether from an underground source or from the surface) and 'Prepared Waters', which may come from any water supply. For Waters Defined by Origin, the Standard requires that precautions are taken at source to prevent contamination by any means, that the conditions of collection are controlled in order to maintain the original microbiological purity and essential composition, and that they are fit for consumption at source. Inspection and monitoring of the source of Waters Defined by Origin is also specified, though no source recognition process of the type applicable to NMWs is required. Furthermore, they can only receive treatment (as in the case of NMWs) for the elimination or reduction of unstable elements and the removal of undesirable elements present in excess of maximum concentrations or (in this case) maximum levels of radioactivity. Antimicrobial treatments may only be used 'to conserve the original microbiological fitness for human consumption, original purity and safety of Waters Defined by Origin'. On the other hand, Prepared Waters, which need not meet any specific standards at origin, can be treated in any manner deemed necessary to make them microbiologically and chemically wholesome.

The microbiological, chemical and radiological quality of packaged waters is not included in the Codex standards, but is defined in the 'Guidelines for Drinking Water Quality' published by the WHO. These standards are not reproduced in their original form here, but are included in Table 3.10.

The Code of Hygienic Practice, which also states that it is to be read in combination with the *Recommended International Code of Practice – General Principles of Food Hygiene*, gives general guidance on precautions in selecting a resource site, protection of groundwater and surface water supplies, hygienic extraction or collection of water and storage and transport of water. It also gives basic advice on design of facilities, processes and packaging, with further sections dealing with maintenance and sanitation, personal hygiene, training, product information and consumer awareness.

Having published standards for both NMWs and other packaged waters, Codex Alimentarius continues to promote them, particularly in those areas where local standards or regulatory requirements do not exist; sometimes they are fully adopted, while in other cases, only the health-based limits required for food safety are applied.

3.5 Latin America

The Latin American market, which includes South and Central America and also Mexico, shows a diversity of approach, in that some countries have adopted the European approach and others are closer to the US model. Some consistency has developed in those countries (particularly in Central America) in which bottlers have become members of the Latin American Bottled Water Association (LABWA), itself a member of the independent International Council of Bottled Water Associations (ICBWA). However, there remain some significant differences across the region.

MEXICO

All water for human consumption in Mexico is regulated by the Ministry of Health, through the 'Regulation of the General Law of Health in the matter of Sanitary Control of Activities, Establishments, Products and Services'. The sanitary authorities are responsible for monitoring public and private suppliers and for granting certificates of Sanitary Condition of Water. This principally applies to waters provided through public supply, but also encompasses packaged waters, regardless of denomination. In Chapter 1 of the third part of the Regulations, entitled 'Water and Ice for Human Consumption and for Refrigeration', as defined by Article 209, water is considered fit for human consumption if it does not cause injurious effects to health and is free of pathogenic germs and toxic substances. The standards by which this is measured are specified in subsequent articles: Article 210 specifies the microbiological standard as ≤ 2

coliform organisms per 100 ml, and absence of faecal organisms. This is the same standard for coliform required by Codex. Articles 211 and 212 deal with the aesthetic characteristics of pH, flavour, scent, colour and turbidity, while Article 213 provides maximum concentrations for health-related and other chemical substances (see Table 3.8).

Table 3.8 Health-related physical and chemical limits for some Latin American waters

	Mexico: all bottled waters	Brazil: Mineral Water and Natural Water	Argentina: Mineral Water
Physical characteristics			
pH	6.9–8.5		4.0–9.0
Flavor	Characteristic	Characteristic	Characteristic
Odor			Characteristic
Color	< 20 units on Pt Co scale	5 hazen units	5 units on Pt Co scale
Turbidity	< 10 units of the silica scale	< 3 units Jackson	< 3 units Jackson
Chemical parameters	(mg/l)	(mg/l)	(mg/l)
Total alkalinity as $CaCO_3$	400		600
Aluminum	0.20		
Antimony		0.005	
Arsenic	0.05	0.05	0.2
Barium	1.0	1.0	1.0
Boron		5	30 (as HBO_3)
Cadmium	0.005	0.003	
Chloride			900
Chromium		0.05	
Cyanide as CN	0.05	0.07	0.01
Copper	1.50	1.0	1.0
Free chlorine	0.20		Absent
Chlorine in chlorinated water	1.0		
Hexavalente chromium	0.05		0.05
Chromium total		0.05	
Calcium hardness as $CaCO_3$	300		
Phenolic compounds	0.001		Absent
Iodide			8.5
Iron	0.30		5.0
Fluoride as F	1.50		2.0
Magnesium	125		
Manganese	0.15	2.0	2.0
Mercury	0.001	0.001	0.001
Nitrate as N	5.0	50 as nitrate	45
Nitrite as N	0.05	0.02 as nitrite	0.1
Ammoniacal nitrogen	0.10		0.2
Permanganate oxidizability	3.0		3.0
Lead	0.05	0.01	0.05
Selenium	0.05	0.05	0.01

Table 3.8 cont.

Sulphate as SO$_4$	250	600
Zinc	5.0	5.0
Active substances on methylene blue	0.5	
Substances extractable in chloroform	0.3	
Substances extractable in alcohol	1.5	
Hydrocarbons		
Dry residue at 180°C		50–2000
Products of contamination		Absent
Pesticide residues		Absent

The definitions for packaged potable water within Article 788 include purified water, NMW and artificially mineralised (carbonated or uncarbonated) water. As a minimum, the microbiological quality and the physical and chemical specifications laid out in Articles 789–792 correspond to those for potable water, the only difference being that NMW and mineralised water must contain from 500 to 1000 mg/l of sodium bicarbonate or sulphate, or an innocuous mixture of salts. Article 793 contains a detailed specification for the carbon dioxide used for artificial carbonation of prepared water.

ARGENTINA

Applicable legislation is included in the Argentine Food Code, Chapter 12, some of which covers Water-based Beverages, Drinking Water and Carbonated Water. Article 985 (Resolution N° 209 of 7.03.94) of the Code contains specifications for NMW and Article 983 deals separately with Carbonated Water. Here, the definition is almost identical to that in the Codex standard for NMW, including source protection, controls during exploitation and limits on allowable treatments. The microbiological requirements are similar, in that the Code requires the absence of parasites, *E. coli*, faecal streptococci, sulphite-reducing anaerobes and *Pseudomonas aeruginosa* in 250 ml of water. However, the limits for health-related substances differ in some instances, and there is a maximum limit of 2 litres on pack size. A variety of descriptions can be used, including 'natural spring mineral table water' or 'natural mineral spring water' or 'natural mineral table water' or 'natural mineral water'.

BRAZIL

In Brazil, Resolution RDC No 54 of 15 June 2000 of the Directorate of the National Agency for Sanitary Enforcemant contains a Standard of Identity and Quality for Natural Mineral Water and Natural Water. This standard refers both to the compositional requirements of the Codex for NMWs and to the GMPs and hygiene requirements laid out in the draft International Code of Hygiene Practice for the Manufacture and Marketing of Bottled Waters (other than NMWs). In the case of both NMWs and natural waters, they must be of subterranean origin, collected under conditions that guarantee maintenance of original composition and absence of pollution, with the only difference between them being mineral content. Further permitted treatments include decanting and filtration, but again, no alteration of mineral content is permitted.

Table 3.8 provides a comparison between the health-related limits for Latin American countries.

3.6   Australia and New Zealand

Legislation governing bottled water in Australia and New Zealand is set out in the Australia New Zealand Food Standards Code. Various parts of this Code, which is written by Australia New Zealand Food Authority (ANZFA: currently FSANZ), deal with labelling requirements, substances added, contaminants and residues, and microbiological limits, but the principal requirements for bottled waters are laid out in Standard 2.6.2 – 'Non-Alcoholic Beverages and Brewed Soft Drinks' – reproduced in Table 3.9. In addition, there are further standards covering food safety requirements, which are applicable to Australia only. As in other markets, the industry is also self-regulating – in this case by the Australasian Bottled Water Institute (ABWI), who have drafted a Model Code against which bottlers can be certificated, based upon mandatory third-party inspections and audits. There are no limits on the treatments permitted, including microfiltration, UV treatment, reverse osmosis, distillation and ozonation.

Table 3.9 Principal health-related limits for packaged water from the Australia New Zealand Food Standard 2.6.2

Substance	Max limit (mg/l)
Arsenic	0.05
Barium	1.0
Borate	30 (calculated as H_3BO_3)
Cadmium	0.01
Chromium VI	0.05
Copper	1.0
Cyanide	0.01 (calculated as CN^-)
Fluoride	2.0 (calculated as F^-)
Lead	0.05
Manganese	2.0
Mercury	0.001
Nitrate	45 (calculated as NO_3^-)
Nitrite	0.005 (calculated as NO_2^-)
Organic matter	3.0 ($KMnO_3$ digested as O_2)
Selenium	0.01
Sulphide	0.05 (calculated as H_2S)
Zinc	5.0

3.7 Asia

The Asian bottled water market, which extends from Jordan to Japan, includes several thousand bottlers in at least 40 countries. Of these, many operate in a relatively unregulated framework, but a growing number are becoming members of the Asia Bottled Water Association (ABWA), which was formerly the 'Far East Chapter' of the IBWA, but which became a founder member of the independent ICBWA. In addition, there are more than a hundred other members, comprising suppliers, distributors, affiliates, as well as new candidates for membership. The ABWA has developed a Model Code, based on the Codex Standard for Bottled/Packaged Drinking Waters other than NMWs, enhanced by the ICBWA model. All members are subject to mandatory inspections and annual unannounced third-party audits to confirm compliance with the Code, and thus enabling use of the Association logo and certification on the label.

All bottled water offered for sale must be safe for consumption. Bottled water originating from natural sources can be labelled 'spring' or 'mineral' water on the same basis as in the USA. Mineral water is SW with a larger amount of dissolved mineral solids, usually above 500 mg/l, and neither SW nor mineral water may have their composition modified through the use of chemicals. The terms 'well water' and 'artesian water' are also commonly used. These must all originate from sources that are well protected and monitored by both the exploiter and the State, but may all be subject to treatments including reverse osmosis, UV, microfiltration, distillation and ozone for the purpose of disinfection. Mineral water or SW must not contain any coliform bacteria or harmful substances at the source. Other bottled waters may undergo a variety of treatments and should meet the regulatory requirements for coliform and aerobic bacteria. Where the national standards are more stringent, the standard of identity is enhanced to meet the national requirement. Pre-packaged ice is also expected to comply with the regulations.

However, the market for purified water from various original sources, including municipal, well water and even sea water, is growing tremendously, and for these there are no limits on the types of treatment available, the only aim being to manufacture bottled water fit for consumption. These waters must not be labelled in such a way as to be confused with SW or mineral water and the label on these water containers must show how they have been treated, for example 'carbonated', 'demineralised', 'distilled', 'de-ionised', 'desalinated'.

When the source for bottled water comes from a community water system and does not undergo any treatment, the product label must state that the bottled water is 'from a community water system' or 'from a municipal source'. However, if the water is subject to distillation, de-ionisation or reverse osmosis, the bottled water product can be legally defined as purified water, demineralised water, de-ionised water, distilled water or reverse osmosis drinking water, and does not have to state the above phrases on its label.

In some Asian countries, such as China, there are national standards only for mineral and purified water. The other categories are either regulated by subnational bodies or by 'company standards' that (in the case of China) must be submitted to the National Technology Supervision Bureau for approval, and then complied with.

A comparison of ABWA Guidelines with WHO drinking water standards and IBWA standards, also showing standards for individual Asian countries, is shown in Table 3.10.

3.8   South Africa

The African market for bottled waters has generally been confined to those countries connected geographically to Europe and to the European tradition. Hence Egypt, Algeria and Morocco, through their Mediterranean link, have a strong and growing market based upon water coolers, most commonly using processed and purified waters. The most notable exception to this is the Republic of South Africa, where a young but growing market has developed within the last few years.

As in other parts of the world, bottled water has been treated simply as a foodstuff, regulated in this case by the Department of Health and the Department of Water affairs. To date, legislation directed specifically at bottled water has been somewhat limited; the draft Regulations Governing Natural Mineral Water (published in 2000 by the Department of Health under the Foodstuffs, Cosmetics and Disinfectant Act 1972) deal only with NMW, the definition of which is based very much on the European Directives. These regulations are undergoing review and it is anticipated that a revised version will be sent out for consultation by the industry within the next year.

Meanwhile, at the industry level, the principal producers in 1997 formed the South African Natural Bottled Water Association (SANBWA). Membership is voluntary, but more than 90% of producers have joined. Their 1999 Standards and Guidelines for Bottling Lines apply to and define NMW and natural spring water, both of which must meet the criteria specified in the draft regulations proposed by the Department of Health. The only difference between them is that mineral water may be abstracted via a spring, well or borehole, whereas an SW must flow naturally to the surface. According to Section 2 of Appendix 3 of the Guidelines and Standards, the following 'Special Requirements' apply.

Natural Spring or Mineral Water shall

(i) be characterised by its contents of certain mineral salts and their relative proportions and the presence of trace elements or of other constituents
(ii) be obtained directly from natural or drilled sources from underground water bearing strata

Table 3.10 Comparing the WHO Drinking Water Standards, IBWA and ABWA Guidelines and some standards from the regulations applicable in some Asian countries

Parameter	WHO (drinking water)	IBWA	ABWA	Indonesia	Singapore	SASO	Oman (NMW)	Malaysia (NMW)	Malaysia (drinking water)	Vietnam	India (NMW)	India (drinking water)	PRC (NMW)	PRC (drinking water)
	mg/l	mg/l	mg/l	mg/l	mg/l	mg/l	mg/l	mg/l	mg/l	mg/l	mg/l	mg/l	mg/l	mg/l
Inorganic constituents														
Aluminum	0.2***	0.2***	0.2***						0.2			0.03		0.2
Ammonia	1.5***	1.5***	1.5***	0.15					0.5	1.5				
Antimony	0.005P	0.006	0.005P				0.005				0.005			
Arsenic	0.01P	0.01	0.01P	0.05	0.05	0.05	0.05	0.05	0.05	0.01	0.05	0.05	0.05	0.01
Barium	0.7	1.0	0.7	1.0	1.0	1.0	1.0			0.1	1.0	1.0	0.7	
Beryllium	NAD	0.004												
Boron (as borate)	0.5P		0.5P				5.0	30			5.0		30	
Cadmium	0.003	0.005	0.003	0.005	0.01	0.01	0.003	0.01	0.005	0.003	0.003	0.01	0.01	
Chloride	250***	250***	250***	250	250	250			250	250	200	200		6.0
Chromium	0.05P	0.05	0.05P		0.05	0.05	0.05	0.05	0.05	0.05	0.05	0.05	0.05	
Copper	2P.1***	1***	2P.1***	0.5		1.0	1.0	1.0	1.0	2.0	1.0	0.05	1.0	1.0
Cyanide	0.07	0.1	0.07	0.05	0.01	0.05	0.07	0.01	0.1	0.07	0.07	0.05	0.01	
Fluoride	1.5	0.8–2.4	1.5	1.0	2.0	001	> 1 and > 2	2.0	1.5	1.5	2.0	1.0	2.0	
Hydrogen Sulfide	0.05***		0.05***				0.05	0.05	0.05	0.05	0.05			
Iron	0.3***	0.3***	0.3***	0.3		0.3			0.3	0.03		0.1		
Lead	0.01	0.005	0.01	0.005	0.05	0.05	0.01	0.05	0.05	0.01	0.01	0.01	0.01	0.01
Manganese	0.5P.0.1***	0.05***	0.5P.0.1***	0.05	2.0	0.05	2.0	2.0	0.1	0.5	2.0	0.1		
Mercury (total)	0.001	0.001	0.001	0.001	0.001	0.001	0.001	0.001	0.001	0.001	0.001	0.001	0.001	
Molybdenum	0.07		0.07											
Nickel	0.02P	0.1	0.02P				0.02				0.02	0.02		
Nitrate (as NO₃)	50	10	50	45	45	45	50	45	10	50	50	45	45	
Nitrite (as NO₂)	3P.Acute 0.2P. Chronic	1.0	3P.Acute 0.2P.	0.005	0.005		0.02	0.005	0.02		0.02	0.02	0.005	0.002
Nitrate/Nitrite	Sum of (conc; GV) ≤ 1	10	Chronic Sum of (conc; GV) ≤ 1			Sum of (conc; GV) ≤ 10								

(Continued)

Table 3.10 (*Continued*)

Parameter	WHO (drinking water)	IBWA	ABWA	Indonesia	Singapore	SASO	Oman (NMW)	Malaysia (NMW)	Malaysia (drinking water)	Vietnam	India (NMW)	India (drinking water)	PRC (NMW)	PRC (drinking water)
Selenium	0.01	0.01	0.01		0.01	0.01	0.05	0.01	0.01	0.01	0.05	0.01	0.01	
Silver	U***	0.025***	U			0.05			0.05	0.01	0.01	0.01	0.05	
Sodium	200***	250***	200***						200	250	150	200		
Sulfate	250***	250***	250***	200		250			400		200	200		
Thallium	0.002	0.002												
Zinc	3***	5***	3***			5.0		5.0	5.0	3.0	5.0	5.0	5.0	
Organic constituents	μg/l	μg/l	μg/l	μg/l	μg/l	μg/l	μg/l	μg/l	μg/l	μg/l	μg/l	μg/l	μg/l	μg/l
Chlorinated alkanes														
Carbon tetrachloride	2.0	5.0	2.0											0.001
1,1-Dichloroethane	NAD													
1,2-Dichloroethane	30	2.0	30											
Dichloromethane	20	3.0	20											
1,2- Dichloropropane		5.0												
1,1,1-Trichloroethane	2000P	30	2000P											
1,1,2-Trichloroethane		3.0												
Chlorinated ethenes (or ethylenes)														
1,1-Dichloroethene	30	2.0	30											
1,2 Dichloroethene	50		50											
cis-1,2-Dichloroethylene		70												
trans-1,2-Dichloroethylene		100												
Trichloroethene	70P	1.0	70P											
Tetrachloroethene	40	1.0	40											
Vinyl chloride	5.0	2.0	5.0											
Aromatic hydrocarbons														
Benzene	10	1.0	10											
Benzo(a)pyrene	0.7	0.2	0.7											
Ethylbenzene	300; 2–200***	700	300; 2–200***											
Phenols		1.0				1.0			2.0	0.5	ND	0.001	0.002	0.002
Styrene	20; 4–2.600***	100	20; 4–2.600***											
Toluene	700; 24–170***	1000	700; 24–170***											

Parameter							
Xylenes	500; 20–1800***	1000	500; 20–1800***				
Polycyclic aromatic hydrocarbons			Free of	Free of	Free of	Free of	Free of
Benzo (b) fluoranthene							
Benzo (k) fluoranthene							
Benzo (ghi) perylene							
Inceno (1, 2, 3–co) pyrene							
Chlorinated benzenes							
Morochlorobenzene	300; 10–120***	50	300; 10–120***				
1,2-Dianbiobenzene	1000; 1–10***		1000; 1–10***				
1,3-Dichlorobenzene	NAD		NAD				
1,4-Dichlorobenzene	3CO; 0.3–30***		3CO; 0.3–30***				
1,2,4-Trichlorobenzene		9.0					
Trichlorobenzenes (Total)	20; 5–50 ***		20; 5–50***				
Miscellaneous organics	μg/l	μg/l	μg/l	μg/l	μg/l	μg/l	μg/l
Acrylamide	0.5						
Dialkylins	NAD	400					
Di-(2-ethylhexyl)adipate	80		80				
Di-(2-ethylhexyl)phthalate	8.0	6.0	8.0				
Edetic acid (EDTA)	600						
Epichlohydrin	0.4P		0.4P				
Hexachlorobutadiene	0.6		0.6				
Methyl tertiary butyl ether (MTBE)							
Micrcystin-LR	1P						
Cyanobacterial Toxin							
Nitrotriacetic Acid	200						
Polychlorinated Biphenyls (PCBs) as decachlorobiphenyl	0.5			Free of	n.d.	n.d.	Free of
Tributylin Oxide	2.0						
Pesticides							
Total Pesticides		0.5		Free of			Free of
Alachlor	20	2.0	20	0.05			

(Continued)

Table 3.10 (*Continued*)

Parameter	WHO (drinking water)	IBWA	ABWA	Indonesia	Singapore	SASO	Oman (NMW)	Malaysia (NMW)	Malaysia (drinking water)	Vietnam	India (NMW)	India (drinking water)	PRC (NMW)	PRC (drinking water)
Aldicarb	10	3.0	10								b.l.			
Aldicarb sulfone		3.0												
Aldicarb sulfoxide		4.0												
Aldrin/Deldrin	0.03		0.03						0.03					
Atrazine	2	3.0	2.0								b.l.			
Bentazone	300		300											
Carbofuran	7.0	40	7.0								b.l.			
Chlordane	0.2	2.0	0.2						0.3					
4-chloro-2-	2.0		2.0											
Methylphenoxyacetic acid (MCPA)														
Chlorotoluron	30													
Cyanazine	0.6		0.6											
Dalapan	1.0	200	1.0											
1,2-Dibromo-3-Chloropropane (DPCP)	0.4–15P	0.2	0.4–15P											
1,2-Dibromoethane	2.0		2.0											
Dichlorodiphenyltrichloroethane (DDT)									1.0					
2,4-Dichlorophenoxyacetic acid (2,4-D)	30	70	30			100			100		b.l.			
1,2-Dichloropropane	40P	5.0	40P											
1,3-Dichloropropane	NAD													
1,3-Dichloropropene	20		20											
Dinoseb		7												
Diquat	10P	20	10P											
Dioxin (2,3,7,8-TCDD)		3×10^5												
Endothall		100												
Endrin		2.0				0.2								
Ethylene D/bromide	NAD	0.05												
Glyphosate	U	700	U											
Heptachlor	0.03	0.4	0.03							0.1				
Heptachlor Epoxide	Total of both	0.2	Total of both							0.1				
Hexachlorobenzene	1.0	1.0	1.0							0.01				

Compound				
Hexachlorocyclopentadiene	9.0	50		
Isoprofuran	2.0	0.2	2.0	3.0
Lindane	20	40	20	30
Methoxychlor	NAD			
4(2-Methyl-4-Chlorophenoxy)butyric acid (MCPB)				
Melolactica	10		10	
Molinate	6.0	200	6.0	
Oxamyl (Vydate)	20			
Pendimethalin	9P	1.0	9P	
Pentachlorophenol	20		20	
Permethrin	20	500		
Picloram	20		20	
Propanil	100		100	
Pyridate	2.0	4.0	2.0	
Simazine	7.0		7.0	
Terbuthylazine	20	3.0	20	
Toxaphene				b.l.
Trifuralin				
Chlorophenoxy herbicides other than 2,4-D and MCPA				
4(2,4-Dichlorophenoxy) butyric acid (2,4-DB)	90	90	100	
Dichloroprep	100	100		
Fenoprep	9.0	9.0		
Necoprop	10	10		
Silvex (2,4,5-7P)	10	9.0	10	
2,4,5-Trichlorophenoxyacetic acid (2,4,5-T)	9.0	9.0		
Disinfectants and disinfection by-products (D-DBPs)				
Bromate	25P	25P	10	
Morochloramine	3000	3000		

1 as bromide

(*Continued*)

Table 3.10 (*Continued*)

Parameter	WHO (drinking water)	IBWA	ABWA	Indonesia	Singapore	SASO	Oman (NMW)	Malaysia (NMW)	Malaysia (drinking water)	Vietnam	India (NMW)	India (drinking water)	PRC (NMW)	PRC (drinking water)
Chloral hydrate (Trichloroacetaldehyde)	10P		10P											
Chloramines (total)		4000												
Chlorate	NAD													
Chlorine	5000; 600–1000***	100	5000; 600–1000***	0.1								0.2		0.005
Chlorine dioxide		800												
Chlorite	200P	1000	200P'	0.1										
Chloractone	NAD													
3-Chloro-4-dichloromethyl-5-hydroxy-2(5H)-furanone (MX)	NAD													
Chiropicrin	NAD													
Cyanogen chloride (as CN)	70		70											
Dichloramine	NAD													
Formaldehyde	900		900											
Trichloramine	NAD													
Chlorophenols														
2-Chlorophenol	0.1–10***		0.1–10***											
2,4-Dichlorophenol	0.3–40***		0.3–40***											
2,4,6-Trichlorophenol	200; 2–300***		200; 2–300***											
Halogenated Acetic Acids														
Monochloroacetic acid	NAD													
Dichloroacetic acid	50P		50P											
Trichloroacetic acid	100P		100P											
Haloacetic Acids (HAA-s), includes mono-, di-, and trichloracetic acid and mono-and dibromoacetic acid		60												
Halogenated Acetonitriles														
Bromochloroacetonitrile	NAD													

Parameter									
Dibromoacetonitrile	100P		100P						
Dichloroacetonitrile	90P		90P						
Trichloroacetonitrile	1P		1P						
Trihalomethanes									
Bromodichloromethane	60		60						
Bromoform	100		100						
Chloroform	200		200						
Dibromochloromethane	100		100						
Total THMs	Sum of (conc: GV) ≤ 1	10	Sum of (conc: GV) ≤ 1			30			0.02
Other chemical/physical parameters									
Color	15TCU***	5TCU***	15TCU***	5TCU	15TCU	15	2	15	5
Odor	Acceptable***	3 T.O.N.	Acceptable***	No smell	Unobjectionable		Agreeable	None	None
PH	< 8 for effective disinfection w/ chlorine***	5–7 purified water	6–8 for effective disinfection w/ chlorine***	6.5–8.5	4.5–6.5 for carbonated 6.5–8.5 for non-carbonated	6.5–8.5	6.5–8.5	6.5–8.5	5.0–7.0
Taste	Acceptable***		Acceptable***		Unobjectionable		Agreeable	Agreeable	
Total dissolved solids (TDS)	1000 mg/l***	500***	1000 mg/l***	500	100–700	1000	150–700	1000	
Turbidity	5 NTU***	0.5 NTU***	5 NTU***	5 NTU	5 NTU	5 NTU	2.0	2.0	1.0
Oxidisability				1.0	3.0	5.0			
Total organic carbon (TOC)									
Radiological constituents									
Alpha activity, gross	0.1 Bq/l	15 pCi/l	0.1 Bq/l	1.0 pCi/l	10 pCi/l	0.1 Bq/l	0.1 Bq/l	0.1 Bq/l	
Beta activity, gross	1 Bq/l	50 pCi/l	1 Bq/l	30 pCi/l	1000 pCi/l	1 Bq/l	1 Bq/l	1 Bq/l	
Combined radium-226 and radium-228		5 pCi/l			3 pCi/l		5.0	5.0	1.5 Bq/l 1.1 Bq/l for 226
Tritium and man-made radionuclides					10 Bq/l for cesium 134 and 137				
Uranium	0.002P mg/l								
Strontium 90		8 pCi/l							

(Continued)

Table 3.10 (*Continued*)

Parameter	WHO (drinking water)	IBWA	ABWA	Indonesia	Singapore	SASO	Oman (NMW)	Malaysia (NMW)	Malaysia (drinking water)	Vietnam	India (NMW)	India (drinking water)	PRC (NMW)	PRC (drinking water)
Microbiological constituents														
E. coli or thermotolerant coliform bacteria	0/100 ml	0/100 ml	0/100 ml	0/100 ml	0/250 ml	0/100 ml	0/100 ml	Nil	Nil	Nil	Nil in any 250 ml sample	Nil in any 250 ml sample	Nil in any 100 ml sample	3 MPN
Total coliform bacteria		01/100 ml No validated total coliform detectable in a 100 ml portion/sample as substantiated by resampling		01/100 ml		4/100 ml and arithmetic mean	1/100 ml 100 ml for second examination	10 MPN/4 MF	10 MPN/4 MF	Nil	Nil in any 250 ml sample	Nil in any 100 ml sample	Nil in any 100 ml sample	
Clostridium perfringens				01/100 ml	0/50 ml									
Colony count at 22°C (72 h)						100 CFU/ml tested within 3 days of production					100/ml	100/ml		
Colony count at 37°C (24 h)				10/ml	10/ml						20/ml	20/ml	50/ml	
Enterococci														
Streptococci					0/250 ml		2/250 ml fecal streptococci			Nil	Nil in any 250 ml sample			Nil
Staphylococcus Aureus				0/250 ml	0/250 ml						Nil in any 250 ml sample			
Sulphite-reducing anaerobes							0/50 ml				Nil in any 250 ml			

Pseudomonas aeruginosa	0/250 ml	0/250 ml	Nil in any 250 ml sample	Nil in any 250 ml sample	Nil
Yeast and mold				Nil in any 250 ml sample
Salmonella and Shigella	0/250 ml		Nil in any 250 ml sample	
Vibrio Cholera and V. parahaemolyticus			Nil in any 250 ml sample	

Key P = Provisional value

Values shown as (for example) 5000; 600–1000*** are to be interpreted as follows: 5000 = health risk limit, 600–1000 = range for aesthetic unacceptability threshold

*** = acceptability aspect (not a health related limit)

(conc:GV) ≤ 1 = Gross value must be ≤ 1   SASO = Saudi Arabian Standards Organization

                                              PRC = Peoples' Republic of China

NAD = No available data

Printed with permission from the Asia Bottled Water Association (ABWA) – excerpt taken from ABWA's Technical Manual.

   (iii) be constant in its composition and have stability of discharge and of temperature, with due account being taken of cycles and natural fluctuations
   (iv) be collected under conditions which guarantee the original bacteriological purity
    (v) be bottled directly at the source, with particular hygienic precautions, by means of a pipeline connecting the source with the bottling facility

There are general requirements for source protection, equipment, materials, containers and processes. Authorised treatments are limited to those necessary for removal of unstable constituents, using coarse filtration, decantation and, where necessary, pre-aeration. The addition and/or removal of carbon dioxide is also permitted. The use of UV radiation is also authorised but this, like the others, may only be carried out provided that it does not alter the essential composition.

The microbiological criteria are laid down in Section 5.1 of the Guidelines:

At source, and thereafter, up to and including the point of sale, an NMW and SW shall be free from:

   (i) parasites and pathogenic micro-organisms
  (ii) *Escherichia coli* and other coliforms and faecal streptococci in any 250 ml sample examined
 (iii) viable *Clostridium perfringens* spores
  (iv) *Pseudomonas aeruginosa* in any 250 ml sample examined

For total plate count:

   (i) At source and immediately after bottling, the TVC shall not exceed 100 organisms per 1 ml when tested at $35°C \pm 1°C$ for 48 hours. The TVC shall be measured within 24 hours after sampling or bottling, the water sample being maintained at $4°C \pm 1°C$ during the period.
  (ii) Thereafter, up to and including the point of sale, the TVC shall be no more than that which results from the normal increase in the bacterial content which the water had at source.

A list of the limits for toxic substances is given in Section 6 and is reproduced in Table 3.11.

It is noteworthy that in South Africa, the large majority of bottlers produce NMW or SW; nonetheless, it is anticipated that future revisions to the industry Guidelines and Standards will also incorporate guidance for the estimated 10% producing purified and distilled products.

Table 3.11 List of limits for toxic substances in the South African Natural Bottled Water Association Guidelines and Standards

II Substance	II Maximum limit (mg/l)	III Test method
Copper	1.0	SABS method 203
Arsenic	0.01	SABS method 200
Cadmium	0.003	SABS method 201
Chromium	0.05	SABS method 1054
Lead	0.05	SABS method 208
Zinc	5.0	SABS method 214
Mercury	0.002	SABS method 1059
Selenium	0.01	SABS method 1058
Fluoride	1 calculated as F	SABS ISO 10304-1
Nitrate	10 calculated as N	SABS ISO 10304-1
Cyanide	0.05 calculated as CN	SABS ISO 204

SABS, South African Bureau of Standards.
Source: Reproduced by permission from The South African Natural Bottle Water Association.

3.9   Conclusions

By examining the various views on food safety and the different legislative requirements in certain markets, an effort has been made in this chapter to establish the criteria by which bottled waters are categorised in different parts of the world. New markets have opened up, often founded on the success of products imported from the major European producers, and the way that they regulate their products often has much to do with the prevailing views, not only of their own regulators, but also of the countries from which their imports originate. In many of the 'old' economies outside Europe but where the European influence has been prevalent, the traditional concept of natural, untreated waters has survived. On the other hand, the parallel adoption by the newer markets, such as Australasia and Asia, of the principles laid down in the IBWA Model Code has continued the trend set in the USA. Meanwhile, the Codex Alimentarius Commission has continued the process of formalising worldwide standards for different types of bottled waters, and increasingly, the newer markets look to these for guidance.

There has also been continued growth in the range of 'added value' products, some of which are simple flavoured waters (NMW or others with added essence) and others containing artificial sweeteners, acids and preservatives, which are effectively soft drinks. However, for the wider bottled water industry, the focus remains on water as the principal commodity.

Throughout the world, the consumption of bottled water is becoming more prevalent and as it does, consumers have a justified expectation that the product will continue to meet the highest standards. At the same time, as technology improves, lower and more precise detection limits for analysis become possible, while higher and higher standards are demanded. This will continue to place a premium, not only on water that undergoes extensive treatment in order to ensure compliance but perhaps more importantly on water which, by a combination of natural protection and good environmental and resource management, can continue to prove to be of the highest quality.

Acknowledgements

The following organisations provided invaluable information through their respective websites:

Asia Bottled Water Association (ABWA): *www.asiabwa.org*
Australasian Bottled Water Institute Inc. (ABWI): *www.bottledwater.org*
British Soft Drinks Association: *www.britishsoftdrinks.com*
Canadian Bottled Water Association (CBWA): *www.cbwa-bottledwater.org*
European Bottled Water Cooler Association (EBWA): *www.ebwa.org*
International Bottled Waters Association (IBWA): *www.bottledwater.org*
International Council of Bottled Waters Associations (ICBWA):
www.icbwa.org
Latin American Bottled Water Association (LABWA): *www.labwa.org*
South African Natural Bottled Water Association (SANBWA):
www.sanbwa.org.za

I would also like to acknowledge the assistance provided by the following people:

Sharon Bergman – Canadian Bottled Waters Association
Dr Bob Hargitt – British Soft Drinks Association
Gwen Majette Haynes – International Bottled Waters Association
Charlotte Metcalf – South African Natural Bottled Water Association
Nadia Six – Director of Nutrition, Packaging and Regulatory Affairs, Nestlé Waters MT
Ita Thaher – Asia Bottled Water Association
Jo Jacobius – Natural Mineral Water Information Service UK,
*www.naturalmineral*water.org

Bibliography

Regulations
Argentine Food Code (1994) Water-based Beverages, Drinking Water and Carbonated Water, Chapter 12. Article 985 (Resolution N° 209 of 7.03.1994).

Australia New Zealand Food Standards Code (2000) Non-alcoholic Beverages, Chapter 2, Part 2.6. Food Standards Australia New Zealand.

Brazilian Directorate of the National Agency for Sanitary Enforcemant (2000) Standard of Identity and Quality for Natural Mineral Water and Natural Water. Directorate of the National Agency for Sanitary Enforcement. (Resolution RDC No 54 of 15 June 2000)

Canadian Bottled Water Association (2003) *Model Bottled Water Code Draft Revision, September, 2003.* Canadian Bottled Water Association, 70 East Beaver Creek Road, Suite 203-1, Richmond Hill, Ontario, Canada.

Commission Directive 2003/40/EC of 16 May 2003, *establishing the list, concentration limits and labelling requirements for the constituents of natural mineral waters and the conditions for using ozone-enriched air for the treatment of natural mineral waters and spring waters.* Official Journal of the European Communities. (No L 126/34)

Department of Health and Human Services (U.S. Food and Drug Administration) (1995) *Beverages: Bottled Water; Final Rule,* Federal Register. Department of Health and Human Services.

Department of Health and Human Services (U.S. Food and Drug Administration) (2002) *Code of Federal Regulations, Title 21 – Food and Drugs, Part 129.35.* US Government Printing Office, Washington, DC.

Department of Health and Human Services (U.S. Food and Drug Administration) (2002) *Code of Federal Regulations, Title 21 – Food and Drugs, Part 165.110.* US Government Printing Office, Washington, DC.

European Council (1980) Directive 80/777/EEC *on the approximation of the laws of the Member States relating to the exploitation and marketing of natural mineral waters.* Official Journal of European Communities (No L 229/1).

European Council (1980) Directive 80/778/EEC (Revoked) *on the approximation of the laws of the Member States relating to the quality of water intended for human consumption.* Official Journal of European Communities (No L 229/11).

European Council (1998) Directive 98/83/EC of 3 November 1998 *relating to the quality of water intended for human consumption.* Official Journal of the European Communities (No L 330/32).

European Parliament and Council (1996) Directive 96/70/EC *amending Council Directive 80/777/EEC on the approximation of the laws of the Member States relating to the exploitation and marketing of natural mineral waters.* Official Journal of the European Communities –(No L 299).

IBWA (2002) *Model Bottled Water Regulation, October 2002.* International Bottled Water Association, 1700 Diagonal Road, Suite 650, Alexandria, VA, USA.

Health Canada (2003) *Food and Drug Regulations, Division 12.* Bureau of Food Regulatory International and Interagency Affairs, Food Directorate, Health Canada.

Mexico Ministry of Health (2002) Regulation of the General Law of Health in the matter of Sanitary Control of Activities, Establishments, Products and Services, Chapter 1. Water and Ice for Human Consumption and for Refrigeration, Mexico Ministry of Health.

Department of Health, Republic of South Africa (2000) Foodstuffs, Cosmetics and Disinfectant Act 1972 (Act No. 54 of 1972) – *Regulations Governing Natural Mineral Water.* Government Gazette Vol 415, No 20823, 28 January 2000 (No R 50).

The Natural Mineral Water, Spring Water and Bottled Drinking Water Regulations 1999. Statutory Instrument 1999 No. 1540.

The Natural Mineral Water, Spring Water and Bottled Drinking Water (Amendment) (England) Regulations 2003. Statutory Instrument 2003 No. 666 for England. (Statutory Instrument 2003 No. 182 for Northern Ireland and Statutory Instrument 2003 No. 139 for Scotland.)

Other publications

American Public Health Association, the American Water Works Association and the Water Pollution Control Federation (1998) *Standard Methods for the Examination of Water and Wastewater*, 20th edn. American Public Health Association.

Codex Alimentarius Commission (1985) *Recommended International Code of Hygiene Practice for the Collecting, Processing and Marketing of Natural Mineral Waters*, CAC/RCP 33-1985. Food and Agriculture Organization of the United Nations.

Codex Alimentarius Commission (1997) *Recommended International Code of Practice – General Principles of Food Hygiene*, CAC/RCP 1-1969, Rev.3-1997. Food and Agriculture Organization of the United Nations.

Codex Alimentarius Commission (1997) *Standard for Natural Mineral Waters*, Codex Stan 108. Food and Agriculture Organization of the United Nations.

Codex Alimentarius Commission (2001) *Code of Hygienic Practice for Bottled/Packaged Drinking Waters (other than Natural Mineral Waters)*, CAC/RCP 48-2001. Food and Agriculture Organization of the United Nations.

Codex Alimentarius Commission (2001) *General Standard for Bottled/Packaged Drinking Waters (other than Natural Mineral Waters)*, Codex Stan 227-2001. Food and Agriculture Organization of the United Nations.

Finlayson, D. M. (unpublished) *Chemical Standards Applied to Natural Mineral Waters and Packaged Waters*. SCI Symposium Natural Mineral Waters and Packaged Waters, 1992..

Miller, R. W. (1993) *This is Codex Alimentarius*. Food and Agriculture Organization of the United Nations.

World Health Organization (1984) *Guidelines for Drinking-Water Quality*, vol 2, World Health Organization.

World Health Organization (1993) *Guidelines for Drinking-Water Quality*, 2nd edn, vol 1, Recommendations, World Health Organization. World Health Organization (2003) *Guidelines for Drinking-Water Quality*, 3rd edn (Draft).

World Health Organization (2003) *Guidelines for Drinking-water Quality*, 3rd edn (Draft).

4 Hydrogeology of bottled water

Mike Streetly, Rod Mitchell and Melanie Walters

4.1 Introduction

Growth in the market for bottled waters has shown a consistent and substantial increase over the last decade. Accompanying this increase have come significant changes in both the legislation and regulations that govern production of bottled waters, and in public opinion that drives the market. Guidelines have also been developed to ensure that the product maintains the highest possible standards required by the consumer.

Arising from this change is the increasing need for the water bottler to understand more about the nature of the raw resource – the factors that govern the occurrence and hydrochemical characteristics of underground water. And yet not only are the processes that create their product natural and largely out of their control, but they also occur below ground, at depths of tens or even hundreds of metres, and so are out of sight, and sometimes, unfortunately, out of mind.

Although the processes may not be directly controllable, they can be influenced for good or ill. Careful development and management of a groundwater source can ensure a steady and reliable supply of the highest quality water. Hasty or ill-judged development can damage the crucial raw source on which bottled waters ultimately rely. In some cases, irreversible damage can be done.

This chapter sets out to do two things: first, to explain the science of the flow and chemical evolution of underground waters so that, second, we can look at how the factors involved interact with each other and can be influenced when managing or developing a source of underground water. Those who are already familiar with the science may use Sections 4.2 and 4.3 for occasional reference but may find Sections 4.4 and 4.5 on source development and management of groundwater sources respectively, more relevant to their present needs.

Once the science is appreciated, it will be seen to be quite straightforward. What is not so straightforward is the combination of the large numbers of factors involved, the degree of interaction between them and the distribution of all of this over large volumes of space and, more worryingly, over time (by the time a change in water quality is detected at a source, the reason for it may have been happening for years). What is needed here is a person whose experience in a broad range of hydrogeological situations enables them to diagnose and predict with confidence. Section 4.5 provides a general framework for applying the science of hydrogeology to the management of an underground source, including some guidance on the commissioning of expert help.

4.2   Understanding underground water – hydrogeology

4.2.1   Underground water – a key part of the water cycle

The concept of a global water cycle has been appreciated for over 300 years. The cycle is driven by solar energy that evaporates sea water to form clouds which in turn fall as rain, supplying the rivers with water which flows back to the sea. Only part of this cycle – that proportion of water which, after falling as rain, percolates (recharges) into the ground before emerging sometime later from springs and boreholes – is of direct relevance to those interested in the development of bottled water sources.

Unlike rainfall, which directly affects our daily lives, underground water is largely invisible and thus is rarely appreciated or understood. This is despite the fact that at any one time, worldwide, there is over 30 times as much fresh water stored underground as in rivers and streams, and that much of our public water supply is sourced from underground water rather than from much more conspicuous surface water reservoirs.

The few places in which groundwater flow may be visible, for instance in large caves, are often dramatic and this fact gives rise to the common misconception that underground water, like surface water, flows in rivers and streams. In fact, such occurrences are largely atypical.

4.2.2   Recharge to underground water

In this section we outline how water enters, flows through and is stored underground, and how it may flow to wells, boreholes, rivers and streams. At present, the average precipitation over the whole of the Earth is estimated to be around 900 mm/year. In the UK, the long-term average rainfall varies from 500 to 4300 mm (precipitation usually falls as rain, but other forms such as snow and dew are included in these figures). As a result of a number of processes, not all precipitation infiltrates into the ground and recharges the body of underground water that supplies streams and boreholes (Fig. 4.1). Recharge is the volume of rainfall that remains after all these processes have occurred. Thus,

$$\text{Recharge} = \text{rainfall} - \text{interception} - \text{runoff} - \text{evapotranspiration} - \text{interflow}$$

Recharge is the primary resource for all supplies sourced from underground water. All the water that is recharged to underground water eventually emerges in springs, streams and rivers. This component of flow to rivers is referred to as baseflow, and because there may be a long period between recharge of the water table and emergence in springs and rivers, baseflow is vital to the support of river flows during dry periods. Human intervention now means that some of this baseflow may be intercepted by wells or boreholes.

The porous nature of most soils enables them to hold a certain volume of water against the force of gravity. This volume is referred to as the field capacity.

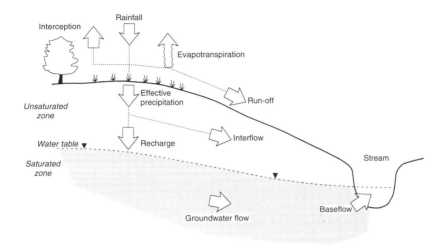

Fig. 4.1 Groundwater flow in the water cycle.

If the soil has less moisture than this (i.e. has a soil moisture deficit) owing to the effects of plants, no water will percolate downwards. Thus, summer rainfall does not usually result in any recharge to groundwater. Consequently, a hot, dry summer (an agricultural drought) may have little effect on groundwater sources. Most groundwater recharge occurs during winter months and a dry winter may lead to a hydrological drought. Soil moisture deficits cannot increase indefinitely, as at some point the process of evapotranspiration will cease to function and plants will wilt.

4.2.3   Groundwater occurrence

Almost all rocks occurring in the surface layers of the Earth's crust contain some void space that may be filled with fluids or gases. Below a depth of about 10 km, the stress caused by the overlying rocks usually causes these to close. Voids of fairly uniform shape (such as the spaces between sand grains) are referred to as pores, whereas voids that are predominantly linear are referred to as fractures or fissures. The ratio of void space in the total rock volume is referred to as porosity. The porosity of clean sands or gravels may be from 25% to 50%, whereas for granite and other fractured rocks, it is usually less than 10%. Different types of porosity are illustrated in Fig. 4.2.

Water occurs in two distinct zones underground: in the upper, unsaturated zone, the voids are only partially filled with water (held in the necks of the pores by capillary forces) and the remainder is filled with air; below this, in the saturated zone, the voids are entirely filled with water. The boundary between these two zones is referred to as the water table and can be observed as the level of water in wells and boreholes where these penetrate through the unsaturated zone to the saturated zone.

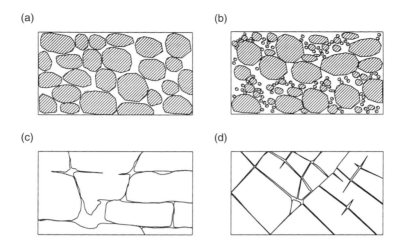

Fig. 4.2 Different types of porosity in rocks: (a) well-sorted sedimentary deposit with high intergranular porosity; (b) poorly sorted sedimentary deposit with low intergranular porosity; (c) rock rendered porous by solution; (d) rock rendered porous by fracturing.

Porosity is not a measure of the ability of a rock to transmit water; a rock may be very porous but if the voids are not connected it will not transmit water easily. Permeability (measured in m/day) is the usual measure of the ability of a rock to transmit water. Typical values of permeability for different rock types are shown in Fig. 4.3. A related parameter is transmissivity (m^2/day), which is a measure of the ability of a formation to transmit water (i.e. transmissivity = permeability × thickness of formation).

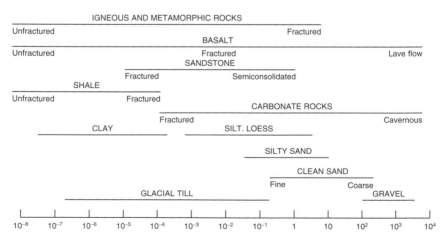

Fig. 4.3 Permeability of selected rocks (from Heath, 1984).

A formation that can store groundwater and that is sufficiently permeable to transmit to wells and springs is referred to as an aquifer. A formation that is effectively impermeable to water is referred to as an aquiclude. Even the most impermeable materials that occur in nature are, in fact, slightly permeable to water, particularly on a long timescale. Where these minor flows are of relevance, these formations may be referred to as aquitards rather than aquicludes. An unconfined aquifer (see Fig. 4.4) is one that contains an unsaturated zone and a water table (one in which the position of the water level in wells and boreholes lies within the aquifer). If an unconfined aquifer is filled to the top by recharge, it will start to overflow when the water table intercepts ground level (i.e. at a spring). However, often aquifers are partially overlain by aquitards or aquicludes (confining layers). If water fills to the top of the aquifer, it cannot overflow (is confined) and the water is stored under pressure. This is referred to as a confined aquifer. If a borehole penetrates through the confining layers into a confining aquifer, the water level will rise up the borehole above the top of the aquifer. The amount of rise depends on the pressure at which the water is stored. The water level may rise above ground level, whereupon the borehole will overflow naturally. This is referred to as an artesian borehole. Where a thin confining layer occurs within the unsaturated zone of an unconfined aquifer, recharge may accumulate on this layer forming a perched aquifer.

Aquifers in which the pores are relatively evenly distributed and interconnected (such as sand and gravels) are referred to as intergranular aquifers. Flow in such aquifers is usually evenly distributed throughout, and the water is slowly filtered through the pores. In fractured aquifers, flow is concentrated in the fractures or fissures, and water often flows more rapidly than in intergranular aquifers. As a result, contamination can travel through fissured aquifers much more rapidly and with lesser retardation than in intergranular aquifers.

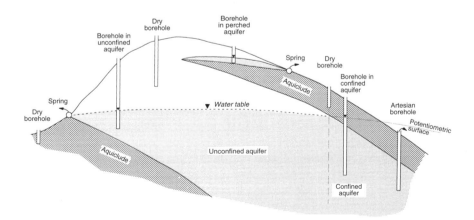

Fig. 4.4 Types of aquifer.

4.2.4   Water levels and groundwater flow

Groundwater will flow from areas with high water levels to those with low levels. This potential energy that drives groundwater flow may be measured by the water level (or groundwater head) in wells and boreholes, measured as height above a standard datum (such as sea level).

The direction of such flows can be determined by constructing a contour map of groundwater levels; groundwater flow will be perpendicular to the contours of groundwater levels (just as flow of water along the ground will be perpendicular to ground level contours). The surface defined by the water level in wells and boreholes is referred to as the potentiometric or piezometric surface. In unconfined aquifers, this is the same as the water table. In confined aquifers, the potentiometric surface lies within the confining layer or above ground level.

The slope of the potentiometric surface is called the hydraulic gradient, and this, together with permeability, controls the rate at which groundwater flows. Groundwater head differences may also occur vertically, and hence groundwater may flow vertically as well as horizontally, e.g. through an aquitard separating an upper and lower aquifer. Groundwater flow, particularly at depth, can also be driven by thermal energy, density differences or in some cases by geochemical factors.

The rate of groundwater flow is dependent on both the hydraulic gradient and permeability. Darcy's law describes this quantitatively:

$$Q = KiA$$

where Q = flow   of   groundwater   (m^3/day), K = permeability   (m/day), i = hydraulic gradient (difference in water levels divided by separating distance) and A = area of rock through which groundwater is flowing (m^2) (see Fig. 4.5). This law is directly analogous to Ohm's law that describes the flow of electricity through a conductor. Darcy's law is valid only for laminar (nonturbulent) flow. As most groundwater flows very slowly, the flow is almost always laminar. However, in fissured aquifers and near boreholes, flows are much faster and may become turbulent, in which case a steeper hydraulic gradient is required to force the same rate of flow through the rock.

4.2.5   Storage of water in aquifers

Storage in unconfined aquifers. When water flows out of an unconfined aquifer, the level of the water table falls. The volume of water that flows out is equal to the volume of water that has drained out of the zone where the water table has fallen.

A preliminary assessment might suggest that this volume is the volume of rock that has drained multiplied by the porosity. However, this is not the case, as not all the water in the voids drains freely under gravity – some of the water is held back by capillary effects. The actual volume of water that drains by gravity

CROSS-SECTION

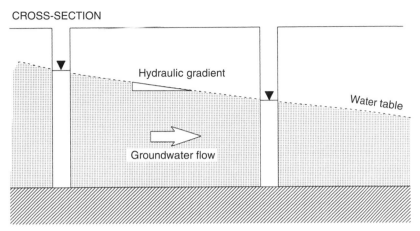

Rate of groundwater flow = Hydraulic gradient x area through which flow is accurring x permeability

PLAN

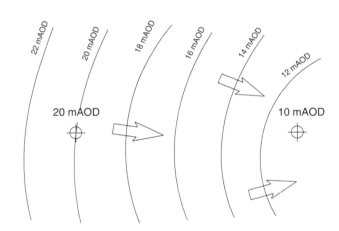

Direction of groundwater flow is perpendicular to groundwater level contours

Fig. 4.5 Groundwater flow and Darcy's law.
*mAOD = metres Above Ordnance Datum.

out of a given volume of rock is referred to as the specific yield (see Fig. 4.6). Thus,

Volume of water released from storage = volume of rock drained × specific yield

The specific yield is always less than the porosity, sometimes significantly – the chalk aquifer in the UK has a porosity of 40% but a specific yield of only 1–2% (also written as a ratio 0.01 : 0.02). This is because the very small necks of the voids in the chalk matrix restrict water from being drained from them.

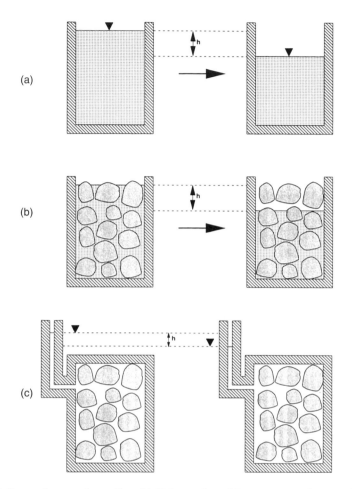

Fig. 4.6 Types of storage in aquifers. (a) Volume released from storage = h × area. Volume released from storage per unit area = h. (b) Volume released from storage = h × drained porosity. Drained porosity is called specific yield or unconfined storage coefficient. (c) Most of the storage has been released by water expanding as the pressure falls. The pores in the aquifer will also compress slightly as the pressure falls.

Storage in confined aquifers. This concept of storage can be best illustrated by an analogy: when air is let out of a bicycle tyre, the volume of air released is much greater than the volume of the tyre when fully inflated. This is because the air in the tyre has been compressed, and expands as it is released. The tyre also contracts a little as the air is let out, contributing a small amount to the volume released. Water is much less compressible than air, but the effect is the same, and slight compression of the aquifer acts in the same way as the contraction of a bicycle tyre. Both of these effects are very small but are still sufficient for confined aquifers to release significant quantities of water from storage when they are pumped.

The coefficient that describes the volume of water released from a confined aquifer by this pressure release mechanism is called the storativity (or confined storage coefficient: see Fig. 4.6). Thus,

Volume of water released from confined storage = area over which water
levels have fallen × fall in water levels × storativity

The value of storativity is much smaller than specific yield (typically 10^{-4}, but may vary from 10^{-3} to 10^{-5} depending on the type of aquifer). Once the confined storage has been used up, the aquifer will become unconfined and water will come from the specific yield.

Natural fluctuations in water levels. Water levels in boreholes are rarely static. Natural fluctuations may occur due to several causes.

- *Seasonal variations in recharge.* Recharge to groundwater occurs mostly during winter, leading to a rise in groundwater levels. During summer months, groundwater levels gradually fall as water is discharged without being replenished. Longer-term fluctuations in groundwater levels may be caused by a succession of dry or wet years.
- *Short-term variations.* These are often observed in boreholes on a timescale of hours to several days. These fluctuations occur most often in deep, confined aquifers and may be linked to variations in atmospheric pressure or to gravitational influence of the moon (earth tides). In aquifers near to the coast, ocean tides may also cause water level fluctuations.

4.2.6   Wells, springs and boreholes

Groundwater flows naturally either to static water (ponds, marshes and wetlands) or to flowing water (streams and rivers). In either case, groundwater is often essential for maintaining the surface hydrological system, particularly during dry periods. Groundwater may discharge either over a broad zone (seepage) or at a distinct point (spring). Springs are more commonly associated with fractured aquifers, which tend to concentrate flows in discrete zones, but springs can also occur in intergranular aquifers depending on how the aquifer intersects the ground. A variety of possible spring configurations is illustrated in Fig. 4.7.

A well is essentially a man-made static source of groundwater, in contrast to a spring, which is a naturally flowing feature. However, many famous 'wells' are in fact springs that have had protective structures built over them. Wells are usually subdivided into two groups according to their method of construction: hand-dug wells or shafts, and machine-drilled wells or boreholes. Hand-dug

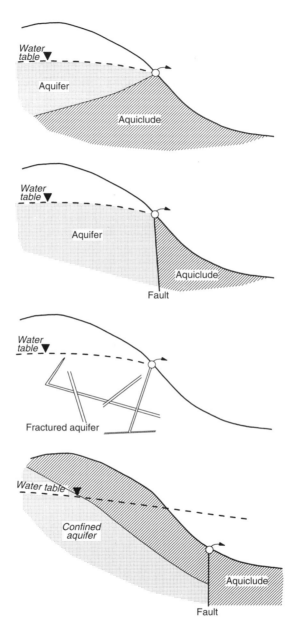

Fig. 4.7 Types of springs.

wells are rarely constructed in developed countries now, as they are slow, labour-intensive and often dangerous to dig. As hand-dug wells usually only penetrate the aquifers nearest to the surface, they are particularly prone to contamination.

A common feature in deep hand-dug wells that have been constructed in competent (i.e. firm) rock is the construction of horizontal collecting galleries referred to as adits. These structures are very effective at increasing the yield of wells, particularly in less permeable aquifers. Drilling techniques are discussed in Section 4.4.4.

4.2.7 Flow to wells and boreholes

When we start to pump a borehole, water is first removed from storage in the borehole. This creates a hydraulic gradient across the sides of the borehole, and water starts to flow from the aquifer into the borehole (Fig. 4.8a). Water is now being taken from storage in the aquifer and the piezometric surface starts to fall (Fig. 4.8b). This creates a cone of depression around the borehole. The radius of this cone is called the radius of influence (r_0). Eventually, water levels stop falling, which is called steady state. Until this point is reached, the situation is transient (unsteady state), with water being taken from storage, causing the levels to fall.

There are three different ways of describing the yield of a borehole depending on the timescale over which the yield is being assessed.

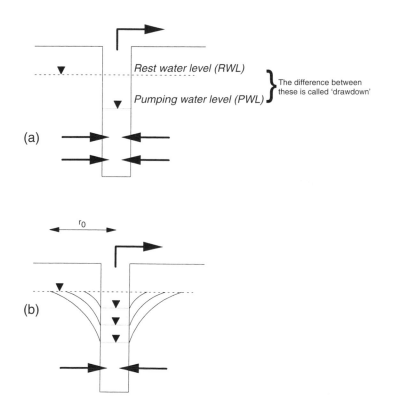

Fig. 4.8 Flow to a borehole. (a) Flow components; (b) Unsteady State pumping, showing increases in water level and radius of influence as water is taken from storage.

- *Borehole yield.* This is a short-term measure of how much can be pumped from a particular borehole. This yield is often measured by the specific capacity. The specific capacity is defined as:

$$\text{specific capacity} = \frac{\text{pumping rate}}{\text{steady-state drawdown at that pumping rate}}$$

  Specific capacity is usually the same order of magnitude as the transmissivity of the aquifer.
- *Aquifer yield.* This is a more general medium-term measurement of the ability of the aquifer to yield water to springs, wells and boreholes.
- *Sustainable yield.* This is a long-term measure of how much recharge entering the aquifer can be drawn out through boreholes. If the sustainable yield is exceeded, water levels will fall year on year as water is drawn from groundwater sources. This is called groundwater mining.

It is also useful to distinguish the potential yield of a borehole (the maximum theoretically available) from the deployable output, which takes into account practical considerations such as pump level, processing capacity and licensed capacity. The yields listed above can be determined by testing the borehole in the following ways.

Step test. As water flows towards a borehole, it converges, and as the same volume has to flow through a smaller and smaller area, the velocity increases. If the velocity increases beyond a certain value (experience suggests 0.03 m/s), turbulent flows may occur and this requires greater hydraulic gradients to drive the same flow of water through the aquifer. Consequently, where turbulent flows occur, drawdown in a pumping borehole will be greater. The extra drawdown is referred to as well loss in contrast to aquifer loss, which is the inevitable drawdown in a borehole for an aquifer with those properties. Well efficiency is a useful guide to the condition of a borehole and is defined as:

$$\text{well efficiency} = \frac{\text{aquifer loss}}{\text{total drawdown}}$$

Well efficiencies are typically in the range 60–70% but may be as low as 10–30%. They usually fall at higher pumping rates and may change with time. Step tests are used for determining both borehole yield and well efficiency. In a step test, drawdowns are monitored while a borehole is pumped for steps of sequentially increasing pumping rate (i.e. 100 min at 50 m^3/day, followed by 100 min at 100 m^3/day and followed by 100 min at 150 m^3/day). At least three steps are performed to define the yield–drawdown curve. Step tests are used to distinguish aquifer losses (which increase in proportion to the pumping rate) and well losses (which increase in proportion to the square of the pumping rate). This is usually expressed as:

$$\text{Total drawdown} = BQ + CQ^2$$

where B = aquifer loss coefficient (BQ = aquifer loss), C = well loss coefficient (CQ^2 = well loss) and Q = pumping rate.

Constant rate test. This test is used for defining the aquifer yield and aquifer parameters. Pumping water levels are monitored while a borehole is pumped at a constant rate. Drawdowns change with time (and with distance if water levels are being measured in surrounding boreholes) can be used to calculate aquifer parameters such as transmissivity and storativity, and may also indicate the presence of aquifer boundaries and interconnection between layered aquifers. Constant rate tests may also be used to determine whether pumping a borehole will have any impact on surrounding water features (lakes, ponds, streams) or other water users.

Long-duration pumping. Preliminary estimates of sustainable yield can be made by studying aquifer geometry, recharge rates and constant rate pumping test results, but the long-term sustainable yield can only be determined by a long period of pumping. It is therefore very important that records of operational pumping rates and water levels are continuously maintained for a borehole.

4.3   Groundwater quality

4.3.1   Hydrochemistry – the history of groundwater
Hydrochemistry is the most distinguishing signature of any groundwater source. While there is relatively little variation in the chemical composition of water falling as rain or snow, the hydrochemistry of groundwaters varies enormously from very lightly mineralised waters that are essentially filtered rainwater to highly mineralised brines. The precise composition of a groundwater source is a record of its history since falling as rain and in particular, a record of the strata through which it has passed and the temperature and pressures to which it has been subjected.

4.3.2   Terms, definitions and concepts
Water is very effective at dissolving salts and some types of organic matter owing to the electrically polarised nature of the water molecule and its tendency to dissociate into constituent ions:

$$H_2O \rightarrow H^+ + OH^-$$

These positively and negatively charged ions are extremely effective at binding with oppositely charged mineral ions to form hydrated ions, thus taking them into solution. This makes water one of the most effective solvents in nature.

The total amount of dissolved material in a sample of water is referred to as the total dissolved solids (TDS). The TDS is measured by heating the sample to 180°C so that all the water evaporates, leaving a dry residue that can be weighed. As these dissolved constituents are mostly in ionic form, water solutions are good conductors of electricity (measured as electrical conductivity (EC) in microsiemens per centimetre (μS/cm)). As the EC of water increases with increasing ionic content, it is approximately proportional to TDS. EC is a useful field measure of how 'mineralised' a water is.

The constituents of groundwaters can be classified as organic (carbon-based) or inorganic. Inorganic constituents are positively charged (cations), negatively charged (anions) or non-ionic. They have been classified as major ions, minor ions and trace elements, according to the frequency with which they occur in groundwater (see Table 4.1). Major ions normally comprise at least 90% of the TDS. As virtually all elements are soluble in water to some degree, only the most common trace elements are shown in Table 4.1. Dissolved organic matter is ubiquitous in natural groundwater. The total organic carbon (TOC) concentration in groundwater is typically in the range 0.1–3 mg/l.

Dissolved gases are commonly present in groundwater and may contribute significantly to the character of a mineral water. Nitrogen and oxygen are the most common constituents of the Earth's atmosphere and are present in most recharge waters. Carbon dioxide occurs in equilibrium with carbonate and bicarbonate ions common in most groundwaters. Methane, hydrogen sulphide and nitrous oxide are products of biologically related processes that can occur in confined aquifers. A variety of units is used in the water industry to express the concentration of dissolved constituents.

Table 4.1 Classification of dissolved inorganic constituents of groundwater

	Cationic	Anionic	Non-ionic
Major ions	Calcium (Ca^{2+}) Magnesium (Mg^{2+})	Carbonate (CO_3^{2-}) Bicarbonate (HCO_3^-) Sodium (Na^+) Potassium (K^+)	Silicon Chloride (Cl^-) Sulphate (SO_4^{2-}) Nitrate (NO_3^-)
Minor ions	Iron (Fe^{2+}, Fe^{3+}) Strontium (Sr^{2+})	Fluoride (F^-) Nitrite (NO_2^-)	
Common trace elements/species	Ammonium	Borate	
	Aluminium Copper		Bromide Iodide Phosphate
		Lead Manganese Zinc	

Mass concentration (mg/l or μg/l). This is the most commonly used unit of concentration and is an expression of the actual mass of that constituent dissolved in a litre of water. This is straightforward when the ion consists of a single element (i.e. Na, Cl, etc.), but confusion can arise where the ion consists of a multi-element species (i.e. HCO_3^-, NO_3^-). The concentrations of these are sometimes reported as the mass of the species in solution (i.e. nitrate as NO_3^-), and sometimes as the mass of the key element in solution (i.e. nitrate as N). It is therefore essential to be clear about the system that has been used for reporting concentrations. Conversion factors for concentration units commonly reported are listed in Table 4.2.

Measure of the number of molecules per litre (mol/l or meq/l). Molarity (mmol/l) is the most fundamental measure of the actual number of molecules in a litre as opposed to the mass of those molecules in a litre (mg/l).

$$\text{Molarity (mmol/l)} = \frac{\text{mg/l}}{\text{molecular mass}}$$

A more common notation is milliequivalents per litre (meq/l), which takes into account the valency of the ion (Na^+ valency one, Ca^{2+} valency two) so that:

$$\text{meq/l} = \text{molarity (mmol/l)} \times \text{valency} = \frac{(\text{mg/l}) \times \text{valency}}{\text{molecular mass}}$$

This is useful because a groundwater sample must have a balanced charge. Thus, the sum of cations (positively charged) in milliequivalents is equal to the sum of anions (negatively charged). This provides a useful check for results from laboratory analyses as shown in Table 4.3.

In that case there is an ion balance (meq/l cations = meq/l anions; within measurement error), which gives us confidence that the analysis has been correctly reported. Usually a difference of less than 5% of the total meq/l (anions plus cations) is considered acceptable.

Alternatively, where one major ion has not been analysed, its concentration can be estimated by the number of meq/l missing from one side of the balance. The mass balance may be affected by minor and trace constituents in some cases and these can be included in the calculation if required.

Table 4.2 Conversion of mass concentration units

Original units	Multiplied by	To give required units
mg/l nitrate as N	4.43	mg/l nitrate as NO_3^-
mg/l ammonia as N	1.29	mg/l ammonia as NH_4^+
mg/l bicarbonate as $CaCO_3$	1.22	mg/l bicarbonate as HCO_3^-
mg/l carbonate as CO_3^-	2.03	mg/l bicarbonate as HCO_3^-

Table 4.3 Ion Balance Calculation

Ion species	Reported value (mg/l)	Molecular mass	Molarity (mmol/l)	Valency	meq/l	Sum (meq/l)
Cations						
Calcium	510	40.08	12.72	2	25.4	
Magnesium	51	24.31	2.10	2	4.2	
Sodium	7.9	22.99	0.34	1	0.34	
Potassium	5.5	39.10	0.14	1	0.14	30.08
Anions						
Bicarbonate	77.5	61.01	1.27	1	1.27	
Chloride	0.6	35.45	0.02	1	0.02	
Sulphate	1383	96.06	14.40	2	28.8	
Nitrate	3	62.01	0.05	1	0.05	
Fluoride	0.1	19.01	0.005	1	0.005	30.12

Parts per million (ppm) or parts per billion (ppb). This is another measure of mass concentration similar to mg/l and μg/l. However, instead of being the mass of an ion dissolved in a litre of water, it is the mass of an ion dissolved in a kilogram of solution (water plus ions). As a simple estimate, ppm can be taken to be equivalent to mg/l and ppb, equivalent to μg/l.

4.3.3   Hardness and alkalinity

Hardness is the property of water that causes it to form scum rather than lather when mixed with soap. Many hard waters also precipitate scale in equipment that heats water, such as kettles and boilers. Hardness is caused by polyvalent cations, mainly calcium and magnesium, which form insoluble carbonates on heating. (Iron and manganese also contribute to hardness.) Hardness that disappears on boiling is *temporary hardness* (sometimes called carbonate hardness); that which remains is *permanent hardness*.

  Total hardness can be measured directly or, more commonly nowadays, the individual ions are measured and the hardness calculated. For concentrations in mg/l, total hardness in $CaCO_3$ is:

$$\text{Total hardness} = [(Ca/20.04) + (Mg/12.16)] \times 50.04$$

Alkalinity is a measure of the capacity of a solution to neutralise an acid. As such, it does not reflect the concentration of a single parameter. However, in almost all natural waters, alkalinity is produced largely by bicarbonate ions and, to a lesser extent, by carbonate ions. The concentration of bicarbonate (and hence the alkalinity) may change rapidly on being exposed to air, as the bicarbonate breaks down to precipitate carbonate and release carbon dioxide. As alkalinity can be measured conveniently in the field, it is a useful measure of bicarbonate at source. Laboratory determinations of bicarbonate and alkalinity have invariably been affected by processes occurring in the sample during transit. Alkalinity is usually reported as an equivalent amount of calcium carbonate; it can also be reported as bicarbonate concentration or as milliequivalents (Table 4.2).

In most groundwaters, if the total hardness is greater than the alkalinity (in equivalent units), then the temporary hardness is the alkalinity, and the permanent hardness is the difference. If sulphate concentrations are low, alkalinity should be similar to hardness.

4.3.4 Evolution of groundwaters

As water flows through the ground, it will gradually acquire a chemical signature from the rocks. Many of the dissolution processes involved are very slow and a groundwater may still be evolving chemically after many thousands of years underground. However, as groundwater flow is also a very slow process, it is possible to track many of the evolutionary sequences by testing the quality of water at various points in the aquifer between the recharge areas and the discharge area. Increasing the rate of flow through an aquifer (e.g. by pumping) may disturb the equilibrium of such a system and lead to discernible changes in groundwater quality.

The chemical evolution of all groundwaters starts with rainwater. There is a common perception that this is pure H_2O, but in coastal areas, rainfall may have a TDS of several tens of mg/l (mainly Na^+, Cl^- and SO_4^{2-} derived from sea water). Even in inland areas, rainwater may have a TDS of several mg/l due to the dissolution of carbon dioxide (CO_2) to form bicarbonate (HCO_3^-). This process releases hydrogen ions (H^+), making rainwater slightly acidic (pH 5–6). In industrial areas, the pH of rainwater may be even lower than this owing to the dissolution of sulphur dioxide/trioxide to form sulphuric acid and nitrous/nitric oxide to form nitric acid. This effect is called acid rain. Rainwater also contains dissolved nitrogen and oxygen. The combination of oxygen and salt gives rainwater a corrosive effect familiar to most car owners living near the sea.

Almost all water that recharges to aquifers passes through a soil zone, and the nature of the processes occurring in this zone exerts a strong influence on the character of the resultant recharge water. These processes may be chemical, biological or influenced by human activity.

Most soils have a capacity to generate acids and to consume most of the oxygen dissolved in the percolating rainwater. The most important acid generated in the soil is carbonic acid by the dissolution of carbon dioxide in water. The main source of carbon dioxide is the decay of organic material in the soil, a process that consumes oxygen. Some organic acids are also generated in the soil zone (particularly in peaty soils), and these may contribute significantly to the TOC of the resultant groundwater. Because this process is linked to biological activity, there may be marked seasonal variations in the composition of recharge waters.

Water infiltrating through the unsaturated zone commonly encounters minerals that are soluble under the influence of carbonic acid to produce recharge water which is often of the calcium bicarbonate type. The following are the main processes that drive the hydrochemical evolution of groundwaters in the saturated zone.

Mineral dissolution/precipitation. Water is a very effective solvent and, given enough time, will dissolve most naturally occurring minerals. Mineral availability is thus the primary control on whether a particular ion is present in groundwater. However, the maximum concentration of the ion that can be held in solution (the solubility) varies with temperature and pressure, which may change as the groundwater flows deeper into the aquifer. Thus, a mineral may be being dissolved at one point in an aquifer while being precipitated at another.

Oxidation/reduction. Recharge water is usually slightly oxidising but, as the groundwater evolves, this oxygen can be consumed by oxidation of organic material so that in confined aquifers (and deep unconfined aquifers), reducing rather than oxidising conditions occur. The oxidising zone is often referred to as the aerobic zone and the reducing zone as the anaerobic zone. The transition from aerobic to anaerobic can appreciably affect the composition of groundwaters. A common series of reactions that may occur in the anaerobic zone includes:

- Reduction of nitrate to nitrogen and/or ammonium ion
- Reduction of the relatively insoluble Fe^{3+} ion to the more soluble Fe^{2+} ion
- Reduction of sulphate ion and hydrogen sulphide gas
- Bacterial fermentation of organic material to produce methane

Ion exchange. It is the process used in commercially available water softening systems; calcium in solution is exchanged for sodium bound on the surface of minerals called zeolites. These occur naturally in basalts, some clays and glauconitic sands and can cause the commonly occurring calcium bicarbonate–type recharge waters to evolve to sodium bicarbonate–type waters.

Processes on discharge to surface. As a groundwater is discharged at surface, either at a spring or from a borehole, rapid changes may occur in the temperature and pressure that may affect the solubility of minerals dissolved in the water. Bicarbonate may break down as it depressurises on discharge to precipitate insoluble carbonates and release carbon dioxide. Fe^{2+} may be oxidised to Fe^{3+}, precipitating a rusty-red, insoluble hydroxide. Dissolved gases are often released on discharge and may be potential hazards: methane can accumulate in buildings, creating a risk of explosion; hydrogen sulphide has an unpleasant smell and at concentrations over 1 mg/l renders water unfit for human consumption; radon is a common constituent of groundwaters in areas underlain by granite, and the by-products of its radioactive decay are harmful to health. These processes may combine in different ways to produce groundwaters of markedly different qualities.

Anionic evolution. A common sequence of evolution of the anionic composition of groundwater is:

Recharge zone: Active groundwater flushing through well-leached rocks. Low TDS with bicarbonate (HCO_3^-) as the dominant anion.

Intermediate zone: Less active groundwater circulation. Higher TDS with sulphate dominant.

Deep zone: Very slow groundwater flow with high TDS and highly soluble minerals (e.g. chloride) present.

This sequence can be explained in terms of two factors: mineral availability and mineral solubility. In the recharge zone, the most soluble minerals such as halite (rock salt, NaCl) have been flushed out long ago so that the groundwater now consists of rainwater modified by dissolution of calcium carbonate. Further into the sequence, there has been less flushing, and minerals such as gypsum and anhydrite (forms of $CaSO_4$) may still be present and contributing sulphate to the system. Evolution to chloride-rich brines occurs only if the water penetrates into deep sedimentary basins. (Chloride-rich groundwaters also occur in coastal aquifers, which may contains slightly modified sea water.)

Groundwater evolution in carbonate rocks. Carbonate-based rocks (limestone and dolomite) are particularly soluble with respect to slightly acidic recharge waters. This explains how caves commonly develop in this type of rock. A consequence is that groundwaters in such aquifers evolve relatively rapidly to calcium or calcium/magnesium bicarbonate–type. Such groundwaters may contain subsidiary quantities of other ions such as sulphate and iron, depending on which other minerals are present in the aquifer. The pH of most of these waters lies between 7 and 8.

Groundwater evolution in crystalline rocks. Most crystalline rocks (such as basalt and granite) contain a high proportion of quartz (SiO_2) and aluminosilicate minerals such as feldspar and mica. Dissolution of these minerals by slightly acidic recharge water usually releases silica and cations (such as sodium, magnesium, calcium and, to a lesser extent, potassium) to groundwater and leads to a rise in pH and bicarbonate concentration. Because this process is slow, whereas flow in fractured rocks tend to be rapid, young groundwaters in crystalline aquifers tend to be dominated by soil zone processes and have relatively low TDS and pH (< 7). Silica concentrations are typically in the range 10–30 mg/l. Chloride and sulphate, when present, can usually be attributed to atmospheric sources.

Groundwater evolution in mixed sequences. Where groundwater flows through a number of different strata, it may follow a complex evolutionary sequence, as minerals are first dissolved and may be then precipitated later in the sequence. Ion exchange is often an important mechanism in this process.

4.3.5   Human influences on groundwater

Up to the twentieth century, groundwater resources had been developed slowly, and generally on a quite localised basis, so that the effects were able to be absorbed by natural processes. Within the last 100 years, and particularly in the last few decades, human activities have progressively modified natural conditions, resulting in significant impacts on groundwater composition in many parts of the world.

Agricultural practices. The widespread agricultural application of NPK (nitrogen, phosphorus, potassium) fertilisers since the 1940s has had probably the greatest impact on shallow groundwater quality. Almost all phosphate in fertilisers is consumed in the soil zone, but significant concentrations of nitrate, chloride and potassium are released to groundwater.

Nitrate concentrations in excess of a few milligrams per litre almost invariably indicate waters derived from shallow aquifers that are polluted, e.g. by septic tanks or from nitrate residues from farming. Boreholes in many catchments that are intensively farmed have experienced rising nitrate levels over many years, and many now exceed relevant drinking water standards. The presence of elevated concentrations of nitrates can have significant implications for the development of bottled water sources. Since elevated nitrate concentrations can be taken as an indicator of surface pollution, and therefore the aquifer not being fully protected from contamination, the recognition of certain categories of bottled waters, such as Natural Mineral Water under European legislation, may be compromised.

Much less information is available on the concentration of pesticides in groundwater owing to their much lower levels of occurrence. However, with more accurate analytical techniques available, this is rapidly becoming recognised as a significant issue.

Sewage. Leakage from sewers and cesspits may contaminate groundwater with elevated levels of nitrate, organic compounds and bacteria. Recent developments in analytical techniques use the presence of compounds such as surfactants and caffeine to prove the effect of human sewage effluent on groundwater.

Acid rain. Burning of sulphur-rich fossil fuels has led to the generation of significant concentrations of sulphur dioxide in the atmosphere in many industrial areas. This forms sulphuric acid in rainwater. The main effect has been on surface ecosystems, but there has also been a discernible effect on some shallow groundwaters.

Solvents. Many industrial processes use a variety of chlorinated organic compounds such as trichloroethene (TCE) and volatile organic compounds (VOCs) as solvents. These compounds are very resistant to biodegradation, and once released into the ground, can be remarkably persistent.

Hydrocarbons. Leakage from underground tanks at petrol stations has recently been recognised as a major source of groundwater pollution (benzene, toluene and some polyaromatic hydrocarbons (PAHs)). Oils and greases are less mobile but may still cause an appreciable problem.

Tritium. Atmospheric hydrogen bomb tests in the 1960s led to significant increases in the concentration of tritium (a stable and harmless radioisotope of hydrogen) in rainwater. In many places, this can still be detected working its way through the groundwater system. The presence of elevated tritium concentrations is often used as an indicator of relatively young groundwaters.

4.3.6 Hydrochemical classification of bottled waters

As groundwaters contain a large number of constituents, classification can quickly become very complex. A useful initial approach is to present the information simply and visually. In doing so, it is important to use meq/l rather than mg/l as the latter is biased towards heavier ionic species. Figure 4.9 shows two pie charts that have been used to compare the concentrations of the major elements

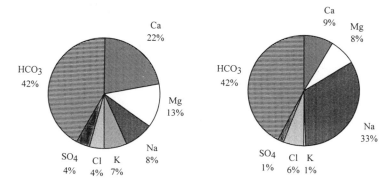

Fig. 4.9 Pie charts to illustrate different groundwater qualities.

of two different waters. A variety of other plots has been developed to illustrate composition and evolution of groundwaters, the most common being the Piper diagram and the Durov plot (see Fig. 4.10).

The simplest classification of bottled waters is in terms of TDS: very lightly mineralised (0–50 mg/l), lightly mineralised (50–500 mg/l), medium mineralisation (500–1500 mg/l) and high mineralisation (1500–10 000 mg/l); the exact boundary between these categories could be selected slightly differently. Most other classifications are based on the dominant cation and anion in terms of meq/l (i.e. calcium bicarbonate–type waters). A classification of well-known brands based on these two approaches is given in Table 4.4.

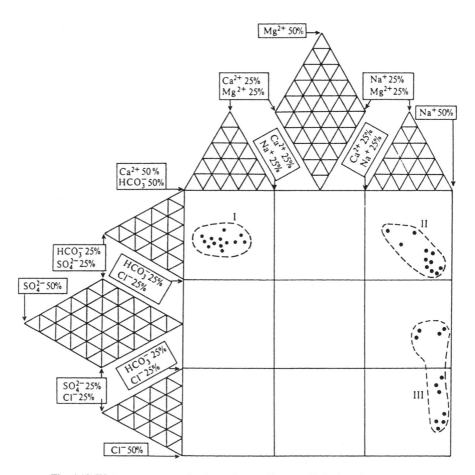

Fig. 4.10 Water types grouped using a Durov diagram: (I) fresh recharge water; (II) ion-exchanged water; (III) old, brackish water (Lloyd & Heathcote, 1985).

Table 4.4 Classification of bottled waters

Types of bottled waters	Mineralisation			
	Very light	Light	Medium	High
Calcium bicarbonate	San Bernado	Buxton Evian Font Vella Levissima Panna Perrier Vera	Boario Ferrarelle	Gerolsteiner Sprudel St Gero
Calcium/magnesium bicarbonate	Spa Reine	Highland Spring San Benedetto Volvic	Badoit Vittel 'Grande Source'	
Calcium sulphate			San Pellegrino	Contrex
Sodium bicarbonate				Appolinaris Hassia Sprudel Uberkinger

4.4 Groundwater source development

4.4.1 Stages of development

This section sets out the stages involved in developing a protected groundwater source that is both chemically and bacteriologically safe.

Groundwater is naturally better protected from contamination than surface water but it can be susceptible to mismanagement: it can become chemically and microbiologically contaminated either via inadequate borehole construction or via contamination in the source zone; it can be affected by overabstraction; and it can have unacceptable effects on other local hydrological features (streams and wetlands). The bottled water manufacturer needs to be aware of how to avoid problems as a result of the activities of others in the area of the borehole and also of their own exploitation of groundwater. The main stages of developing a resource are as follows:

(1) *Resource evaluation*: identification of the physical characteristics of the source that govern the quantity and quality of water delivered to it. This should include assessment of the catchment area and the hydrogeology.
(2) *Source definition*: measurement of the physicochemical and microbiological characteristics of the water. Also defining the variability of the source in terms of flow and composition.
(3) *Source construction*: borehole drilling and testing or spring works construction.

(4) *Source management*: continuous monitoring to identify any changes occurring and to provide information for managing the source. This must
include designing a strategy for protecting against any factors in the source
or associated works that represent risks to the source water quality.

4.4.2   Resource evaluation
The catchment area of a source is the area in which any rain falling on the
ground may find its way to the source. The boundary of a surface water
catchment is defined by the ridges – surface water divides – shown by the
topographic contours on the maps. Definition of catchment areas of groundwater sources is more subtle. They are ideally defined by a contour map of the
water table (see Fig. 4.11). However, the position of the water table is not always
known accurately enough to allow this; instead, the groundwater catchment
must be defined indirectly, using information such as the extent of the aquifer
and the nature and position of the water table as indicated by other boreholes
and springs. If the recharge to the aquifer is known, the area of a groundwater
catchment can be estimated from the following relationship.

$$\text{Groundwater catchment area} = \frac{\text{average flow}}{\text{average recharge}}$$

Where the source is tapping a confined aquifer, the catchment area may be a
considerable distance away, for example, the recharge area for the Great Artesian
Basin in Australia is hundreds of kilometres away from the centre of the basin.
    Once the catchment area has been defined, the next stage in the evaluation of
the resource is to develop a hydrogeological conceptual model of the processes
occurring between the catchment zone and the source. The aim of this model is
to provide a qualitative understanding of how the aquifer system behaves. The
conceptual model should define inputs to the system (usually recharge plus any
surface water inflows) and outflows (which may include spring flows, river
baseflows and any borehole abstraction). The initial conceptual model may
identify areas on which little information is available and which require further
investigation.
    A water balance calculation is a useful first step in moving from this qualita-
tive appreciation of the system to a semi-quantitative understanding. The water
balance works on the following principle:

$$\text{Inputs} - \text{outputs} = \text{change in storage}$$

Estimating how much storage there is in the system is important in assessing the
vulnerability of the system to drought: a source with a large storage component
will be less vulnerable than a system with small storage. The quantification of the
system can be made further by constructing a computer groundwater model of
the aquifer, although this should only be attempted if such a detailed under-

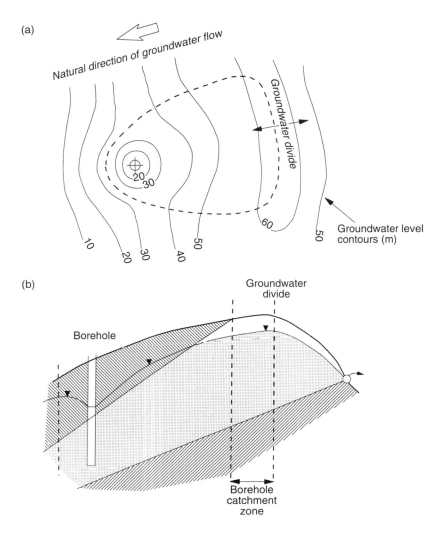

Fig. 4.11 Borehole catchment zones: (a) Unconfined aquifer; (b) confined aquifer.

standing is absolutely necessary, as it requires a large amount of detailed information for calibrating the model.

When developing a source, it is essential to include an impact assessment in the hydrogeological study in order to predict the existing water features that may be affected. In many cases, the potential effect is not significant in terms of environmental degradation, but at some sites even a small increase in abstraction effects a large environmental impact. A formal environmental impact assessment for the proposed development may need to be submitted to the planning/licensing authorities before permission is granted.

4.4.3 Source definition

There is some degree of variability in the quality and quantity of most ground-water sources. As a consequence, it is necessary to monitor the source for a reasonable period in order to define the quality and the reliable yield. For more stable sources, a period of one year may be sufficient, but shallow sources may show a degree of variability from year to year and will need several years of monitoring. There is a variety of different systems operating worldwide for the registration of mineral waters and spring waters, and the appropriate details should be ascertained for the target market.

4.4.4 Source construction

4.4.4.1 Springs

Protection. As springs discharge at ground level, there is often significant potential for contamination from the surrounding soil. The first step in developing a spring must therefore be to secure the spring head against contamination. As each spring is likely to be unique, there are no standard constructions for this, but great care should be taken (1) to avoid accidentally blocking or diverting the spring and (2) to separate the spring flow from any other potentially contaminated sources of water such as shallow soil water.

In some cases it may be impossible to prevent all shallow waters from joining the spring flow, and in these cases it is essential to cordon off the area in which shallow flows may be contributing to flow and strictly prevent any potentially contaminating activities in this area. To avoid turbidity, a filter pack of washed and graded gravel and sand is often included in the spring head construction, as well as a settlement tank with a washout pipe within the spring box. A standard spring box construction is illustrated in Fig. 4.12. Water should be piped directly from the spring head to the intended point of use to prevent contamination; however, a sampling point for water quality testing should be included together with draining and access points for future use.

Yield. Because springs are a natural occurrence, it is not usually possible to increase their yield significantly without drilling a borehole. In some cases, several springs may occur in the same vicinity and it may be possible to collect all these flows into one chamber either by channelling or by cutting the spring back into the water table. As flows cannot be increased, it is important to measure the natural flows accurately to determine the maximum yield available and its variability through the year.

Variability. Deep-sourced springs may have remarkably constant yields and quality; the Bath Hot Springs in England have been monitored since the early days of hydrochemistry and appear not to have changed significantly in quality or quantity for several hundred years. However, shallow-sourced springs are

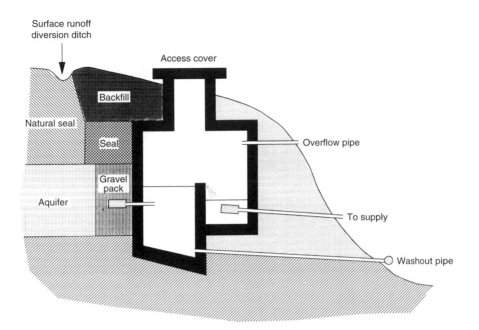

Fig. 4.12 Typical spring chamber construction.

more likely to have variable quality and quantity, which may make them less suitable as natural mineral water sources.

4.4.4.2 Boreholes

The drilling of boreholes is usually carried out by a specialist drilling contractor, and it is recommended that this work should be supervised by an experienced hydrogeologist or engineer. Some of the steps that may be required in drilling and testing a borehole are discussed here.

Site selection. In many cases, the flow of water through an aquifer is fairly evenly distributed. This does not mean that all boreholes in an aquifer will yield the same amount of water, only that the precise yield can never be predicted accurately in advance; there is always an element of risk in drilling boreholes. As a result, the precise siting of boreholes is dictated primarily by practical considerations such as finding a level spot with sufficient room for a drilling rig near to the point at which the water will be required.

In aquifers in which borehole yields are particularly variable, the chances of success may be enhanced by drilling near to faults (which may be indicated on geological maps, detected by aerial photography or in some cases by spring lines). Surface geophysical surveys may also be of some assistance in locating suitable sites. Many drilling contractors will also recommend water diviners or dowsers. There is no consensus within the water industry as to the effectiveness of water divining; all statistical tests for the success of water diviners for siting boreholes (as opposed to locating pipes etc.) show no significant advantage, but their continued popularity cannot be denied.

Drilling techniques. Percussion drilling is the simplest but usually the slowest form of drilling. A heavy chisel is continually raised and dropped, breaking up the formation below. The chisel is then withdrawn and a bailer with a flap bottom is dropped down the hole to pick up the debris. As the drilling progresses, temporary steel casing is allowed to drop down, thus sealing the sides of the hole against collapse.

In rotary drilling, a rugged drill bit is attached to the end of one or more lengths of drill pipe. The top of the drill pipe is rotated by the drilling rig while downward pressure is applied. This causes the drill bit to grind against the rock formation. While this is happening, drilling fluid is continuously pumped down the drill pipe and returns up the borehole, carrying the drilling chippings with it. The drilling fluid may consist of air, foam, water or a mixture of water and clay referred to as 'mud'. The choice of drilling fluid depends on the depth of hole and type of strata encountered. A variant on rotary drilling is reverse circulation drilling in which the drilling fluid passes down the hole between casing and drill pipe and then comes up the drill pipe. Rotary drilling is usually quick and is widely applicable in most formations.

Rotary drilling rigs can often be equipped for downhole hammer drilling. This uses compressed air to fire a percussive drill head repeatedly (similar to a pneumatic drill) and to flush the chippings up the side of the borehole. This is particularly useful in hard formations, but the potential depth of penetration below the water table is limited by the size of the compressor used to pressurise air.

Drilling a borehole can contaminate an aquifer. It is therefore essential to ensure that a drilling rig is cleaned before starting and that great care is taken to prevent spillage of oil or diesel during drilling and testing: fuel bowsers should ideally be double-bunded and kept away from the borehole location when not in use. Specification of biodegradable greases and drilling fluids is also recommended. Drip trays and absorbent matting should be placed beneath the drilling rig and other machinery.

Borehole logging. Where the geology of an area is known, the required depth of a borehole can be predicted in advance. Where drilling is more investigative or the aquifer is more variable, it is often useful to make actual measurements of the rock properties down a borehole. To make these measurements, a specialist contractor will be required to run a suite of geophysical tools down the borehole. A wide range of properties can be measured by borehole geophysics, but it is most often used to detect the presence of clay-rich layers, fissures and inflow zones (where the temperature and EC of the inflowing water may change). Closed-circuit television (CCTV) surveys may also be run to examine the nature and extent of fissures.

Borehole construction. Where a borehole has been drilled in a reasonably competent (firm) formation, the borehole sides may stand up without any support. However, it is necessary to install casing (pipes) in the top section of the borehole (usually from surface to water level) and to secure this with cement (sulphate-resistant) in order to prevent near-surface contamination from entering the borehole. Where the formation is less competent and shows signs of collapse, it is also necessary to install slotted casing or screen against the water-yielding sections so that the aquifer can be supported while still yielding water to the borehole. The size of slots or screen is determined by the size of grains in the aquifer formation. Where the aquifer is very fine-grained, it may be necessary to install a filter pack (gravel pack) of graded sand or gravel in the annulus between screen and aquifer face to prevent the aquifer grains from passing through the screen slots. All materials used in borehole construction should be resistant to corrosion. Stainless steel (e.g. grade 316) casing and screen are normally used for natural mineral water and spring water boreholes.

Borehole development. The process of drilling a borehole creates much fine material that can clog the water-yielding horizons in an aquifer. After the casing and screen have been installed, the borehole should be developed to flush out this loose material. There is a variety of techniques including airlift pumping, over-pumping and surge pumping. Boreholes in limestone aquifers may be developed by acidisation (the introduction of concentrated hydrochloric acid into the borehole). The acid reacts with the limestone, generating carbon dioxide, and enlarges the fissures around the borehole by dissolution. The acid is consumed in a few hours and the borehole is then flushed again to remove loose debris and the residual chloride from the acid–limestone reaction.

Test pumping. After drilling and development, a step test and constant rate test should be carried out to determine the borehole yield (see Section 4.2.7 for more details on pumping tests).

Water quality testing. Water quality should also be monitored during this process as the quality will be affected by drilling at the start of testing and will only equilibrate towards its long-term value after a prolonged period of pumping. Values of pH, EC, temperature and alkalinity should be measured daily at the wellhead. Samples for analysis of major ions should also be collected daily with samples for minor ions, perhaps every two or three days during testing. At the end of the test, a sample should be collected for analysis of a comprehensive suite of determinants. Hydrochemical analysis and classification are discussed in Section 4.3.

Borehole redevelopment. The condition of a borehole may deteriorate with time (often indicated by a drop in well efficiency), and it may become necessary to redevelop it. A CCTV survey should be run to examine the condition of casing and screen and to detect the presence of clogging caused by chemical precipitation or by biofouling (e.g. by iron bacteria). Boreholes can be redeveloped by brushing, by airlift pumping or by acidisation to remove encrustation. Boreholes can be sometimes relined if the existing casing is leaking.

Pumps. Pumps are selected on the basis of borehole diameter, required yield and pumping water level. Where the pumping water level is within 7 metres of the surface, it is possible to use a surface-mounted suction pump. Where the pumping water level is deeper than this, it is necessary to use an electrical submersible pump or a surface-mounted engine that drives a downhole rotor.

4.5 Management of groundwater sources

4.5.1 Record keeping

Like all manufacturing processes, production of spring water and natural mineral water requires constant vigilance to ensure quality and to provide advance warning of the onset of problems. The collection and preservation of records on all aspects of a groundwater source, including its construction, operation, maintenance and abandonment, is an essential activity. If neglected, it will generally not be possible to identify and correct problems economically, possibly leading to declining yield or deterioration in water quality.

Every groundwater source should have a regular monitoring scheme. A file should be established when initial development plans are formulated, and it should be maintained right through to the time when the borehole is finally abandoned or otherwise disused. The data collected should include:

(1) Details of the initial design of individual boreholes, and of the wellfield if multiple boreholes are used in exploitation of the aquifer.
(2) Borehole construction records
(3) Pumping test data, including tests conducted at the time of construction and subsequent step tests to monitor borehole yield characteristics
(4) Records of the type and capacity of the pump and the depth at which it is installed
(5) Operating records, which should include pumping rates and static and pumping water levels
(6) Records of any maintenance that is carried out on the borehole
(7) Records of abandonment

The type of forms used and the format of the data collected are not of critical importance. What is of most importance is that the records are collected on a routine basis and that the data are regularly reviewed. If the data show any signs that parameters (such as water levels, yield or water quality) are moving away from their long-term values, a period of more detailed monitoring and assessment should be implemented.

Consideration should be given to installation of data loggers which can automatically collect water level, water quality (e.g. EC) and pumping data at a preset frequency, i.e. at intervals ranging anywhere from minutes to hours to days.

4.5.2 Sampling and water quality analysis

The chief distinguishing feature of most bottled waters is their hydrochemical quality, particularly for natural mineral waters that are required by legislation to have the stable chemical composition displayed on the label of the bottle. It is therefore particularly important to have accurate measurements of the individual constituents that contribute to that unique character. Because of the nature of individual parameters, measurements of some should be made in the field, while others must be determined from samples collected and sent to the laboratory for analysis.

There are two main reasons for measurement of parameters in the field: (1) it allows rapid assessment of changes in the sampled water and (2) it provides a control for measurements taken later in the laboratory, particularly for those constituents that may change in transit. Key determinants that should be measured in the field include EC, temperature, pH, dissolved oxygen and alkalinity. Most of these can be determined using hand-held meters or probes linked to data loggers (to allow regular monitoring). Alkalinity can be measured using a field titration kit.

The key consideration when collecting a water sample for laboratory analysis is to ensure that it is representative of conditions in the aquifer. Water that has been standing in a borehole or spring tank for several days will have changed in composition, and it is necessary to flush out the standing water before sampling. The sample should be collected in a clean bottle and immediately sealed with no air space remaining in the bottle. Samples should be refrigerated and then transferred to the laboratory in a cool box. For determinants that may change during transit, the laboratory may supply bottles with a particular preservative. For instance, when analysing for dissolved iron, the sample should first be filtered (to remove any solid iron particles) and then preserved by adding concentrated nitric acid, which prevents iron from precipitating during transit.

A variety of analytical techniques are available for most determinants. The laboratory may suggest the most appropriate technique, taking into account the accuracy and limit of resolution required. The frequency of sampling and analysis will depend on the size of the source, and the particular requirements of the regulatory body that implements the relevant legislation. For a moderate-sized source (e.g. 10–30 Ml/year), the following frequencies may be appropriate:

Water levels, pumping rates	Daily
pH, EC, temperature	Daily
Taste, odour, visual check	Daily
Major and minor ions, toxic elements and metals	Monthly or quarterly
Full suite of analysis	Annually

4.5.3 Control of exploitation

One of the most important aspects of source management is to ensure that the abstraction is sustainable in the long term, and that it does not cause detrimental impact upon the aquifer or on other abstractors. This requires

- that flow rates are maintained at a level that is commensurate with the amount of water actually required for the production process, ensuring that the total volume of abstraction does not exceed the available resources (i.e. avoid overabstraction);
- that the source is adequately equipped and protected so as to exclude the risk of pollution or contamination;
- that the quantity and quality of the abstraction is continuously monitored and that any deficiencies which may occur in the course of exploitation are promptly rectified;
- that routine analyses are undertaken as required by local or national regulations.

A decline in yield and/or a deterioration in water quality is likely to occur unless the abstraction is kept within the level that is sustainable and the source is properly managed.

4.5.4 Risk assessment and definition of protection zones

The consequences of contamination of a bottled water source are often far greater than that for a public water supply source, and the controls on catchment areas should therefore be proportionally stricter. This is particularly true for natural mineral waters, which may not be treated in any way that affects their natural composition. The catchment to the source must be identified so that potential risks associated with activities within the catchment can be understood and, where possible and practical, controlled and managed.

The area in which rainfall enters the aquifer is the most vulnerable part of the system because contamination can also enter at this point. The aquifer itself may have strong filtering capacity, but this is not effective against all forms of contamination. During its passage through the soil zone and shallow geological strata, groundwater may pick up anthropogenic contaminants that have come from a number of sources. Some of these are:

- Dry deposition of atmospheric contaminants
- Regional application of fertilisers and pesticides

- Changes in land use (e.g. ploughing up grassland or pasture may result in a pulse of nitrate being leached from the soil zone)
- Urban development, including sewerage, industry, roads (oils and salt), railtracks (herbicides) and airport runways (de-icing agents, oils and hydrocarbons)

Although the general characteristics of an aquifer may be well known, the sensitivity of borehole water sources to various types of land use is often poorly known. The vulnerability of the source will depend on the nature of the aquifer and the land use activities within the catchment to the source. For unconfined aquifers, the catchment area is often relatively simple to define, flow times are often relatively short and the effects of activities in the catchment area are easier to identify. Where an aquifer is confined, it is often assumed that the source will be well protected from human activities. Although this is often the case, it is not always true, particularly where groundwater storage and flow occurs primarily within fractures, such as in limestone and basalt. In such cases, the definition of recharge areas can be very difficult, and it can be hard to assess the risk posed by activities that may be at some distance from the borehole source. It is therefore strongly recommended that a protection zone be defined around the source, in order that the level of potential risk arising from various land use activities can be identified and managed.

A simple but robust approach to defining protection zones is to say that the whole source catchment should be designated a protection zone, since contamination entering the ground anywhere in this area could reach the water table and hence the source. Whilst this may be a reasonable approach for a small source, it may not be possible to exert control over the whole catchment for a large source. In such a case, the usual approach is to prioritise areas according to their travel time, i.e. the time it would take groundwater to travel from that point to the source. This approach is standard for large public water supplies sourced from boreholes and springs. In the UK, three zones are defined:

- *Zone I*: Inner Source Protection. A 50-day travel time from any point below the water table to the source (minimum 50 m radius). Strict controls are maintained in this area.
- *Zone II*: Outer Source Protection. A 400-day travel time from any point below the water table to the source or, in high-storage aquifers, the larger of this and the catchment area calculated using 25% of the long-term abstraction. Some activities may be discouraged in this area.
- *Zone III*: Source Catchment. The complete catchment of the groundwater source. Management policies such as reduction of fertiliser inputs may be appropriate for this area.

In the case of confined aquifers, the catchment zone may be a considerable distance from the source. However, in these cases it is worth defining an intermediate transit zone. In this zone, the aquifer is naturally protected from contamination by the impermeable confining layer but can be contaminated where this is breached, e.g. by other boreholes (abandoned boreholes are a common source of contamination in confined aquifers).

The level of control that can be exerted over land use activities, and in particular, the degree to which potentially contaminating activities can be controlled, will depend on land ownership within the catchment area. The ideal situation is one in which the abstractor has complete control over land use and so can minimise the risk of pollution of the source. Where the aquifer is unconfined and hence vulnerable to contamination, and where the abstractor has limited or no control over land use, very careful consideration needs to be given to the potential for pollution of the aquifer and to the type of bottled water source that is developed. For example, it may be preferable to develop a source that can be treated as opposed to one in which the scope of treatment is limited by legislation, if there is a possibility that the aquifer is, or could in future become, polluted by agricultural nitrates.

4.5.5 Monitoring, maintenance and rehabilitation

The bottled water manufacturer is affected by one single factor that separates him from the rest of the water industry – that he is tied to a single, often named, source. There is therefore usually no freedom to obtain water supplies from another location, should something go wrong with the borehole or spring. Proper care and maintenance of the source, coupled with a detailed knowledge of its history and operational use, are therefore essential to the long-term viability of the source.

Often the borehole or spring are the least well known of the bottled water manufacturer's assets, and yet they are the most important. While large amounts may be spent in marketing, or in upgrading the bottling factory, comparatively minor sums are often spent on the water source itself. This partly stems from the fact that there is relatively little that can go wrong with a borehole, or the borehole pump, but also from the fact that, being below ground, they are 'out of sight and out of mind'. However, the results of a breakdown can be catastrophic and could potentially result in major disruption to production; this is most likely to occur during times of peak production when the aquifer and borehole supply are under greatest stress.

Awareness and understanding of a particular problem is as important as its cure, yet proper diagnosis requires a range of hydrogeological and operational information, which is all too often not available because appropriate monitoring has not taken place. Maintenance and rehabilitation have traditionally lacked glamour and can often be tainted with the embarrassment of something gone wrong. However, economic advantages can be gained from provision of adequate and proper management of the source. The three aspects of borehole management, namely, monitoring, maintenance and rehabilitation, should therefore be regarded as integral parts of a cohesive programme, rather than as separate functions.

- Monitoring is required to show the need for active maintenance measures.
- Maintenance covers two fields: pump and borehole structure, and involves a programme of routine actions to maintain performance and to prevent borehole deterioration.
- Rehabilitation is the action needed to repair a borehole or a damaged pump that has failed owing to inadequate monitoring and maintenance.

The maintenance regime for the borehole will depend on the local conditions in the particular aquifer. Both the frequency of maintenance and the action needed to maintain borehole performance vary, but in all cases should be based on data from the monitoring programme.

4.5.6 Monitoring borehole yield

The yield of a water supply borehole depends on three factors: the aquifer, the borehole and the pump, and a decline in yield can be related to a change in one of these factors. The routine monitoring of water levels and pumping rates is the most important aspect of borehole source management, since a problem can be identified only if data are available on the water level/discharge rate relationship in the pumping borehole. Measurements of specific capacity and available drawdown are neither difficult nor time-consuming and are based on the following:

(1) Static (nonpumping) water level – this should be measured at least weekly, at end of the longest nonpumping period (often the end of the weekend).
(2) Maximum pumping level – as with static water level, this should be measured at least weekly, at the end of the longest pumping period (normally the end of the working week).
(3) Pumping rate – measured at the same time as the maximum pumping level.

Water level and pumping rate data should be collected and reviewed on a more-or-less continuous basis to develop an understanding of the way in which water levels in the borehole respond to pumping. The onset of problems can then be easily identified and then either halted or reversed through proper maintenance. For example, if the discharge rate is maintained at a fairly constant level, a slight but continuous fall in water level may indicate a depletion of resources within the aquifer, or possibly deterioration of the pump, while a more marked fall in water level in the borehole that is not seen in nearby observation boreholes is most likely to be due to deterioration in the structure of the borehole itself.

A more quantitative assessment of borehole behaviour can be obtained by carrying out a carefully planned step drawdown test (see Section 4.2.7). Such tests should be undertaken at least annually. The timing of the test should take into account natural seasonal changes in the aquifer, so that, for example, a comparison is not made between conditions at the end of summer when aquifer water levels are at their lowest, and in spring when levels are correspondingly high.

Although the condition of a borehole can be monitored through changes in the water level/discharge rate relationship in the pumped borehole, the conditions within the aquifer can only be determined through monitoring water levels in observation boreholes that are unaffected by pumping. The monitoring of water levels in both the production borehole and in one or more observation boreholes should therefore be seen as an integral part of the monitoring programme.

Some of the more common reasons for declining yields, and some of the possible solutions to the problem are summarised below.

The aquifer	Where the observed decline in yield is accompanied by a fall in water level, but with no change in the specific capacity, the cause of the problem can usually be attributed to a change in conditions within the aquifer.
	A decline in yield may occur as a result of falling water levels in the aquifer, possibly caused by a decline in recharge (drought) or excessive withdrawals (overabstraction).
	The most immediate solution to the problem would be to reduce the rate of abstraction, although this may be practical only if the problem relates to a short-term occurrence, such as a summer drought. Longer-term solutions may be to spread the abstraction over a larger area of the aquifer (more boreholes, each pumping less water) or even implement measures to artificially enhance recharge to the aquifer.
The borehole	A fall in yield that is due to problems within the borehole is manifested by a reduction in specific capacity, but no change in water level. In particular, a marked fall in water level in the pumping borehole that is not seen in nearby observation boreholes is most likely due to deterioration in the borehole structure.
	This often results from blockage of the well screen by rock particles, deposition of carbonate or iron compounds, build-up of biofilm inside the borehole or through reduction in the length of open hole by movement of sediment in the borehole.
	The solution is usually to redevelop the borehole through physical (airlift pumping and surging, scrubbing) and/or

chemical (acid to dissolve encrustations, disinfectant to remove biofilm) means.

The pump   Where the cause is attributable to the pump, no change is usually seen in water level or specific capacity.

The cause may be due to wear of pump impellers and other moving parts or to loss of power from the motor.

The solution almost always requires that the pump is removed from the borehole and is either reconditioned or replaced.

4.5.7   Changes in water quality

A change in water quality can have fundamental impact on the viability of a bottled water source, and so water quality monitoring is perhaps the most important aspect of source management.

Regulations governing bottled water production almost always require that water quality is assessed on a regular basis to ensure that the water continues to meet set standards.However, analyses done for regulatory purposes, although probably comprehensive in terms of the parameters assessed, will generally not be frequent enough to determine whether changes are occurring, so that correct-ive action can be undertaken if required. The routine monitoring programme therefore needs to focus on key parameters that are indicative of changing conditions.

A change in water quality can occur as a result of changes in the aquifer or in the borehole and can affect the biological, chemical or physical quality of the water. Variations in the quality of water pumped from the borehole are observed when changes occur in the quality of water moving through the aquifer, and may involve almost any substance soluble in water. Changes in chemical quality may occur in a variety of ways.

- Natural changes in the hydrochemistry of the aquifer in response to sea-sonal fluctuations in rainfall and recharge. Such changes will generally have few concerns for the bottled water producer, but must nevertheless be recognised. Some regulations, such as those governing natural mineral waters, require that the composition of the water is stable. This requirement must take into account the limits of natural fluctuations, which can vary from virtually no change in confined aquifers in hard rock areas to sub-stantial changes in shallow, unconfined, unconsolidated aquifers. What is of particular importance is that the source water is monitored on a suffi-ciently regular basis so that deviations from the 'norm' can be identified.
- Changes in water quality can be induced by pumping and can occur in several ways. As the pumping rate increases, water is drawn from areas increasingly further away from the borehole. This can result in water from different geological strata or different fracture systems entering the bore-hole. In areas close to the sea, brackish water can enter the bottom of the

borehole through upconing of the saline interface. The solution to such changes in water quality is usually to pump either at uniform rate so that water is always obtained from the same catchment area or to reduce pumping to a rate that ceases to pull water of undesirable quality into the borehole.

- Accidental or intentional release of potential pollutants within areas of influence of water supply boreholes. Of particular concern to bottled water producers are long-living substances in groundwater, and which may travel unchanged over long distances within the aquifer, such as herbicides, pesticides and other complex organics, petroleum products and substances that contain trace concentrations of metals.

Deterioration in the biological quality involves the occurrence in the water of bacteria, viruses and/or parasites associated with human or animal wastes. Such deterioration indicates, in nearly all cases, a connection between ground surface and the open section of the borehole. This could occur via pollution of the ground in the vicinity of the borehole, or via the borehole itself, perhaps if the integrity of the casing has deteriorated or if the wellhead has been flooded. Protection against pollution in the immediate vicinity of the borehole can be provided through identification of catchment protection zones and restriction of land use activities within the catchment area (see Section 4.5.4). Of particular importance in an unconfined aquifer is exclusion of livestock and other potentially polluting activities from the Inner Protection Zone. Proper sanitary design of the borehole and headwork will also help to prevent leakage of polluted water into the borehole.

Changes in physical quality involve the appearance, taste and temperature of the water. Taste and temperature changes are likely to be associated with changes in water quality, and will occur for similar reasons. A change in appearance or colour may occur as a gradual presence of rock particles in the water, indicating that the strata in the zone adjacent to the borehole still contain fine-grained material that was not adequately removed during borehole development (see Section 4.4.4). Several solutions are possible, depending on the nature of the problem. Some of them are:

- Removal of the pump and redevelopment of the borehole
- Decreasing the pumping rate to reduce inflow velocities and prevent mobilisation of fine particles
- Installation of a course filter at the wellhead to remove particles before they enter the distribution system – but note that allowing the problem to continue could lead to excessive wear and damage to the pump and possibly to the borehole itself.

The sudden appearance of suspended solids in the water is more likely to indicate the failure of well screens or the rupture of the well casing, perhaps as a result of corrosion. The solution in this case may involve either the removal of the old casing and screen and replacement of new material or the installation of a smaller-diameter well screen inside the original casing. The former is often not possible and the latter is likely to restrict the size of the pump that can be installed, and hence the maximum yield that can be obtained. Careful monitoring of physical parameters is therefore an essential aspect of routine water quality monitoring, so that the onset of problems can be identified and remedial measures taken before catastrophic failure occurs.

In planning the development of a new source and in determining the scope and frequency of the water quality monitoring programme, due account must therefore be taken not only of the natural changes that may occur within the aquifer but also of the artificial changes in water quality that may occur as a result of the abstraction, and the potential risks to the source that could arise due to both historic and future pollution incidents. Failure to implement a thorough and regular monitoring programme is likely to result in unforeseen problems, most probably at a time when the source is under greatest threat and when a reliable water supply is most urgently needed.

Bibliography

British Soft Drinks Association (1995) *The Guide to Good Bottled Water Standards.* A useful guide to technical aspects of developing and managing a bottled water source.

Domenico, P. & Schwartz, F. (1990) *Physical and Chemical Hydrogeology.* A comprehensive textbook on general hydrogeology. Wiley.

Driscoll, P. (1986) *Groundwater and Wells.* Johnson Filtration Systems. An invaluable reference for all aspects of well drilling and development.

Freeze, A. & Cherry, J. (1979) *Groundwater.* A comprehensive textbook on general hydrogeology. Prentice Hall.

Green, M. & Green, T. (1994) *The Good Water Guide.* A guide for general information on a wide variety of bottled waters. Rosendale Press.

Heath, R. C. (1984) *Basic Groundwater Hydrology.* US Geological Survey Water Supply Paper 2220. United States Department of the Interior.

Hem, J. (1985) *Study and Interpretation of the Chemical Characteristics of Natural Water.* A useful reference on hydrochemistry. USGS Water Supply Paper 2254.

Kruseman, G. & de Ridder, N. (1990) *Analysis and Evaluation of Pumping Test Data.* A reference for a comprehensive analysis of pumping tests. IILRI.

Lloyd, J. & Heathcote, J. (1985) *Natural Inorganic Hydrochemistry in Relation to Groundwater.* A useful reference on hydrochemistry. Clarendon Press.

Price, M. (1996) *Introducing Groundwater.* A book for a detailed introduction to hydrogeology. George Allen & Unwin.

Stumm, W. & Morgan, J. (1981) *Aquatic Chemistry.* A valuable reference on hydrochemistry. Wiley.

5 Water treatments

Jean-Louis Croville and Jean Cantet

5.1 Why and when water must be treated

The bottled water industry employs many different types of water from various sources, including naturally flowing springs, boreholes and previously treated municipal supplies. Water may be bottled for human consumption or may be used for some other essential part of the process such as cleaning but, regardless of its origin or ultimate purpose, it may be necessary or desirable to treat it further before use. Depending on the incoming quality, there is now a wide range of treatments available, and this chapter attempts to provide some basic information to assist bottlers in choosing which methods to use. Water treatment may be used for various reasons, some of which are discussed below.

5.1.1 Compliance with local regulations

Local regulations may impose limitations on the quantities of some microbiological and chemical elements in the bottled product. Most of the time, these regulations are based on international legislation or recommendations or guidelines such as those provided by the WHO and Codex (see Chapter 3). Depending on the nature and objective of the treatment, the treated water might be bottled under a different status (Mineral Water, Spring water, Natural Water, Prepared Water, Drinking Water, Purified Water, etc.) according to local definitions. The main constituents are the microbiological indicators of contamination (coliform, *Pseudomonas aeruginosa*, etc.) and chemical elements of 'health significance' (inorganic and organic elements).

5.1.2 Quality reasons

For some specific quality reasons, it is necessary to remove some elements from the water before bottling, even if their quantities are below the local regulation limits. A very practical example is in the case of unstable elements such as dissolved iron, which may give brown deposits of oxidized iron in the bottled product after a few weeks of storage. Such treatments can be applied to any water, and are usually accepted by local regulations for natural waters, as they do not modify the composition in terms of the characteristic constituents.

5.1.3 Marketing reasons

For marketing reasons, some modifications may be made to the original composition of the water, such as removal of minerals or addition of valuable constituents (salts, vitamins, fibers, etc.), to be adjusted to market needs and preferences, and the target product. In the case of a treatment that strongly modifies the original composition of the water, the product can no longer be classified as a 'natural' water.

5.2   Water treatment objectives

5.2.1   Removal of undissolved elements
Undissolved elements found in water can be in particulate or colloidal form, and may originate from the aquifer, or from one of the process steps themselves, such as a mass media filter. Removal is generally done using a retention process, which is preceded by oxidation and/or coagulation, if necessary. The main retention processes include mass media filters and bag and depth filters.

5.2.2   Removal/inactivation of undesirable biological elements
Undesirable biological elements found in water may originate from the source (in the case of surface or superficial water), or from contamination during the water process, and must be removed or inactivated before bottling. They include (ranked in size from larger to the smaller) protozoa, mold and algae, bacteria and finally viruses. Removal of these elements can be achieved by physical retention (adsorption/screen effect). Inactivation is generally obtained by destruction of the cell structure and genetic material using chemical oxidation (chlorine, ozone, peroxide, etc.) or irradiation (ultraviolet (UV)). In many cases, such microbiological treatments are used as safety barriers just before bottling to prevent any potential accidental contamination.

5.2.3   Removal of undesirable chemical elements
Undesirable chemical elements, in stable or unstable form, are of natural origins (due to the geological characteristics of the aquifer), or from contamination linked to human activities (industry, urban and agricultural). Specific removal of these elements can usually be achieved by one (or a combination) of the standard water treatments based on oxidation, filtration, adsorption, ion exchange or bioelimination.

5.2.3.1   Iron and manganese
Iron and manganese are dissolved in their stable form in the water whilst it is underground, but as soon as they contact the outside air, they may begin to oxidize and precipitate out naturally (though manganese oxidation with air takes somewhat longer). Removal of iron and manganese consists of a forced oxidation with an oxidant (principally air or ozone in the bottled water business), followed by mass media filtration (sand, anthracite, etc.).

5.2.3.2 Ammonium

Ammonium is not identified as a 'chemical of health significance' by WHO; however, it has to be strictly controlled in nonsterile water, considering its ability to be biologically oxidized into nitrite, which is recognized to be toxic. Removal of ammonium can be done by chemical oxidation with sodium hypochlorite or by bio-oxidation. Due to the risk of by-product formation, bio-oxidation is the recommended process in bottled water.

5.2.3.3 Arsenic

Arsenic is identified as a 'chemical of health significance' by WHO; it is widely distributed throughout the earth's crust and very often found in underground water originating from volcanic strata. Removal of arsenic can be done by coprecipitation with ferric hydroxide, or adsorption onto activated alumina, granular ferric hydroxide or manganese dioxide sand.

5.2.3.4 Fluoride

Fluoride is identified as a 'chemical of health significance' by WHO; it is a fairly common element found in a number of minerals. Very often, fluoride is associated with highly mineralized and naturally carbonated water. Removal of fluoride can be achieved by adsorption onto activated alumina.

5.2.3.5 Organic matter

Although not considered as a toxic element, organic matter may have to be removed from water to limit the risk of uncontrollable biological growth in the process that may lead to biomass accumulation and off-odor/taste issues. Organic matter may also develop toxic by-products when oxidized with chlorine or ozone. Removal of organic matter can be done by adsorption onto activated carbon or biological degradation on media filter. Membrane filter processes such as nanofiltration (NF) and ultrafiltration (UF) can also significantly reduce the level of organic matter.

5.2.3.6 Volatile organic compounds and pesticides

Volatile organic compounds (VOCs) and pesticides are identified as 'chemicals of health significance' by WHO; they usually result from contamination by human activities, though some VOCs are of natural origins, particularly in the case of ancient sedimentary and volcanic reservoirs bearing naturally carbonated water. Removal of VOCs and pesticides can be done by adsorption onto activated carbon.

5.2.4 Addition of valuable 'elements'

A wide range of constituents are added to water for commercial reasons. The different categories of water or beverage available in the market are:

- Flavored (flavors, flavor enhancers, etc.)
- Nutritional (mineral salts, vitamins, etc.)
- Functional (plant extract, fibers, etc.)

Addition of these elements can be done by traditional batch preparation, or by in-line dosing of a concentrate. Specific chemical and microbiological treatment may be necessary to stabilize such products. Carbon dioxide addition is also a very common water treatment; it can be done by dosing natural or industrial carbon dioxide directly into the water flow (in-line injection or tank saturator).

5.3   Water treatment processes

5.3.1   Filtration

Filtration is a physical process widely used in water treatment to remove different types of particles, such as suspended solids (turbidity), colloidal compounds, metalloid complexes and biological species. Depending on the application, the filtering part is a screen, a membrane or granular material: running conditions such as flow rate and size of the particles requiring removal may necessitate adjustment of the process (see Fig. 5.1). Typically, the size of particulate material ranges from approximately 0.001 μm to 100 μm, in highly variable shapes. In each case, the particulate matter is accumulated in or on the filter, and the pressure drop across the filter medium is one of the key parameters of the process monitoring.

5.3.1.1   Mass media filters

Mass media filters are common in water treatment processes. The most frequently used are classified as follows:

- Rapid sand filters
- Slow sand filters
- Pressure or gravity filters
- Diatomaceous earth filters

Mass media filters are used for removal of particles down to 20–30 μm. Depending on the material used as filtration media, they have very high retention capacities, and are therefore well adapted to the treatment of water with heavy particle load.

Pressure filters, which are very common in the rapid sand filter category, are widely used in the bottled water business. The vessel is normally cylindrical, made from steel, stainless steel or plastic materials with a diameter of up to 4 m. The majority are downflow forward, with upflow backwash (for the cleaning

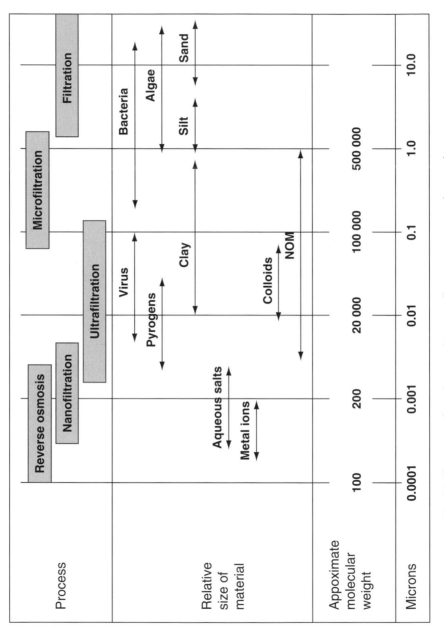

Fig. 5.1 Size ranges of membrane and mass media–type processes and contaminants.

step), as this is the simplest arrangement in terms of hydraulics and protection of filtered water quality. In these running conditions, the upper part of the filter is the effective section involved in the filtration process, as the continuous 'cake' formation (accumulation of the deposit) contributes to the filter efficiency. The main components of a pressure filter are represented in Fig. 5.2.

The minimal equipment installed for such filters comprises:

- Pipes and valves required for the production mode and the backwash mode
- Vessel itself with a floor usually fitted with nozzles
- Media

The vessel is also fitted with an upper air vent system. The sizing is mainly established based on the flow rate calculated on

- a filtration rate, defined as the ratio between flow rate and cross section (expressed in m^3/m^2h). In the bottled water treatment, the normal range for a rapid sand filter is from 8 to 15 m/h, according to the scope.
- an empty bed contact time (EBCT), defined as the ratio between the height of material and the filtration rate (expressed in minutes). Depending on the application, EBCT range is between 5 and 15 min.

The most commonly used media materials are silica sand and anthracite. In many filters, only a single grade of sand (supported on coarser sand and gravel)

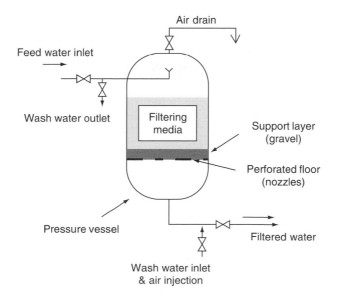

Fig. 5.2 Pressure filter.

is used but dual- and multimedia beds are sometimes employed. For iron removal applications, a standard configuration would be:

- First support layer with a 4–8 mm gravel, 5 cm deep
- Second support layer with a 1.3–1.5 mm coarser sand, 15 cm deep
- Filtration layer with a 0.55 or 0.95 mm sand, 80 cm deep

Efficient filtration during the production mode is achieved through:

- Controlling the filtration flow to within design limits
- Limiting the flow and pressure variations
- Monitoring the filtrate quality and pressure drop
- Maintaining the correct frequency and conditions for the backwash of the filter

Simple instruments, such as flowmeters and pressure gauges, allow continuous monitoring of the process and can be complemented by specific instruments such as turbidity-meters, sensors, etc.

Filtration process control is also achieved through efficient backwashes. Backwashing is done using back flow and should provide a minimum bed expansion of 10–20% to ensure a partial fluidization and adequate cleaning of the media. The required backwash flow is governed by media size, type and water temperature. Dual- and multimedia beds need slightly greater bed expansion than single-media beds to maintain a good stratification. Higher backwash flow may be needed at the start of the washing to overcome the initial resistance of the bed. To prevent this, air injection is often used before backwash to create pathways for better distribution of backwash flow through the bed. Air injection rates are normally between 20 and 40 m/h and backwash is done with a flow rate equivalent to 2–3 times the nominal flow rate. The duration of air injection and backwash varies, but typically would be less than 5 min for air injection and 10–20 min for the backwash. The final step is a rinse to recover the standard quality of the treated water.

5.3.1.2 Other filters
Removal of particles, microparticles or microbiological elements from the water needs different filter types. The three principal categories generally used are:

(1) Depth filters characterized by a fibrous or metallic matrix, which provides a random pore size distribution; the more commonly used materials are polypropylene, cellulose, glass fiber and sintered metal. In this case, particles are trapped mainly within the thickness of the matrix. These filters are cheaper than surface filters and have high retention capacities; they are often characterized by a 'nominal' pore

size, indicating the ability of the filter to retain the majority of particulates (60–98%) at or above the rated pore size. Particles smaller than the nominal size may also be trapped in the matrix (adsorption).

(2) Composite filters with several layers of filtrating materials made of glass microfibers or polymers. Particles larger than the pores inside the filter matrix are in this case trapped mainly at the surface of the filter, while smaller particles are often trapped inside the matrix. These types of filters have both membrane and depth filter properties. The number of pleated layers gives to these filters a high retention capacity, and their 'controlled' pore structure leads to a better efficiency compared with depth filters. They are also cheaper than membrane filters.

(3) Membrane filters can be described as a porous matrix with a controlled pore size. Particles are stopped at the surface or in the upper part of the membrane thickness. All particles larger than the pores are trapped (nominal cutoff). They are often characterized by an absolute pore size, indicating the pore size at which a challenge organism of a particular size will be retained with 100% efficiency under strictly defined test conditions. Cost is higher than for other types of filters.

In order to increase the efficiency of the filtration stage, depth filters (high capacity, low cost) are often combined with membrane filters (high selectivity). The most widely used filters are:

- Bag
- Coreless
- Cartridge

The bag system is a low-cost device that enables treatment at a high flow rate ($60\,m^3/h$ for a 40 in.), but with a low capacity (Fig. 5.3). It can be the right solution for accidental issues. Depending on the materials used to produce the bags (polyamide, viscose, polyester, polypropylene, PTFE, etc.), particle size between 1 and $1000\,\mu m$ can be removed with variable efficiency. Possible application is post filtration, after a mass media filter (e.g. for iron removal) to protect the downstream part of the process against possible release of particles from the mass media.

The coreless filter is a recent development that is adapted for the removal of particles and capable of treating at high flow rates with low cost; each coreless element can treat up to $80\,m^3/h$ (Fig. 5.4). The coreless cartridge is fixed on a stainless steel support mounted in the housing. Existing housings can also be quickly upgraded to coreless technology on-site with a conversion kit that includes a core as a permanent part of the filter assembly. Elements between 10 and $40\,\mu m$ are now available. So this technology is well adapted to particle filtration and can be used as the first step in a water treatment process.

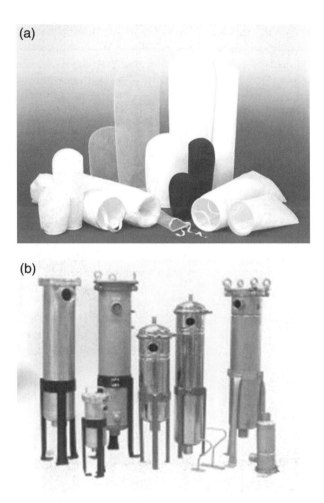

Fig. 5.3 Bags, filters and housings: (a) bags; (b) filters and housings.
(Photo courtesy of Pall Corporation.)

Filter cartridges (10, 20, 40 in.), which are also widely used, are more expensive and have a lower capacity than bag and coreless, but this is compensated for by a higher efficiency. They can be used for particle removal and microbiological filtration with available cutoff size between 0.1 and 100 μm. According to the application (i.e. the cutoff size), the flow rate will be between 0.25 and 1 m³/h for a 10 in. cartridge. Generally, there are three elements in this type of cartridge:

(1) Filtering material, made of nylon, polypropylene, PVDF, PTFE
(2) Support layers, mainly in polyester or polypropylene
(3) Polypropylene core

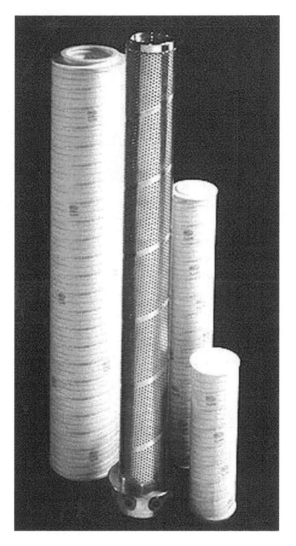

Fig. 5.4 Coreless cartridge. (Photo courtesy of Pall Corporation.)

To increase the filtering surface, the membrane is pleated (see Fig. 5.5), and several cartridges can be installed in the same housing.

Some suppliers also propose metallic filters (stainless steel) with cutoff size between 1 and 30 μm. The choice of filtration systems is dependent on:

- Treatment flow rate
- Objective of the treatment (particles, microbiology)
- Cost of maintenance

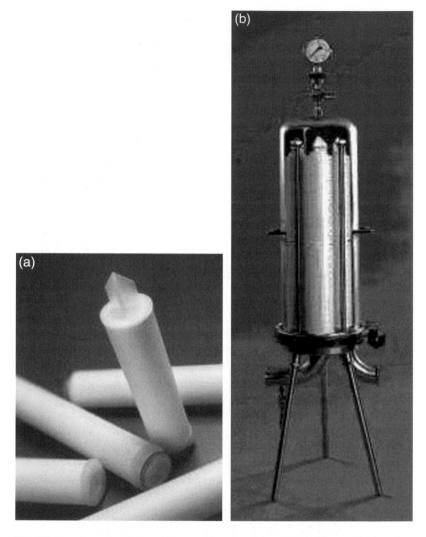

Fig. 5.5 Structure of a filter cartridge and housing: (a) cartridge; (b) cartridge housing;

Monitoring of the process is mainly achieved through observing the pressure drop and checking the outlet water quality; specific integrity tests also allow detection of possible defects in the reliability of the system. In general, cartridges are for one-time use, while bags and coreless filters can be cleaned and reused.

A strict process control is required for microfiltration processes. It is essential to perform accurate monitoring to check that the filter runs continuously and that it is full of water. Otherwise, biomass buildup often develops at the surface of the membrane, and this can lead to quality issues. Sanitation must also, therefore, be done periodically.

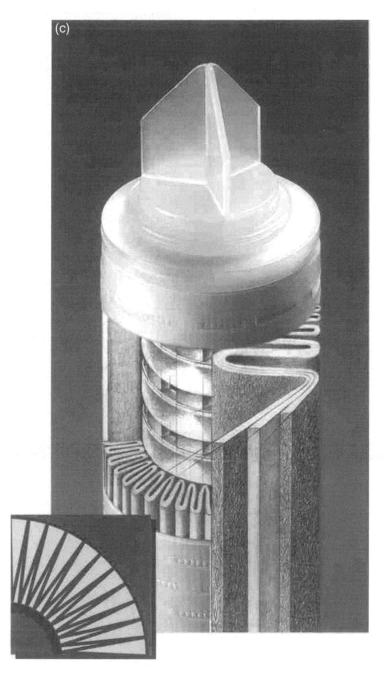

Fig. 5.5 cont'd (c) fluted cartridge.
(Photo courtesy of Pall Corporation.)

5.3.2   Membrane processes

Membranes represent an important new set of methods for drinking water treatment. These processes are: microfiltration (MF), ultrafiltration (UF), nano-filtration (NF), which are filtration-controlled membrane processes; and reverse osmosis (RO), which is a diffusion-controlled membrane process. Electrodialysis (ED) is also sometimes classified in the category of ion exchange processes, as it uses an ion exchange membrane. Filtration processes are used in two possible running conditions:

(1) Dead-end filtration in which all the flow goes through the filter. The main advantage of this mode is that it treats all the water without any loss, but the pressure drop increases with the fouling over time, and it is therefore necessary to increase the working pressure progressively to maintain the process efficiency (Fig. 5.6a). In this case, the lifetime of the filter depends very much on the inlet water quality. This configuration is used in many MF processes.

(2) Cross-flow filtration in which only a part of the inlet flow (up to 98%, depending on the application) goes through the membrane, the other part remaining untreated. In this configuration, the fouling on the membrane is continuously removed by the 'shear-force' engendered by the flow velocity at the membrane surface. This configuration is mainly used in UF, NF and RO processes (Fig. 5.6b).

RO has been primarily used to remove salts from brackish water or seawater, although it is also capable of very high rejection of organic compounds. ED is

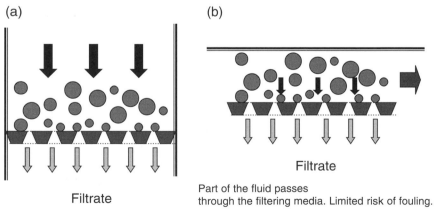

Fig. 5.6 Filtration modes: (a) dead-end filtration; (b) cross-flow filtration.

mainly used to demineralize brackish water and seawater and to soften freshwater. NF is the most recently developed membrane process, used to soften
freshwaters and remove disinfection by-product (DBP) precursors. Finally, UF
and MF are used to remove turbidity, microbiological elements and particles
from freshwaters. Figure 5.1 gives the size range of membrane processes and
contaminants.

Membranes, the key components of the process, are manufactured as flat
sheets, hollow fibers or coated tubes. They must be configured into elements to
manage the flow streams in the membrane machines and support the membrane
under the required hydraulic pressures.

Flat sheet configurations include the plate-and-frame and spiral-wound
design; the latter predominates all forms including hollow fiber and tubular.
The plate-and-frame design allows a variety of feed- and permeate-channel
designs, but is a high-cost approach (see Fig. 5.7a). The spiral-wound design
affords the best 'all-around' characteristics, of high packing density, low cost and
rugged high-pressure operation (see Figs 5.7b and 5.7c). With the recent advent
of specialized feed-channel spacer materials, a wider range of applications such
as RO and NF now employs the spiral design.

Hollow fibers (0.5–2 mm diameter) are mainly used in UF processes
(Fig. 5.7d). They can handle high solid loading without plugging and can be back-
flushed to remove fouling layers. Since they are self-supporting homogeneous
fibers, they are limited by the tensile, compressive and flexural strengths of the
membrane material, which is porous. This limits the operating pressure and flow
rates to less than those of spiral wounds.

Coated tubes are large-scale versions of hollow fibers (0.6–2.5 cm), with the
membrane coated on the inside wall of a support porous material. This support
material gives the tube its strength so that higher operating pressures are possible
(Fig. 5.7e).

Principal materials used in membrane manufacturing are: cellulose acetate
(CA), polyamide (PA), polysulfone (PS), polyvinylidene fluoride (PVDF) and
acrylonitrile (AN). All these are suitable for organic membranes, whereas metallic oxides (AlO_2, TiO_2, ZrO_2) or carbon-based coatings are best for inorganic
membranes. In addition to the shape and the material used for membrane
production, pore size and 'cutoff' – expressed in daltons (the unit for molecular
weight) – are of course key parameters for the choice of membrane processes.
The membranes are installed in pressure-bearing housings, often with several
membranes in the same housing.

A typical membrane process design for RO, NF and UF includes a pump to
provide the necessary pressure and cross-flow velocity, the filter elements and
housings, connecting parts, control valves and instruments (pressure gauges,
flowmeters, conductimeter, etc.). The applied pressures are between 2 and 4
bar for UF, 4 and 8 bar for NF and above 8 bar for RO application (depending
on the initial salt level). Prefilter cartridges or other pretreatments (e.g. MF or

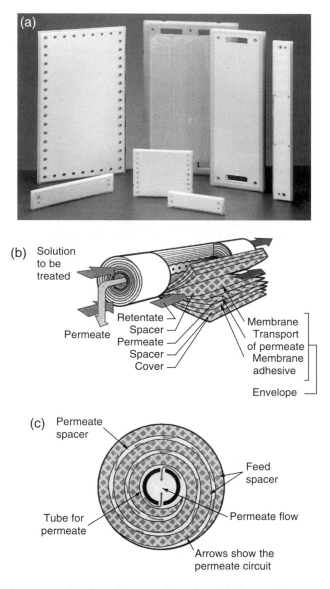

Fig. 5.7 Some examples of membrane configurations: (a) plate-and-frame membrane (photo courtesy of Pall Corporation); (b & c) spiral-wound membrane;

(Continued)

UF prior to RO) are usually included to protect the expensive membrane against physical, chemical or biological fouling. In some cases, mainly for RO, chemical injection (antiscaling agents) is used. Figure 5.8 gives a schematic view of a membrane process, which is generally assembled on a skid.

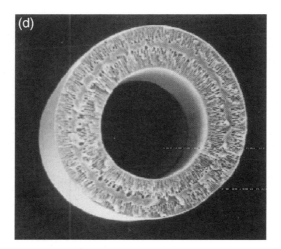

Fig. 5.7 (*Continued*) (d) hollow fiber membrane
(photo courtesy of Pall Corporation)

The key indicators for the efficiency of the process are:

- Pressure drop from the upstream to the downstream side
- Recovery, defined as the ratio of permeate to feed volume
- Outlet water quality
- Rejection rate, defined as the percentage of elimination for a component

NF and RO applications usually have a 75–85% recovery, and UF and MF can reach up to 98%. Efficiency is directly proportional to effective pressure, water temperature and characteristics of the water (e.g. fouling index).

To maintain high efficiency for RO, it is sometimes necessary to include a pretreatment such as UF and iron removal. Biofouling is a major issue for RO membranes because of the difficulty in cleaning and sanitizing them efficiently. This is due to the spiral-wound configuration of the membrane and the nature of the membrane itself; only few newly developed membranes are compatible with hot water and strong chemical sanitation.

5.3.3   Adsorption

Adsorption is the physical and/or chemical process in which a substance is accumulated at an interface between two phases (e.g. solid/liquid). The adsorbate is the substance being removed from the liquid phase, and the adsorbent is the solid phase onto which the accumulation occurs. Adsorption of substances onto adsorbents takes place because there are forces at the molecular level that attract the adsorbate from solution to the solid surface. Two operating modes (physical or chemical) can be used in adsorption:

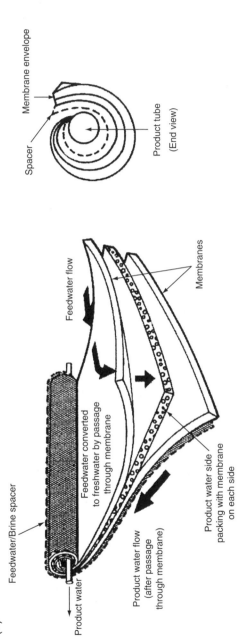

Fig. 5.7 (*Continued*) (e) tubular membrane (American Water Works Association/Letterman, R. D. (1999) *Water Quality and Treatment—A Handbook of Community Water Supplies*, 5th edn. © McGraw-Hill, Inc. Reproduced with permission).

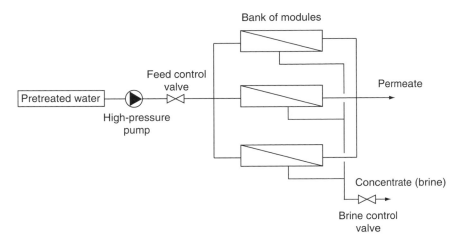

Fig. 5.8 Schematic view of a membrane process.

(1) Regeneration operation: when maximum capacity is reached, adsorbent is regenerated. Such regeneration includes a desorption step (removal of the adsorbate trapped on the adsorbent) using a chemical reagent such as soda or acid followed by a neutralization step. The adsorbent then recovers its full capacity for adsorption.

(2) One-way operation: when maximum capacity is reached, the adsorbent is changed. This happens when the adsorbate is trapped deep inside the porous structure and is very difficult or impossible to extract, or if capacity and cost of the adsorbent are such that regeneration is not economical.

For purposes of water treatment, adsorption from solution occurs when impurities in the water accumulate at the solid/liquid interface. Adsorption plays an important role in the improvement of water quality. The ideal case, of a single adsorbate being selectively removed onto an adsorbent, occurs very seldom in practice; the objective in most real systems is to remove several adsorbates. The heterogeneous mixture of compounds in natural waters reduces the number of sites available for the trace compounds, because of direct 'competition' for adsorption sites or because of pore blockage.

The adsorption equipment is similar in design to pressure filters used for filtration applications. Sizing is also mainly established from the filtration rate and EBCT (up to 20 min). The 'breakthrough curve', and hence the adsorbent capacity, must be assessed beforehand through small-scale column tests (SSCTs) to size the unit. Depending on the application and the adsorptive material, regeneration and/or backwash operations can be adapted in terms of flow rate and duration. For all types of media, the principle of the support layers with gravel and sand is maintained but has to be adapted according to the adsorptive material.

In all cases, process efficiency is determined by monitoring the constituent being removed. Other parameters, such as pressure drop, should also be checked to control the process.

5.3.3.1 Activated carbon

Of all water treatments based on adsorption, activated carbon is by far the most widely used. The raw materials are lignite, coal, bone charcoal, coconut shells and wood charcoal. Pores are developed during activation, partly by burning away carbon layers, the thickness of which depends on the structure of the starting material. The activation step, which produces the pore structure, influences the adsorptive properties. It is accomplished either by heating to 200–1000°C in the presence of steam, carbon dioxide or air, or by wet chemical treatment at lower temperatures with exposure to agents such as phosphoric acid, potassium hydroxide or zinc chloride. The method and temperature of activation strongly influence pore size distribution and chemical properties. Hence, depending on the production method of the activated carbon, the initial conditioning before treating water for bottling becomes very important (because of the risk of pH modification or chemical compound release). In most cases, the activated carbon is a single-use material because regeneration (steam or thermal reconditioning) is not possible on-site. The main 'target constituents' to be removed by activated carbon for drinking water are:

- Specific organic molecules that cause taste and odor, mutagenicity and toxicity
- Natural organic matter (NOM) that causes colors
- Disinfection by-products (DBPs)
- Volatile organic compounds (VOCs and synthetic organic chemicals (SOCs)
- Chlorine or ozone

Activated carbon is available in powdered form (in mixed reactors) or in granular form (using a 'bed' configuration): granular activated carbon (GAC) is the type most commonly used in the bottled water business. Backwashing of GAC filter adsorbers is essential for removal of solids, to maintain the desired hydraulic properties of the bed and to control biological growth. To limit bacterial growth on the media, the activated carbon filter may also require regular sanitation using hot water and steam.

5.3.3.2 Manganese dioxide

Manganese dioxides (either from natural sand or synthetic origins) are being increasingly used for water treatment with the objective of removing manganese and arsenic. Several products are available in the market with different efficiencies, but in each case, the material is placed in a conventional 'sand-type' filter vessel, sized according to the standard parameters (filtration rate, EBCT) and based on previous capacity tests. Backwashing conditions need to be adapted to the materials as some of them are very crumbly and can produce a lot of fine particles during this operation.

For manganese removal, a chemical regeneration can be done when the material is saturated or when the breakthrough point is reached. This regeneration, which is a reoxidation of the media, is done using potassium permanganate or chlorine. For maximum efficiency, a backwash is required before chemical regeneration.

For arsenic removal, manganese dioxide is an effective adsorbent, and high performance can be easily achieved (release of arsenic is $< 2\,ppb$) with high reliability. In this case, adsorption is reversible, which means that regeneration of the material is possible. This chemical regeneration includes caustic soda for desorption, acid neutralization and a final water rinse step. Regeneration is a constraint of this process, and the management of wastewater should also be taken into consideration.

Manganese dioxide has a strong affinity for iron; thus, when a manganese dioxide filter is used for the removal of manganese or arsenic, it is essential that iron be removed first. Iron is a 'competitor' to manganese, and is preferentially trapped on the material, and in this case, adsorption sites become less or not at all accessible for other compounds, and so the efficiency decreases for manganese removal.

5.3.3.3   Activated alumina
Activated alumina is mainly used for the removal of arsenic, fluoride and, occasionally, selenium. As with other adsorbents, preliminary tests (pilot tests) are essential before any industrial design to assess the performance of the process (capacity) and the impact on water characteristics. Efficiency depends on the activated alumina used, the running conditions and the water characteristics. For fluoride or arsenic removal, adsorption is reversible and so chemical regeneration can be applied when the material is saturated or when a fixed breakthrough point is reached.

5.3.4   Ion exchange
Ion exchange is becoming extensively used in water treatment. The process has been primarily used for the removal of hardness and for demineralization, but is now being implemented for other applications. As its name implies, ion exchange describes the physicochemical process by which ions are transferred from a solid to a liquid, and vice versa. All ion exchangers, whether natural or synthetic, have fixed ionic groups that are balanced by counter-ions (anions or cations, which exchange with ions in solution) of opposite charge to

maintain electroneutrality. Figure 5.9 gives a schematic diagram of this exchange.

As the resin initially containing counter-cation A^+ is placed in a solution containing B^+, diffusion is established due to the concentration difference between counter-ions A^+ and B^+ in the resin and in solution. Exchange occurs until equilibrium is reached between the solution within the matrix of the resin and the bulk solution.

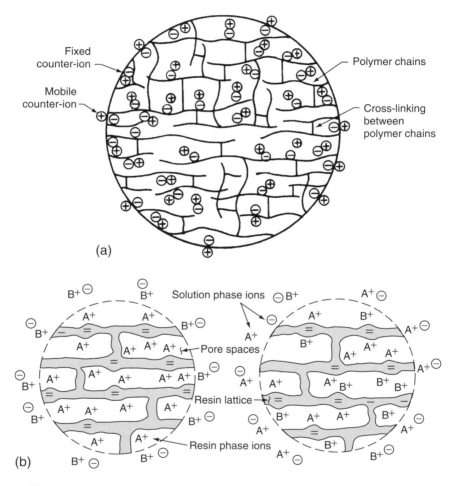

Fig. 5.9 Schematic diagram of a cation exchange resin. (a) Structure of organic cation-exchanger bead (American Water Works Association/Letterman, R. D. (1999) *Water Quality and Treatment—A Handbook of Community Water Supplies*, 5th edn. © McGraw-Hill, Inc. Reproduced with permission). (b) Initial and steady state of a cation-exchange resin (Montgomery, J. M. (1985) *Water Treatment Principles and Design* © John Wiley & Sons, Inc. Reproduced with permission).

Nowadays, most resins used in water treatment processes are synthetic ion exchange resins. Due to their higher capacity, synthetic resins are preferred to natural materials such as greensand, clay, peat, bauxite, charred bone or zeolites. The majority of ion exchange resins is made by the copolymerization of styrene and divinylbenzene (DVB). Styrene molecules provide the basic matrix of the resin, while DVB is used to cross-link the polymers to allow for the general insolubility and toughness of the resin. They are generally available in spherical shapes with particle diameter size between 0.04 and 1.0 mm. Particle size is a key parameter for the rate of exchange (proportional to either the inverse of the particle diameter or the inverse of the square of the particle diameter) and also for the hydraulic design of the column (pressure drop). To produce the various types of cationic and anionic resins, the plastic structure is reacted with either acids or bases. The ionizable group attached to the resin structure determines the functional capability of the exchanger. There are four general types of ion exchange resins based on their functional groups, which are used in water treatment:

(1) Strong-acid cation exchangers
(2) Weak-acid cation exchangers
(3) Strong-base anion exchangers
(4) Weak-base anion exchangers

Each category of resins must be implemented with adapted running conditions and regeneration procedures. Strong-acid exchangers can convert neutral salts into their corresponding acids if operated in the hydrogen cycle (e.g. $NaCl + R\text{-}H \leftrightarrow HCl + R\text{-}Na$). In this case, the functional groups can be derived from sulfonic (HSO_3^-), phosphonic ($H_2PO_3^-$) or phenolic groups. For this type of resin, the chemical regeneration is usually done with HCl or H_2SO_4, or during a sodium cycle where the resin is regenerated with NaCl. The hydrogen cycle removes nearly all major raw water cations and is usually the first step in demineralizing a water. It can be represented by the following reaction:

$$CaSO_4 + 2(R^-.H^+) \leftrightarrow (2R^-).Ca^{2+} + Na_2SO_4$$

The sodium cycle is used for softening waters and also for the removal of soluble iron and manganese in the following form:

$$CaSO_4 + 2(R^-.Na^+) \leftrightarrow (2R^-).Ca^{2+} + Na_2SO_4$$

Weak-acid exchangers differ from strong-acid exchangers in that weak-acid resins require the presence of some alkaline species to react with the more tightly held hydrogen ions of the resin. The exchange is a neutralization with the alkalinity (HCO_3^-) neutralizing the H^+ of the resin. So weak-acid resins will split alkaline salts but not monoalkaline salts (e.g. $NaHCO_3$ but not NaCl). In contrast with strong-acid resin regeneration where a large excess of regenerant is required to create the concentration driving force, weak-acid resins use up to 90% of the acid (HCl or H_2SO_4) regenerant, even with low acid concentrations. Weak-acid resins are favored when the untreated water is high in carbonate hardness and low in dissolved carbon dioxide; they are primarily used for achieving simultaneous softening and dealkalization.

Strong-base exchangers split neutral salts into their corresponding bases if operated on the hydroxide cycle (e.g. $NaCl + R\text{-}OH \leftrightarrow NaOH + R\text{-}Cl$). They are typically used after cation exchangers to remove all anions for complete demineralization, but weakly substances such as silica and carbon dioxide can also be removed. In the chloride cycle (e.g. $SO_4^{2-} + 2(R^+.Cl^-) \leftrightarrow 2(R^+).SO_4^{2-} + 2Cl^-$), strong resins have been used for nitrate and sulfate removal. The functional sites are derived from quaternary ammonium groups. There are two types of strong-base resins: type I, with three methyl groups as the functional group; and type II, with an ethanol group that replaces one of the methyl groups. Type I has greater chemical stability, while type II has a slightly greater regeneration efficiency and capacity. Depending on the use, the strong-base resins are regenerated with NaOH or NaCl.

Weak-base exchangers remove free mineral activity (FMA) such as HCl or H_2SO_4. They are sometimes used in conjunction with strong-base resins in demineralizing systems to reduce regenerant costs and to attract organics that might otherwise foul the strong-base resins. They are regenerated with NaOH, NH_4OH or Na_2CO_3.

Resins are mainly used in fixed-bed columns, which are steel pressure vessels constructed to provide a good feed and regenerant distribution system and appropriate bed support, including provision for distribution of backwash water and enough free space above the resin to allow for the bed expansion that is expected during backwash. Again, the key parameters in designing the process are EBCT, rate and capacity. This is why, before any sizing, SSCTs must be performed to check the capacity of the resins, in order to monitor the possible release of chemical compounds. These tests also give data about the selectivity of the resin, the competitiveness with other elements and the impact on the main characteristic components of the water.

The stability of an ion exchange resin under certain physical, chemical and/or radioactive conditions plays a major role in many applications. Physical stress may change the structure of the resin, the most common example being excessive osmotic swelling and shrinking, which may break the bed. Mechanical compression, abrasion and excessive temperature can also rapidly degrade the characteristics of a resin. Chemical degradation can occur, with the breaking of the polymer network, modification of the functional groups or fouling of the resin by the species in solution. Possible release of by-products is also a high risk for the resins.

Bacterial growth can be a major problem with anion resins because the positively charged resins tend to 'adsorb' the negatively charged bacteria that metabolize the adsorbed organic material – negatively charged humate and fulvate anions. The use of chemical disinfectants or hot water is not recommended. To correct this issue, special resins have been developed, which contain bacteriostatic long-chain quaternary amine functional groups on the resin surface.

5.3.5 Chemical oxidation

Chemical oxidation processes play several important roles in the treatment of drinking water. Chemical oxidants are used mainly for the oxidation of reduced inorganic species to destroy taste- and odor-causing compounds and to eliminate color. Because many oxidants also have biocide properties, they can be used to control biological growth. The most common chemical oxidants used in water treatment are chlorine, chlorine dioxide, permanganate, ozone and air.

Chlorine is the most widely used oxidant in water treatment. Molecular chlorine is typically provided in pressurized cylinders so that it exists as a liquid under pressure. It is then added to the water by reducing the pressure in the tank and releasing it as a gas. Because of the risks involved in transport and handling of hazardous chemicals, liquid sodium hypochlorite (NaOCl), supplied as a concentrated bulk solution, is increasingly used despite its higher cost. The main problem with hypochlorite is the stability of the product, which tends to degrade over time, particularly when it is stored at high temperatures and/or exposed to sunlight. Chlorine oxidation is mainly used for:

- Oxidation of iron and manganese
- Taste and odor control
- Color removal
- Ammonium removal (using NaOCl)

The main disadvantage of chlorine is the production of a variety of chlorine-substituted halogenated compounds, e.g. chloramines are produced when chlorine oxidation is used for ammonium removal.

Chlorine dioxide (ClO_2) is a greenish-yellow gas at room temperature, unstable at high concentrations and potentially explosive upon exposure to heat, light, electrical sparks or shocks. It does not hydrolyze in water like chlorine, remains in its molecular form as ClO_2 and results from the reaction of sodium chlorite ($NaClO_2$) with either gaseous chlorine (Cl_2) or hypochloric acid (HOCl). Its first application in the past was for taste and odor control, although it is also an effective oxidant for reduced iron and manganese. One of its principal advantages is that it does not react with ammonia; also, it does not enter into substitution reactions with NOM to the same degree as chlorine. So it does not produce trihalomethanes (THMs), haloacetic acids (HAAs) or most of the other oxidation by-products that result from chlorination. Chlorite (ClO_2^-) is considered to be the principal oxidation by-product of chlorine dioxide.

Potassium permanganate ($KMnO_4$) has been used as a water treatment oxidant for decades. It is commercially available in crystalline form and is prepared on-site. As an oxidant, $KMnO_4$ is more expensive than chlorine and ozone, but it has been reported to be as efficient for iron and manganese removal and may require less equipment and thus less capital investment. Potassium permanganate solution is injected in the water to be treated; contact time after oxidant addition is typically 5 min at 20°C, followed by mass media filtration. The process is more efficient at pH values above 7.5. It is also used for the chemical regeneration of manganese dioxide in manganese removal filters.

Ozone gas is formed by passing dry air or oxygen through a high-voltage electric field. The resultant ozone-enriched air is added directly into the water by means of porous diffusers or venturi systems. The injection point is generally at the base of a contact tank (in the case of diffusers), or on the infeed supply (in the case of a venturi injector). Subsequent mixing and diffusion in the contact tank over a period of 10–15 min allows dissolution of the gaseous ozone. The most common use of ozone in Europe is for manganese removal, while in the USA, it is used widely as a final disinfection step prior to bottling. Ozone is also sometimes used to transform nitrites to nitrates during a nitrification process. Because of its high oxidizing power, ozone reacts with many constituents of the water. In particular, it can lead to the formation of brominated by-products when applied to waters with moderate to high bromide levels. Some of this oxidized bromine will continue to react to form the bromate ion (BrO_3^-). Other by-products attributed to ozone include hydroxyacids, aromatic acids and hydroxyaromatics. When ozone reacts with organic contaminants in water, including NOM, it partially oxidizes them to lower molecular weight with higher polarity. These biodegradable by-products can contribute to biofouling problems in the water process if not properly controlled. Ozonation is therefore often followed by a biological activated carbon (BAC) process, to remove such biodegradable organic matter (BOM). Of all the previous oxidation processes, ozone is the most widely used in the bottled water business to oxidize dissolved iron or manganese. Ozone is sometimes used in conjunction with hydrogen peroxide or UV irradiation to produce hydroxyl radicals ($OH°$), which have powerful oxidative properties.

Air oxidation is the simplest oxidizing process and is most widely used in the bottled water industry for iron removal. However, air oxidation is not effective for manganese removal when the pH of water is below 9.5, due to its very slow kinetic reaction; hence, a more powerful oxidant such as ozone is required for this application. Several devices are used to inject the air (venturi, porous diffuser, direct injection in a mixing pot), and an oxidation tower is set up to

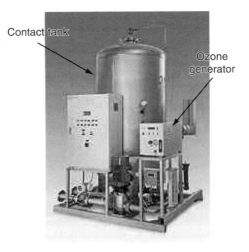

Contact tank

Ozone generator

Fig. 5.10 Typical ozone process in an ozone system.

increase the required contact time between water and air for oxidation reactions. To enhance the mixing between air and water, the oxidation tower is sometimes partially filled with plastic or metallic rings or natural materials such as volcanic pouzzolane. The airflow rate must be properly adapted to the water flow rate and to the initial level of iron. Table 5.1 gives the oxidation potential for the main oxidants.

5.3.6 Biological processes

Biological processes are commonly used in the treatment of tap water. The main benefits from these processes are iron and manganese removal, ammonium and nitrate removal (nitrification and denitrification) and the elimination of BOM. In many cases, biological processes are oxidizing processes in which the oxidation is done by different types of microorganisms. These microorganisms are fixed on a support material, such as sand, anthracite, manganese dioxide or activated carbon, within a conventional filter vessel, and they use the element to be removed as a nutrient for their growth. In general, the efficiency of biological processes is dependent on the running conditions, and more accurately, the possible variations of the running conditions, such as flow rate, temperature, dissolved oxygen level, change in the water characteristics, due to which the process control can be critical, particularly during the starting phase.

Table 5.1 Standard potentials for chemical oxidants used in water treatment

Oxidants	$OH°$	O_3	H_2O_2	$KMnO_4$	ClO^-	$HOCl$	O_2	ClO_2
$E°$ (V)	2.85	2.08	1.78	1.68	1.64	1.48	1.23	0.95

In the bottled water business, the use of a full biological process is generally limited to ammonium removal (biological nitrification) and in some cases to organic matter removal. However, it is known that many physicochemical-based iron and manganese removal processes become in effect 'hybrid' processes (physicochemical + biological) with time, due to the fixing of microorganisms on the support.

Biological nitrification is an oxidation process done in two stages: first, the conversion of ammonium (NH_4^+) to nitrite (NO_2^-) and second, conversion of nitrite to nitrate (NO_3^-). *Nitrosomonas* and *Nitrobacter* are species of bacteria that can be involved, under aerobic conditions, in ammonium removal. Natural nitrifying flora becomes established over time and the biological process starts; the first phase leads to nitrite production, and after a short nitrite buildup, the second part of the reaction starts with the production of nitrates. During this phase, it is necessary to control as well as possible the key parameters involved in the biological reaction:

- Temperature ($> 17°C$)
- Flow rate (as constant as possible)
- Dissolved oxygen (> 2 mg/l)
- pH (optimum range 7.2–7.8)

The typical time to reach a fully efficient nitrification is between 3 and 5 weeks, depending on the running conditions and the water characteristics. Some adjustments (e.g. increasing the temperature by adding a heat exchanger) can be considered during this key step. Of course, parameters such as filtration rate (8–10 m/h) and EBCT (10–12 min) have to be optimum for the process design itself. Concerning the process control, flow rate variations must be avoided and dissolved oxygen must be checked. Periodic backwashes have to be done (based upon pressure drop) maintaining 'gentle' conditions to limit the biomass loss due to abrasion.

In addition to being an efficient adsorbent, activated carbon is also a useful support for bacterial growth. Many BAC beds are used in order to achieve some biodegradation of organics in the water. The total organic carbon (TOC) is not significantly reduced by ozone alone, but the biodegradable proportion of TOC is increased (up to a factor of 10 depending on running conditions). Removal of this assimilable organic carbon (AOC) with BAC may be sufficient to give a significant reduction in TOC.

Biological processes require very strict hygienic conditions. Any uncontrolled microbiological growth or contamination in the biological filter may dramatically affect the efficiency of the treatment and can be source of contamination downstream of the process. Specific microbiological pretreatment might be necessary to protect the biological process, and regular backwashing of the biological filter is recommended to eliminate any accumulated excess of biomass.

5.3.7   Microbiological treatments

Microbiological treatments are designed for a partial or full inactivation or removal of microorganisms. It is important to check if this type of treatment is permitted by local regulations. The microbiological quality of a source water, or the efficacy of a treatment system for inactivation of microorganisms, can be assessed by direct monitoring of pathogens or by use of an indicator system. Because pathogens are a highly diverse group, generally requiring a highly specialized analytical technique for each pathogen, the use of indicator micro-organisms is a more popular method. The main processes used for the control of water microbial quality are:

- Physical treatment, such as MF or UF (removal)
- Chemical treatment, such as ozone, chlorine, chlorine dioxide (destruction)
- Photochemical treatment with UV irradiation (destruction)

Combinations of some of these treatments are also sometimes used, e.g. UV and MF.

5.3.7.1   Microfiltration

The principle of MF is described in Sections 5.3.1.2 and 5.3.2. It successfully removes the vast majority of microorganisms present in a water, with a high removal efficiency (up to 5 log reduction) for cysts and bacteria, where the filter cartridges typically have 0.45 and 0.22 μm pore size. These cartridges are set up in a stainless steel housing (see Fig. 5.5). Microorganisms are trapped on the upstream side of cartridges, which allows high efficiency for their removal. In normal running conditions, the cartridges have to be changed regularly every 3–6 months, or more frequently, if pressure drop increases. Process control is mainly done through the monitoring of the pressure drop and the outlet water quality. Integrity tests can also be done to check the nondegradation of the cartridge.

For better physical removal of microorganisms, UF is also sometimes used; in this case, all the microorganisms, including viruses, are stopped by the membrane barrier. This solution is more complex and costlier than MF because of the equipment required for implementing it (see Section 5.3.2 for more details). The bottled water industry generally uses MF rather than UF.

5.3.7.2   Chemical treatment

Chemical treatments used for bottled water, in particular chlorine and ozone, vary. For example, some countries impose requirements for the maintenance of high hygienic conditions during water transportation from the spring (in the case of tankering) or during water storage, and in this case, chlorine is sometimes added to the water to guarantee water quality. The dose has to be adapted according to the water composition, i.e. NOM , iron, manganese and ammonium

(a)   (b)   (c)

Fig. 5.11 Schematic view of the action of ozone on microorganisms: (a) attack of the ozone molecules onto the microorganisms; (b) first impact of ozone producing 'holes' in the cell membranes; (c) full destruction of the cell membrane by ozone.

levels. When chlorine has been added, a dechlorination step (usually using GAC) is implemented just before bottling.

Due to its high oxidizing power, ozone is also used in the bottled water business to inactivate microorganisms by destruction of the cellular membrane. Two major configurations using ozone are:

(1) A standard microbiological treatment in which running conditions are calculated to reach what is referred to as a CT value (ozone concentration multiplied by contact time), allowing a total inactivation of microorganisms. A standard CT value is 1.6 (0.4 mg/l for 4 min). The ozonation process is immediately followed by a UV device or an activated carbon filter installed after the contact tank to eliminate the residual ozone.

(2) 'Germ free' bottling, where water is bottled with an ozone residual to prevent any microbiological growth in the final product: the ozone residual having the effect of sanitizing the packaging (cap and bottle) at the time of bottling. Considering the time needed for total natural elimination of the residual ozone, a high CT can be reached and bromate formation might be an issue.

Chlorine dioxide is not known to be used in the bottled water business, except (very rarely) for the water used to rinse bottles prior to filling.

5.3.7.3   Ultraviolet

Antimicrobial treatment based on UV irradiation is well known and widely used. Disinfection is achieved through the degradation of the nucleic acids (mainly adenine and thymine) within the bacterial cell. The maximum effect is obtained at a wavelength of around 260 nm. Although all microorganisms are sensitive to UV radiation, the sensitivity of the organisms varies, depending on their resistance to penetration by UV energy. Certain literature mentions a phenomenon of photoreactivation of UV-disinfected microorganisms; however, the operation of these repair processes in microorganisms is not very clear.

UV radiation is obtained from mercury discharge lamps, which can be of two types, low pressure and medium pressure. This depends on the partial pressure of the mercury vapor in the lamp. Low-pressure lamps are more efficient at converting electrical to UV energy, giving an output within a narrower wavelength band than medium-pressure lamps. However, the latter give a much higher output of UV radiation for the same size unit; they are generally more compact and the capital costs are lower. During the first 100–200 running hours, an initial decrease in intensity is detected, but over time the efficiency is stable. Generally, the life of a low-pressure lamp is estimated at 8000 h. It best operates in a continuous mode, as one 'on/off' operation is equivalent to 1 h of operating time for the lamp. Most UV disinfection units are tubular in design, with water flowing parallel to the long axis of a centrally positioned lamp or a bank of lamps, which are housed in quartz tubes to protect them. The inlet and outlet of the UV housing are perpendicular to the lamp; the same UV housing can hold several lamps installed around the diameter of the housing. Figure 5.12 shows a general view of a UV device.

Specific sensors are set up to monitor the UV intensity and allow detection of any defects in the running conditions, e.g. if a lamp is off. An 'hours run' meter also gives an indication of the remaining lamp life. The UV transmissivity of

Fig. 5.12 Schematic view of a UV system. (Reproduced with permission of Hanovia Ltd.)

Table 5.2 Water treatment processes

General processes	Type	Main applications	Efficiency	Advantages	Disadvantages
Filtration	Mass media filter	Particle and iron removal	Moderate to high	Backwashable	Size
	Bag filter	Particle removal	Variable	High flowrate/low cost	Low capacity
	Coreless filter	Upstream particle filtration	High	High capacity and flow	Low range of cut-off size
	Cartridge	Particle removal/ microbiology	High	Low cutoff size	Operation cost/oneway
Membrane	Microfiltration	Particle removal/ microbiology	High	Absolute cut-off	Clogging
	Ultrafiltration	Particle removal/ microbiology	High	Some large organic molecules/viruses	Investment cost
	Nanofiltration	Partial demineralization	Moderate to high	Removal of some dissolved molecules	Investment cost
	Reverse osmosis	Total demineralization	High	High quality of water	Bacterial growth (biofouling)
Adsorption	Activated carbon	NOM, VOCs, DBPs, chlorine, ozone removal	Variable	Wide range of 'pollutants' removal	Bacterial growth/no possible regeneration
	Manganese dioxide	Manganese and arsenic removal	High	Selectivity/capacity/ regeneration	Wastewater from the regeneration
	Activated alumina	Arsenic and fluoride removal	Variable	Selectivity/regeneration	Low capacity/ wastewater for the regeneration
Ion exchange	Resin	Softened water/manganese removal/ demineralization	Variable	Wide range of products	Possible release of chemical compounds/ bacterial growth/low selectivity

Chemical oxidation	Chlorine	Iron and ammonium removal/microbiology control	High	Inexpensive/strong oxidant/persistent residual	Halogenated by-products may contribute to taste & odor issues
	Chlorine dioxide	Taste & odor control/iron and manganese removal/microbiology control	High	No HAA and THM by-products/no reaction with ammonia	Chlorite and chlorate formation/may produce undesirable odors
	Potassium permanganate	Regeneration of manganese dioxide/iron and manganese removal	Variable	No halogenated by-products/easy to feed	Process control (pink water)
	Ozone	Manganese oxidation/microbiology control	High	Very strong oxidant	By-products/process control/impact on NOM
Biological irradiation	Air	Iron oxidation	High	Very simple	Weak oxidant
	Pressure filter	Ammonium removal	High	No by-products	Process control
	UV	Microbiology control	High	No by-products	Impact of turbidity on transmissivity

water decreases with increasing color, turbidity, iron and manganese; so these elements have to be removed before the UV step if the irradiation is to maintain a high efficiency. The applied dose, which is expressed in mWs/cm^2 or mJ/cm^2, depends on the application. The standard dose is in the range 25–$40\,mJ/cm^2$, which is effective for the majority of microorganisms, except for some viruses and for molds and amebae that may need a higher dose (up to 200–$250\,mJ/cm^2$). In addition, UV can be used to destroy residual ozone in the water, and for this purpose, $180\,mJ/cm^2$ is required. To reach its full efficiency, a UV device requires a few minutes (lamp heating) when water is not flowing through the UV chamber. A key requirement to maintain a high efficiency level for a UV device is to clean the quartz tube periodically for full transmissivity. Table 5.2 provides a summary of the general water treatment processes.

5.4 Conclusion

The objective of this chapter was to give a brief overview of the treatments available for use in the bottled water business. Though the range of such treatments is very wide, the following factors must be taken into account when considering which treatments to use.

- It is an expensive affair in terms of investment and operation;
- It may generate liquid/solid wastes;
- It may generate undesirable by-products in the treated water;
- It may itself be at the origin of quality issues (failure of treatment/contamination).

Implementation of any water treatment must be absolutely justified, and the method and sizing must be correctly adapted to the need.

Many constituents in the composition of the water may affect not only the elements to be treated but also the performance of the water treatment itself.

It is vital to have a thorough knowledge of the raw water composition, over a period long enough to understand the natural fluctuation in composition (e.g. during low and high flow periods). This will make it possible to limit as much as possible any inadequacies in definition and rating, which may otherwise lead to extra costs in investment and operation, loss of performance and quality issues.

Water treatment processes should be subjected to a thorough hazard identification process (HACCP) and the results incorporated in the quality management system. Finally, most important is that any treatment used must be authorized by local regulations for the specific status of the water, whether it is ultimately being bottled or used simply as part of the bottling process.

Bibliography

American Water Works Association (1999) *Water Quality and Treatment: A Handbook of Community Water Supplies*, 5th edn, Letterman, R. D. (ed). McGraw-Hill, New York.

Hall, T. (ed) (1997) *Water Treatment Processes and Practices*, 2nd edn. WRc Swindon, Wiltshire

Montgomery, J. M. (1985) *Water Treatment Principles and Design*. Wiley, New York.

6 Bottling water – maintaining safety and integrity through the process

Dorothy Senior

6.1 Nature of water

Water is an amazing substance with anomalous properties that are pertinent to review, as they influence how it is appropriate to handle this very special product in the bottling process.

6.1.1 Physical properties

One of the characteristics of water is that it can exist in all three states of matter – solid (ice), liquid (water) and gas (water vapour) – in normal climatic conditions. Evaporation can take place from water or ice at any temperature. Liquid water is at its maximum density at 3.94°C. In ice, at 0°C and below, the existence of a very open molecular structure causes the solid state to be less dense than its liquid counterpart. As ice melts, the molecular lattice breaks up and the molecules can pack more closely together, resulting in a denser substance. This unusual characteristic has enormous biological and environmental consequences, ensuring survival of underwater life. As liquid, water tends to form a sphere (as in droplets) and it is drawn by gravity to eventually seek a level with the whole. Given energy, water creates rhythm that can be observed in the pattern of waves or in meandering watercourses. Water moves in layers that pass each other at varying speeds, and vortices create the pattern of changing activities in water. Because water has a high heat capacity, it requires a large increase or decrease in the surrounding temperature to affect the temperature of the water, which will only equilibrate with the environment over time. Its high boiling point of 100°C ensures its continued existence as liquid on the Earth.

6.1.2 Chemical properties

The molecular structure of water is asymmetrical, with the bonds to the hydrogen atoms at an angle of 105° (Fig. 6.1). A partial separation of electric charge (as shown in the figure) produces a bipolar molecule, with the highly electronegative element oxygen forming 'hydrogen bonds' with the hydrogen atoms in adjacent water molecules to produce strong intermolecular forces. The strength of this bonding accounts for the large latent and specific heat capacities of water, its cohesive nature and for the inward-acting forces at the surface that enable water to have the highest surface tension of all liquids. This facilitates capillary action and the ability of water to wet surfaces.

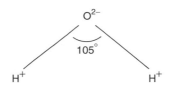

Fig. 6.1 Molecular structure of water.

The bonding within other substances is weakened in the presence of water, resulting in the separation of electrostatically charged cations and anions which, when surrounded by water molecules, become dissolved forming a solution. This powerful solvent action of water is vital for plant and animal life as it enables transport of chemicals and nutrients and facilitates life processes. However, this same characteristic can also work to transport harmful pollutants and toxic substances. It is almost impossible to produce or store pure water since virtually all substances are soluble to some extent in water and almost all chemical changes are dependent on water.

6.1.3   Biological properties
Water provides the medium for all biochemical reactions through four of its characteristics: (1) solvent property, (2) heat capacity, (3) surface tension and (4) properties on freezing. Most biological processes take place within a narrow temperature range; consequently most organisms cannot tolerate wide variations in temperature. The maintenance of a narrow range of temperature as a result of water's very high heat capacity thus makes it ideal for supporting animal and human life. Water is needed to transport food around the body, eliminate waste, regulate body temperature and to control organ functions. It sustains all the processes of life, carrying dissolved elements vital for every form of life from the tiniest bacterial cells to the most complex human organism.

6.2   Influencing factors

It follows from the foregoing brief outline of the nature of water that it is highly susceptible to chemical and organoleptic change as well as being easily contaminated by allochthonous (those not indigenous) microflora. In order to maintain its active balance and integrity, it is necessary to develop a 'water consciousness' – a means of recognising and preserving the qualities of wholesome water. To achieve this it is important to give careful consideration to all the materials that will come into contact with the water and also to the procedures for handling the water as 'product' and to appropriate hygiene practices.

The following section of this chapter provides guidance on the factors to be considered when selecting equipment, materials, and processes, with a view to minimum product degradation and maximum product safety. They will also be an essential part of the prerequisite programme to support a Hazard Analysis Critical Control Point (HACCP) system, as described in Chapter 9.

6.2.1   Materials in contact with water

Consideration of materials in contact with water throughout the bottling process is likely to include

- Plant equipment (lining of boreholes, pipelines and tanks, etc.) (see 6.2.2)
- Bottle-filling equipment (see Chapter 7)
- Filters or treatment equipment (see 6.2.3)
- Carbon dioxide (see 6.2.4)
- Process air (see 6.2.5)
- Packaging formats (see 6.2.6)

6.2.2   Plant equipment

Whenever feasible, it is advisable to have plant equipment that is dedicated to the bottling of water. Residues of sugar, fruit cells and flavours associated with soft drinks are extremely difficult to eradicate and can lead to organoleptic and microbiological problems with water. Even given the transient status of water within the bottling plant, it is worth choosing materials that have as little reaction with water as possible and by design are capable of being maintained to a high standard of hygiene. All materials used in contact with water must be approved for food use.

Stainless steel is the most widely used material since it meets these requirements. Where plastics are selected on economic grounds, it is important to be satisfied that these not only have approval for food contact but that they specifically do not affect the water in any way. Contact surfaces should be smooth to facilitate product flow and also to enable them to be cleaned easily and effectively.Where permanent joints are needed in stainless steel, they should be smooth and continuous, polished to the same standard as the surrounding material and able to augment the mechanical strength of the material. Integrated stainless steel (ISS) fittings should be used on pipework designed for use with cleaning-in-place (CIP) systems. Hygienic, good quality tungsten inert gas (TIG) welds are needed to avoid areas for build-up of contamination, localised corrosion and to facilitate cleaning and disinfection. Orbital welding ensures reproducible quality and a high grade of stainless steel is recommended, e.g. 316 with a low carbon content.

Any pipework should be self-draining to avoid residual liquid that could lead to contamination of product either through cleaning substances or through build-up of bacteria. Equally, deadlegs and air spaces should be avoided while installing the plant. Where valves are used, similar grade stainless steel combined with hygienic design will maximise product safety. Typically, these will be diaphragm valves with a straight-through flow path to minimise turbulence. Servicing of valves should be possible in-line. This standard of stainless steel material is appropriate for use in any equipment, starting with lining of bore-holes through transporting pipework, tanks, fillers, etc. Although 304 stainless steel is appropriate for general use, 316 is better for salty environments. The grade selected would also be based on the action of cleaning and disinfecting chemicals, as well as suitability for product water contact.

Consideration needs to be given to gaskets in pipework joints and valves. Materials are continually developing and improving. The material used in bottled water equipment needs to be approved for food use by local Regulations and compatible with cleaning methods and chemicals used.

Elastomers, such as silicone, which are synthetic, rubber-like materials with compressible, springy properties, can be misaligned. Because of this, protrusion of the gasket into pipework can cause difficulties in fully draining the system, providing sites for cleaning products to remain and thus lead to cross-contamination. To avoid this, the pipework needs to be designed in such a way that it cannot be misaligned.

Polytetrafluoroethylene (PTFE), which is not an elastomer, is an extremely stable material with lubricious properties. The downside of PTFE is its low compressibility and tendency to 'creep', which can lead to a permanent change in shape. Again, the design of the pipework and the gasket material needs to be compatible. In the case of teflon, spring arms to self-centre the seal to the pipework build in a mechanical stop. Specialised technical advice on the selection of suitable joints is recommended, together with a schedule for their inspection and servicing.

6.2.2.1 Tankers

Under ideal conditions, water for bottling is supplied from the source directly to the factory through a pipeline. Sometimes, however, environmental or local planning considerations make it impossible to build a factory close enough to the source to make this possible and, in this case, tankering becomes the alternative procedure. Indeed, there are some factories that operate entirely through the use of tankers, with dedicated fleets delivering tens (sometimes hundreds) of loads each day. Whether operating such a fleet or only a single tanker, the following factors must be controlled:

- The tanker loading station, including valves, hoses, joints and fittings, must be kept secure, as the source is often in a remote location;
- The loading equipment must be properly designed to ensure maximum integrity; this means at best that the hose is permanently fitted to a 'swivel' joint at the loading pump, thus eliminating the need to connect and disconnect with every delivery;
- The delivery end of the hose must be capped at all times when not in use, and great care taken not to contaminate it when connecting it to the

tanker. The loading pump and hose should be regularly cleaned and disinfected, with full records being maintained of all cleaning, disinfection and loading activities.

Tankers should be dedicated to use for water. Many water bottlers have in the past suffered serious organoleptic and other quality defects as a direct result of using tankers that were previously used for other purposes such as milk or fruit juice. Tankers should be maintained clean (not just sanitised) and to this end they should be fitted with appropriate spray ball fittings. The frequency of cleaning will depend to some extent on the frequency of use. Perhaps, paradoxically, those used continuously are in effect 'cleaner' than those used only on an occasional basis: for the former, a monthly CIP may be sufficient; for the latter, cleaning and disinfection may need to be performed before every load. In either case, periodic microbiological analysis will help in deciding the appropriate regime. Tanker unloading stations at the factory require the same design features as the loading stations and must be managed with the same degree of diligence.

6.2.3   Filters

Various applications for filtration involve contact with product water – either directly for the water or indirectly in providing protection for carbon dioxide, process air or the bottling environment.

Direct filtration of product water may be used for a variety of reasons depending on the category of water: e.g. Natural Mineral Water (NMW), Spring Water (SW) or other bottled drinking water. In the case of NMW, mechanical filtration is permitted by European Legislation to remove particulates such as sand and unstable elements, but must be carefully selected to avoid any compositional change and allow any autochthonous (indigenous) organisms to remain in the downstream filtrate. Depending on the amount and size of particulates, such as sand from groundwater supplies, it may be found beneficial to use two or three grades of filter in series. The first 'rough' filter should be as near the source as possible to prevent aggressive action of sand in moving water through pipework. The final 'fine' filter would then be prior to bottle filling. Another reason for filtering in series is to prolong the life of the final filter, which is the most costly.

One important factor to be borne in mind is that the effectiveness (and, by definition, any failure) of such a filter is not detectable by eye, and may only become apparent through deteriorating microbiological results. In order to monitor filter performance and detect failures early, some filter manufacturers have developed nondestructive technology to enable testing the integrity of the filters. Such 'integrity tests' are performed by applying a pressure differential using sterile air across the filter and by measuring the rate of pressure decay. This is a simple exercise that should be performed regularly (recommended weekly) and always following filter installation, cleaning and disinfection.

Several factors need to be considered in filter selection, such as the nature of the water to be filtered, the flow rate, temperature, pressure drop, surface area, the degree of filtration required and whether prefiltration is taking place. The choice of micron rating of a final filter for NMW will need to take into account the diminutive size of autochthonous organisms, permitting these into the filtrate, but at the same time retaining allochthonous species, which are much larger. This fulfils a requirement to exercise due diligence in protecting consumer safety. Each NMW will be different, but typically a nylon medium of 0.2 μm rating will satisfy these requirements.

Carbon dioxide and process air need to be filtered to remove any particulates and to provide microbiological safety. Hydrophobic cartridges, typically of 0.01 μm rating, are employed for this purpose. Compressed process air, however, should pass through several stages of prefiltration such as coalescing filter, desiccant dryer and carbon adsorption. Activated carbon filters are sometimes used to remove taints and odours from carbon dioxide and process air. In this application, carbon may also adsorb volatile organic compounds (VOCs). Where carbon filtration is used, care must be taken that it does not introduce microbiological contamination. For this reason, the carbon filter should be upstream from the final filtration. To reduce the risk of air contamination to product water around the filling operation, the air supplied to an enclosed environment can be filtered and maintained under a positive pressure.

It is advisable to use reputable suppliers of filters, to advise in the first instance on suitable filtration options and to provide supportive technical services in the event of any enquiries. It is important to realise that any filtration is a process and as such it can go wrong, resulting in a greater problem than would occur if the filter had not been there in the first place. It is therefore necessary to monitor results of filtration regularly. In addition, a schedule for changing filtration cartridges at appropriate frequencies is imperative. It is prudent to maintain records of such replacements and of filter efficiency.

6.2.4   Carbon dioxide

Some groundwaters have a natural carbonation, but where a water is still (noncarbonated), carbon dioxide gas may be added to achieve a sparkling, effervescent product. Carbon dioxide is a colourless, odourless and non-poisonous gas readily soluble in water. It can be produced as either a by-product of a fermentation process or chemically as a by-product of fertiliser manufacturing. Carbon dioxide has on occasions been found to impart taste and/or odour taints to bottled waters. There have also been concerns, notably in 1998, regarding levels of benzene in supplies of carbon dioxide to the beverage industry. Purification of the gas is required to ensure its suitability as an additive for bottled waters and other beverages.

In 1999, the International Society of Beverage Technologists (ISBT) produced *Quality and Purity Guidelines for Carbon Dioxide*. These provide valuable technical information on parametric limits and other relevant data for producers of the gas and its users in the beverage industries. The guide was revised in 2001 and entitled *Quality Guidelines & Analytical Procedure Bibliography for 'Bottlers' Carbon Dioxide*.

The British Soft Drinks Association (BSDA) convened a working group in 1998 to address the issue of carbon dioxide safety and suitability and, as well as supporting the ISBT Guidelines, also identified approved laboratories within the UK for undertaking analysis of carbon dioxide through ring testing.

Water bottlers are advised to give high priority to the safety and purity of carbon dioxide used as an ingredient for their products. Supplier audits and HACCP can ensure that all routes liable to cause contamination are covered. Deliveries of carbon dioxide can be analysed through an approved laboratory, and organoleptic evaluation of the gas carried out, using an adapted mini carbonator, similar to the one often used at home to make single-serve carbonated drinks, before offloading from the tanker.

6.2.5   Process air

Compressed air is used at virtually all stages of the water bottling process, depending on the type of equipment installed. It may be used for bottle cleaning equipment prior to filling; however, in instances where polyethylene terephthalate (PET) bottles are blown direct to line, neither water nor air rinsing may be necessary prior to filling. It is also used in the fillers themselves and in cap vibratory bowls. If bottle blow-moulding equipment is used on-site, compressed air is used there also.

In any of these functions, the microbiological and organoleptic quality of the process air must be of a very high standard; this can be achieved by appropriate filtration to 0.01 μm rating. For machinery after capping of bottles, i.e. labelling, packing, palletising, etc., the quality of compressed air is important but less critical. It is sometimes feasible to provide these two qualities of process air from two different generating supplies.

6.2.6   Packaging formats

While water is only briefly in contact with plant equipment during the bottling process, it may be in contact with the primary packaging for several months. For this reason it is even more important to consider packaging formats to maintain the integrity of the water and maximise shelf-life for the product.

Fundamentally, the functions of packaging are to contain the product and to protect it from possible hazards and contamination during transportation and distribution, and throughout its shelf-life. Packaging also serves to inform the distributor, retailer and consumer about the product, its identity, volume and characteristics and any other legislative requirements. In addition, it acts as a silent salesperson and marketing tool to present the product and identify the brand.

Various packaging formats and sizes may be used to provide consumer choice in different retail outlets. Sizes range typically from 20 cl to 2 l, though 3 and 5 l containers are also available. Larger containers, typically 19 or 22 l, are used for water coolers (see Chapter 10). Although the bottle or any other container is the main part of the packaging, the closure must also be considered carefully as an integral component.

6.2.6.1 Glass

Of all packaging materials, glass is the most suitable for containing water for a variety of reasons:

- It is chemically inert and does not affect the quality, odour or taste of the water;
- It is usually a clear material and therefore enhances the clarity of the water itself;
- It is hygienic – a major consideration;
- It is rigid, which gives it strength and enables it to be run efficiently on high-speed bottling lines;
- It is impermeable and has the best ability of all bottle materials to retain carbonation, enabling maximum shelf-life to be prescribed;
- It is suitable for still and sparkling water;
- It is resealable, recyclable and is perceived as high quality.

To be set against the above advantages, glass is the heaviest packaging material for bottles. Great care is needed in its handling, both during the filling and distribution process and in consumer use, owing to its breakability. Although it is an inherently strong material, it can be weakened by impact, and this can lead to spontaneous explosion at a later time if the water in it is carbonated.

Glass has been used for making bottles for hundreds of years. It is made from readily available natural ingredients: sand (silica, SiO_2), soda ash (Na_2CO_3) obtained by chemically treating common salt and limestone ($CaCO_3$). These are the main components, but small amounts of other elements may be added to achieve particular qualities or colours. Glass bottle production is schematised in Fig. 6.2.

Glass is recyclable and large percentages of cullet (recycled glass), typically 85%, are used to advantage in the manufacture of bottles, reducing the process' energy consumption. Furnaces vary but 25% energy savings are an accepted industry standard. The average energy consumption of a furnace is 1380 kWh/t of glass melted, so each tonne of cullet used saves 345 kWh of (gas) energy. The incorporation of cullet also reduces carbon dioxide emission resulting from the process.

Both single-trip and multi-trip bottles are available in glass. Multi-trip bottles need to be more robust, use more glass and can become very scuffed through many

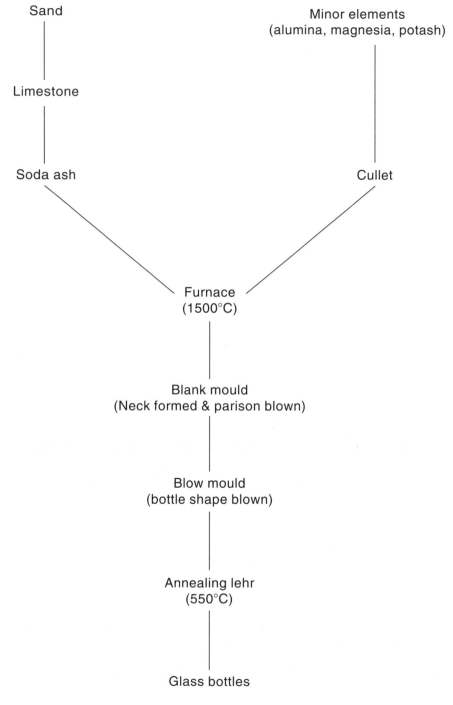

Fig. 6.2 Glass bottle production.

journeys, detracting from the presentation of the product. Other considerations related to multi-trip bottles are the logistics of returning bottles to the production unit, the use of hot caustic soda solutions, and large volumes of water and energy use. These are serious environmental issues and, at the same time, returned bottles also need vigilance to ensure their continued fitness for purpose.

Single-trip bottles are often the preferred choice for bottled water. They provide fitness for purpose, bringing least risk of compromise to product safety and integrity. Single-trip glass bottles eliminate the need for bottle washing, since all that is required prior to filling is rinsing with water or air. They present the product in premium condition. Advances in design and the manufacturing process have also enabled light-weighting of single-trip containers.

6.2.6.2 Polyethylene terephthalate

PET is now by far the most important material used for the packaging of bottled waters. It is a very lightweight material with good clarity. It can be used for still and sparkling products, though it does lose carbonation through the wall of the bottle, which restricts the shelf-life that can be prescribed. There is a risk of organoleptic effects where acetaldehyde content in the bottle is greater than 4 ppm. PET bottles are resealable, unbreakable and recyclable.

When empty, PET bottles lack rigidity, which, together with their light-weight, makes them unstable; there are difficulties in running bottles at high speed unless they are conveyed suspended by the neck ring. PET is a synthetic thermoplastic polyester derived from crude oil. Terephthalic acid, derived from xylene, is reacted with ethylene glycol to produce the esterified monomer. This is polymerised and the material supplied to bottle manufacturers as granules. The production process is schematised in Fig. 6.3.

Few water-bottling companies produce their own glass bottles, but many have plastic bottle manufacturing facilities, which have distinct advantages in saving transport and storage costs. The bottles are made in a two-stage process: first, the injection-moulding of preforms, during which process the neck finish is completed. In the second stage, which may take place at the bottling facility, the test tube–shaped preforms are stretch blow-moulded to their final shape and size, which may take place at the bottling facility. In recent years, blow-moulding PET bottles from preforms directly to the water bottling line has become virtually the norm. The manufacture of the preforms themselves is also increasingly being done by bottlers, with consequent improvements in control over the process and significant cost savings.

PET is recyclable, although plastics reprocessing is still very much in developmental stages; the largest stumbling block is the lack of collection facilities. Technology is developing to enable recycled PET to be reused for food contact, but at present most reprocessed material is utilised as fibre for clothing, sleeping bags, carpets and brushes, and as films for thermoformed packaging and photography.

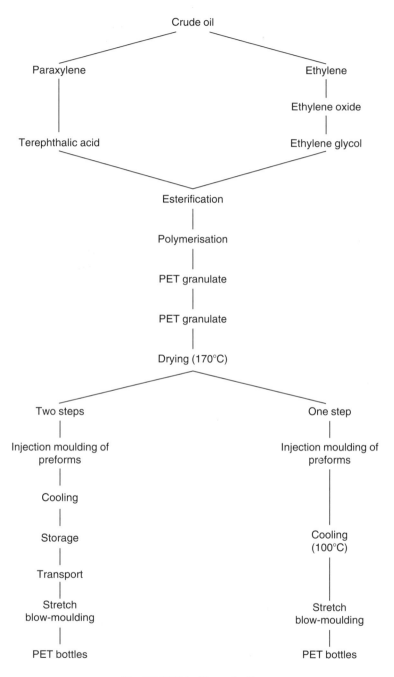

Fig. 6.3 PET bottle production.

Lightweighting of PET bottles, without loss of performance and with better environmental impact, is now attainable with recent advances in design and manufacture.

6.2.6.3 Polyvinyl chloride

For many years, polyvinyl chloride (PVC) has been used for packaging still waters. Of all synthetic materials, PVC was probably the one most researched and the compound was capable of being formulated specifically to be compatible with water, or even a particular water. However, the preference for, and versatility of, PET, together with pressure from consumer groups, has led to the decline and, ultimately, the demise of PVC in the bottled water industry. Over several decades, PVC served the industry well, being light in weight and relatively low cost.

6.2.6.4 Cans

Cans are used for only a very small percentage of the packaged water market. They have the advantage of being lightweight, easily handled and recyclable. They hold carbonation very well and can be used in vending machines. A disadvantage is that the product is not visible. Cans are ideal for single servings but have the disadvantage of not being resealable. Very expensive plant equipment is required to fill this relatively small percentage of product.

Aluminium is preferred to steel as it gives better organoleptic and corrosion test results. Both steel and aluminium cans are internally coated with an epoxy-based lacquer that protects the product from direct contact with the metal. The sealing mechanism between the can body and the end is critical to this packaging format, and precise specifications and monitoring of this are essential throughout the filling process.

6.2.6.5 Cartons

Cartons are used for an even smaller percentage of packaged waters than cans. Cartons, where used, are usually of a multilayer laminate material. A layer of paper provides rigidity to the pack, an aluminium layer forms a barrier to gases and polyethylene (PE) layers top and bottom ensure that the pack is watertight. The resultant packs are lightweight and easily handled. This method of packaging is very space-efficient. Cartons can only be used for still product, and, although major advances have been made in laminate, there may be an effect on the taste of the water that limits its optimum shelf-life.

6.2.6.6 Polycarbonate

Polycarbonate is typically used for large bottles – 11, 19 and 22 litres – for use with water coolers. Considering the size of the bottles, they are lightweight. A collection system is needed to facilitate reuse and a very thorough bottle washing process is required (see Chapter 10 for further information).

6.2.6.7   High-density polyethylene

In some countries, notably the USA, high-density polyethylene (HDPE) is used for packaging water. It is lightweight, low cost, robust and recyclable. Disadvantages are that it is not a clear material, it cannot be used for carbonated water and it may (if the process is not well controlled) affect the organoleptic properties of the water. However, there is a successful market in larger-sized packages, particularly 1.0 and 2.5 US gallons. The use of HDPE was actually an offshoot from the dairy industry, for whom (prior to the development of PET technology) it was the only source of larger containers. In a development somewhat similar to that in the dairies, many US bottlers have developed 'through the wall' arrangements with the HDPE bottle manufacturers, who manufacture their bottles on the same site as the bottling facility.

Considerations in the choice of container materials for bottled water are summarised in Table 6.1.

6.2.6.8   Closures

The efficacy of any packaging format in fulfilling all its functions relies heavily on an efficient and suitable sealing system. There must be compatibility between the bottle or container and the closure. Both must be designed to work together.

Just as with the bottles, the materials used for closures must not in any way affect the product water – organoleptically, physically or chemically. Equally, the water and closure must be compatible so that the materials of the closure and its performance are not impaired. Overall and specific migration testing as well as organoleptic evaluation will provide evidence of a closure's suitability for the task. The size of the closure is significant: the smaller the closure diameter, the less contact there will be between closure and product, and this is particularly important for smaller-volume bottles.

In carrying out suitability trials with prospective packaging formats, it can not be assumed that what is compatible with still water will also be suitable for carbonated water.

Closures require both strength and rigidity for application on high-speed bottling lines and to withstand substantial internal pressures during service. Products carbonated to 3–4 volumes of carbon dioxide can result in an internal pressure within the headspace of up to 10 bar, especially at elevated temperatures, since the solubility of carbon dioxide in water decreases as the temperature increases. Both at the design stage and while filling the bottle, it is important to be aware of the optimum percentage vacuity to minimise internal headspace pressure. This internal pressure must be capable of release prior to complete removal of the closure. In the case of glass bottles, this is usually achieved through irregularities in the finish (the mouth of the bottle). However, in PET bottles, where more precision is achieved in the finish, a venting system has been designed into the thread for this purpose. Such prior venting minimises the risk

Table 6.1 Factors influencing choice of containers for bottled water

Type of bottle/ container	Advantages	Disadvantages
Glass	Chemically inert – will not affect quality/odour/taste Clear Strength in rigidity – can be run efficiently on high-speed bottling lines Hygienic Retains carbonation well Can give maximum shelf-life to product Suitable for still or sparkling product Resealable Recyclable Impermeable High quality	Heavy Care needed in handling, both in filling and consumer use Breakable Impact damage can result in spontaneous explosion with possibility of injury to consumers Could be subject to foreign body contamination from broken glass
PET	Lightweight Clear Suitable for still or sparkling product Resealable Unbreakable Recyclable	Nonrigidity can cause (1) spillage on opening (2) instability of pallet with still product Loss of carbonation Limited shelf-life for sparkling products
PVC	Lightweight Low cost Resealable Unbreakable Recyclable	Not clear Cannot be used for sparkling product unless biorientated Effect on total viable count (TVC) Adverse views of consumer groups
Cans	Lightweight Holds carbonation well Use in vending machines Easily handled Recyclable	Not resealable Expensive plant required for relatively small percentage of business Product not visible
Cartons	Lightweight Easily handled Space-efficient	Only suitable for still product May affect taste Limited shelf-life
Polycarbonate (water coolers)	Lightweight Returnable Reusable	Requires collection system Requires thorough bottle-washing facilities Effect on TVC
HDPE	Lightweight Low cost Robust Recyclable Suitable for large containers	Not clear Cannot be used for sparkling product May affect taste/smell

of closures 'missiling' (blowing off the bottle and becoming a missile), which can cause possible injury.

As well as withstanding internal pressure, closures need to be able to sustain external pressure in stacking, storage and transport.

Much emphasis is now being placed on the ease of opening without mechanical aids, especially since the elderly are becoming a more significant percentage of the population.

Tamper evidence is an important feature in the light of increasing incidence of malicious tampering. All closures used on bottled waters must therefore be tamper-evident to contribute to their safety and integrity. This is a legal requirement in many markets. As far as cap materials are concerned, they may be metal, e.g. aluminium or aluminium alloys and tin plate, or plastics, e.g. polypropylene (PP), low-density polyethylene (LDPE) and high-density polyethylene (HDPE).

The closure may be one-piece (plastics only) or two-piece. Metal closures usually have a flowed-in plastic liner, and two-piece plastic closures have a resilient and compressible liner usually based on PVC or PE. Metal closures are stamped and drawn out of sheet that has been suitably decorated and lacquered. Crown closures are usually made of tin plate as this has the attributes of high strength and rigidity, whereas rolled-on closures need to be softer and here aluminium is more suitable. Plastic closures, introduced in the 1980s, are mainly injection-moulded and any printing or design work is applied after this process. Although crown closures are used on some single-use bottles, most bottled waters are fitted with screw caps. In the case of metal closures, the thread is formed at the point of application to fit the individual bottle, and this type of closure is especially suitable for glass bottles. Plastic closures have their thread moulded in the injection-moulding process.

The chosen closure should be applied immediately after filling the bottle to maximise the integrity and safety of the water. Recommended application and removal torques are provided by the closure manufacturer, and these should be closely adhered to and monitored throughout the bottling process.

In the last few years, sportsbottles have become very popular (see Fig. 6.4). There have, however, been a few unfortunate incidents related to the small components of sportsclosures. A working group, a subcommittee of the technical packaging committee of the BSDA, together with input from closure suppliers, has developed a *Code of Practice for Sports Closures*, which gives advice on assessing the risks of choking hazards, test procedures and labelling recommendations.

Water bottles discussed in this chapter are essentially for single usage. Once empty, they should be discarded appropriately, to facilitate their recycling. It is recommended that bottles should not be refilled or reused by consumers. Contamination can result from this. In addition, other components of the packaging format, e.g. the sports closure, are not designed for repeated use.

Fig. 6.4 Generic sportsbottle. (Photograph courtesy by James Roddick, Gleneagles Spring.)

6.3   Labelling

Labelling performs several functions as part of a packaging format: labels identify the product and proclaim the brand; they act as silent salespeople. The average consumer can scan approximately 1.3 m of supermarket shelf space per second, so the label has about 0.1 s to make an impact, thus providing a challenge to both marketing executives and designers.

There are various requirements of legislation to be incorporated into the label design – for example, within the European Community these include the product description, volume declaration, name and address of the bottler and their country of origin, and indication of Best-Before-End (BBE) duration. Although not a legal requirement in all territories, important element of information declared on the label is a Typical Analysis panel. This provides levels of composition for several parameters enabling the consumer to make a choice relating to taste preferences or dietary needs. Composition is usually given in mg/l. The bottler may wish to include other consumer use advice on the label, such as serving suggestions and storage recommendations. Nutrition facts are often given. With increasing awareness of environmental issues, as well as requirements of legislation, identification of bottle material and recyclability may also be displayed. As well as the name and address of the bottler, it can be helpful to the consumer to give a website address through which they can direct any enquiries they may have in relation to the product.

On single-unit bottles, a bar code is printed on the label; however, in the case of multipacks, the bar coding is usually on the outer sleeve or other means of collating the bottles in the pack. In designing new or revised labels, it is advantageous to liaise with enforcing authorities to ensure that the requirements of legislation have been met.

6.4   Shelf-life, batch coding and traceability

The requirement for a BBE date relates to legislation. In addition, it facilitates stock rotation within the distribution and retail systems and also provides guidance for the consumer. It must be understood that the dates given refer to the bottle in its unopened state. The duration of shelf-life that can be prescribed for a bottled water will be influenced by the bottle and closure materials and by the size of the container. Generally, the smaller the container, the shorter the shelf-life. The shelf-life may be different for still and sparkling products, depending on the nature of the packaging.

Water bottled in glass, either still or sparkling, remains in good condition for several years. Considerations in this case focus on the effects of the closure mechanism and on the outer presentation remaining in pristine condition. Still water in PET bottles remains in good condition for many months, perhaps years. However, sparkling products in PET lose carbonation through the wall of the bottle, showing greater loss from smaller bottles through the relationship of volume of water to surface area of the bottle. This restricts shelf-life to about 12 months for bottles of 1 l or more capacity and considerably less for smaller sizes. Figure 6.5 illustrates typical carbonation loss from 2 l PET bottles.

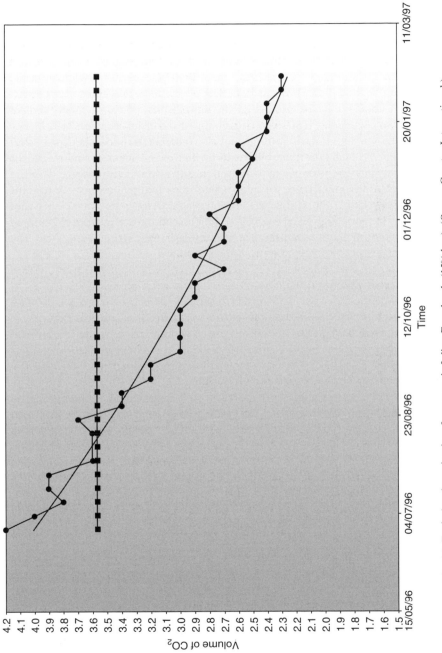

Fig. 6.5 Typical carbonation loss from generic 2-litre Euro bottle (15% loss). (*Source:* Constar International.)

Still and sparkling water in cans can be given a shelf-life of at least 12 months, but it is advisable to be satisfied that there is no increase in aluminium content beyond that period.

Individual waters may behave differently over time depending on their characteristics and composition in relation to the packaging used, and it is wise to monitor performance chemically, microbiologically and organoleptically during and beyond prescribed shelf-life, which can then be reviewed accordingly.

Batch coding may be positioned on the closure, bottle or label, and, in addition to giving the BBE date, it usually shows a lot number, which may refer to the date and time of bottling, and also indicate the particular line or shift of bottling. To facilitate traceability still further in the event of any recall, the bottles themselves may be coded with the date and time of manufacture. This can be located in the area where the label covers it so that it does not confuse the consumer. Where PET bottles are blown direct to line, this marking is not required, though it may be advisable to retain records of preform manufacturing dates relevant to the bottling dates on which they were used.

6.5   Hygiene practices

Bottled waters are classed as food and in many instances receive little or no treatment prior to bottling. Good hygiene practices are therefore an essential prerequisite of a successful operation. These prerequisites, together with HACCP (described in Chapter 9), form the basis of good practice for the industry.

The starting point for protection is the source of water itself (see Chapter 4). Because this is a highly specialised area, it is recommended to have input on developing a groundwater source for bottling from a reputable hydrogeologist. Wherever possible, it is good to have control of activities in the catchment area through specific practices to protect the groundwater from all kinds of pollution. A protocol for the entry of visitors to the bottling plant or its water source will ensure maximum protection to staff, premises, equipment and products.

6.5.1   Buildings and facilities
Many bottled waters, especially NMWs, are bottled at source, and the location of the source and/or the bottling facility is important, which should be selected on a criterion of minimum risk of pollution from the external environment such as factory fumes and industrial and agricultural odours. The buildings need to be sound and well maintained with well-fitted external doors and windows to prevent ingress of pests and insects. A reputable pest control contract is advisable.

A high standard of maintenance within the plant contributes towards achieving good hygiene. Consideration needs to be given to when such maintenance will be undertaken so as to minimise any risk to the product. A planned maintenance schedule is the ideal. Selection of substances used, e.g. lubricants, needs care and these should always be approved for use in food production environments.

When extensive structural repairs or developments are undertaken, these are best tackled during a planned shutdown period. The choice of paint and when it is used is also important since some paints are very odorous and can taint water.

6.5.1.1   *Internal building surfaces*
Floors should be sealed and easily cleanable. In filling rooms, the floor should be coved at junctions with the walls. Walls and ceilings should be light in colour to reflect as much light as possible, as well as to make any soiling visible. The surfaces should be smooth and impervious to facilitate effective cleaning. Junctions between walls and ceilings in filling rooms should be coved. Light fittings should, where possible, be flush with the ceilings and should be shatterproof. External windows should not open into filling rooms. Windows should be close fitting and any sills sloped to discourage their use as shelves. External windows can be fitted with appropriate mesh to prevent insect ingress. The mesh itself must be capable of being effectively cleaned.

6.5.2   *Layout and process flow*
A design enabling continuous process flow – with materials receipt and storage at one end, finished goods and despatch at the other and the processing stages in sequence in between – will optimise good hygiene practice and minimise any cross-contamination (see Fig. 6.6).

Within this design, it is desirable to allocate a particular area for the vulnerable open bottle stages of rinsing, filling and capping. With provision of an enclosure for these stages, it is feasible to control this environment by treating the air inside and imposing a positive pressure.

6.5.3   *Ancillary facilities*
Facilities provided for staff, laboratories, maintenance workshops and chemical storage should be separated from the main production area and be capable of being maintained to the same hygienic standard as the rest of the plant. In addition to washbasins, which must be provided in toilets and canteens, it is recommended to provide further hand-washing facilities at entry points to filling rooms. In all locations, hand washbasins (not to be confused with sinks) must be maintained immaculately and should not be used for other purposes.

6.5.4   *Cleaning and disinfection*
Routine, scheduled cleaning and disinfection is an essential part of running a water bottling plant. This not only protects the safety of the product but also provides a good working environment for employees. It removes extraneous matter that could be conducive to microbial activity and provide a risk of foreign body contamination.

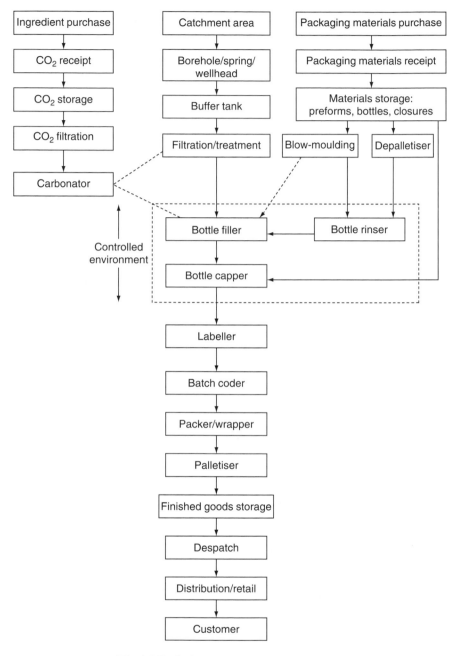

Fig. 6.6 Typical process flow for bottling water.

Schedules will define what is to be cleaned, who will undertake the task, the frequency of cleaning and disinfection and how it will be done. They also prescribe chemicals, their use and precautions and means of monitoring the efficacy of the procedures. Like other potable liquid installations, a water bottling plant is often designed for CIP, i.e. the process is carried out with bottling equipment in its assembled state.

The choice of the cleaning chemical or disinfectant must be made taking into consideration whether the plant is dedicated to water or whether it is also used for soft drinks, the level of soil/biofilm expected, frequency of cleaning, compatibility with equipment and filter materials and its rinsability and removal to leave no trace of odour or taste.

For dedicated water equipment, cold/ambient cleaning and disinfection may be appropriate and effective; for equipment shared with soft drinks, more rigorous routines with high temperatures are required to achieve satisfactory results.

Each plant will be different, so the cleaning programme and its frequency will be planned accordingly to maximise product integrity and safety (see Chapter 8).

6.5.5   Personnel

It is important that the staff involved in the production process or its ancillary functions are appropriately trained in the awareness of the vulnerability of product water and how to protect its cleanliness and safety, as well as to be experts in procedures to carry out their particular roles. A company hygiene policy will outline expected standards of personal hygiene and fitness, specifying any medical screening the company may adopt, and good practices. Such a policy will also instruct on the use of protective clothing and on where this is applicable. The policy will include requirements on reporting sickness or infections and on the suitable covering of cuts and open wounds.

Personal hygiene aspects include the need for clean, short fingernails, clean, covered hair, and a requirement for not using perfumes and aftershaves. A lack of jewellery will also appear as a requirement. The good practices include when and where to wash hands, housekeeping, handling and use of primary packaging components, filling room protocol, locker usage and eating/drinking locations. An endorsement by the company chief executive can be included in the food safety and hygiene policy, which lends credence to top management commitment.

Importantly, as well as being given training, staff need to be appropriately supervised or managed to ensure that the training and the company policy are fully implemented. In this way, each and every employee contributes to the product's safety and integrity.

6.5.6   Good manufacturing practice

Good practices incorporated within personnel training specify that product containers are never used for anything other than product. All too easily they can be viewed as handy containers for a whole range of things – lubricants, cleaners, nuts and bolts, flowers, etc. – and this could pose a serious risk to the product.

Fingers should never be inserted into open bottles as this will lead to microbial contamination.

Any visitors or contractors to the plant need to be briefed appropriately on relevant practices.

The principles of good hygiene practice will be incorporated within an overall quality management system (see Chapter 9) and as such would be subject to audit and review (see Chapter 11) to ensure that such practices are effectively implemented.

Bibliography

Code of Practice for Sports Closures. (2003) The British Soft Drinks Association Ltd, 20/22 Stukeley Street, London WC2B 5LR.

Guide to Good Bottled Water Standards (2002), 2nd edn. The British Soft Drinks Association Ltd, 20/22 Stukeley Street, London WC2B 5LR.

Harwood, M. (1995) The important closure device. Paper presented at PETCORE Conference, Geneva, 31 January.

Industry Guide to Good Hygiene Practice: Bottled Water Guide. (2001) Chadwick House Publishing. Chadwick Court, 15 Hatfields, London SE1 8DJ.

Moody, B. (1997) *Packaging in Glass*, revised edn (first published 1963). Hutchinson Benham, London.

Oliphant, J. A., Ryan, M. C. & Chu, A. (2002) Bacterial water quality in the personal bottles of elementary students. Canadian Journal of Public Health (September–October).

Quality Guidelines & Analytical Procedure Bibliography for 'Bottlers' Carbon Dioxide (2001), revised edn. International Society of Beverage Technologists. (ISBT), 8120 S. Suncoast Boulevard Homoassa Florida 34446–5006.

Schwenk, T. (1965) *Sensitive Chaos*. Rudolf Steiner Press, London.

Schwenk, T. & Schwenk, W. (1989) *Water – The Element of Life*. Anthroposophic Press, New York.

The Glass Container Industry. Energy Consumption Guide 27. Future Energy Solution, Didcot, Oxfordshire.

7 Filling equipment

Fred G. Vickers and John Medling

7.1 Introduction

Developments in the field of small packs and the technology of their production continue to be challenging. Of central importance within these developments is the filling system – one area within the packaging line that can directly affect product quality. The development of container types for bottled waters has been dramatic. Modern filling systems are now required to cope with a wide range of containers and materials. Glass, PET and PVC are now all commonplace. In many instances, special bottle shapes are used as a means of providing some individuality to the marketeer. The container now not only serves as a means of containing the product but has also become an 'art form' in order to attract customers as products compete on the supermarket shelves.

Bottle filling systems have been developed to cater for this wide range of containers: e.g. developments in handling to suit different bottle types, precise control of the filling process and the control of filling accuracy to meet increasingly demanding quality standards. Additionally, the filling environment has evolved to provide the level of cleanliness required for production of sensitive products such as bottled water. Most bottlers are familiar with conventional filling machines utilising cam-activated control of the filling process. Recent developments have introduced the concept of electropneumatic control of the filling function, offering the bottler more precise control.

We will examine two filling options: the filling of carbonated waters and the filling of noncarbonated (still) products. We will also consider the introduction of water flavouring – a recent addition to the bottled water scene. Filling machine speeds have dramatically increased, with 40 000–60 000 bottles per hour (bph) on smaller container sizes now becoming commonplace. To meet these higher speeds, filling system 'blocs' incorporating rinser/filler/capper are now industry standards. In some applications, complete 'super blocs' incorporating flavour dosing/mixing including labelling have been used. Electronic bottle inspection systems (for glass bottles) have also been introduced to the bloc system.

7.2 Filling systems

There is a wide range of filling systems available and it is impossible to provide
here a comprehensive study of all types of machines. For the purpose of this
review, we will examine primarily the development of machines utilising a short
tube filling system, and later in the chapter, new developments in the design of
hygienic volumetric filling systems.

Figure 7.1 illustrates a cross-sectional view through a typical filling machine.
This machine incorporates a product carousel ring bowl. The filling valves are
mounted on the rotating ring bowl element: product is fed to the machine via a
central rotary distributor assembly, and passes to the ring bowl via a series of
radial feed pipes.

Our starting point is the basic short vent tube filling system with a product
carousel ring bowl to fill carbonated water. Figure 7.2 illustrates a typical
filling valve system suitable for filling carbonated water products. Most of
the latest generation of short tube fillers use a valve integrated within the
product ring bowl (internally mounted in Fig. 7.2). This basic system is
capable of filling a whole range of carbonated products and is in common
use within the bottled water industry. Figure 7.3 shows an alternative valve
mounting position, the complete valve arrangement being external to the product ring bowl.

Product is fed to the machine ring bowl via a feed pump or directly from an
in-line carbonation system. Product level within the ring bowl is maintained by
means of control floats, or more usually now by a series of level probes positioned within the ring bowl. Figure 7.4 illustrates the filling sequence for a basic
machine handling carbonated product.

Derivatives of this basic machine concept have been developed. When handling carbonated product, the potential of the system can be increased by an
additional ring channel. This provides for the filling sequence to include a pre-
evacuation/preflushing process (Figs 7.5 and 7.6). This same system also provides a collection channel for the recovery of CIP(cleaning-in-place) fluids. The
pre-evacuation phase (vacuum system) could be applied to any rigid container,
e.g. glass. Alternatively, with the use of lightweight PET bottles, this extended
sequence could be used to provide for a gas(CO_2)-flushing phase prior to filling
(can be fed by a separate CO_2 supply).

These systems were developed for products that are extremely sensitive to
oxygen contamination. Today, with the increasing demand for improved product quality in the bottling of carbonated water products, filling machines with
pre-evacuation or preflushing are becoming more desirable.

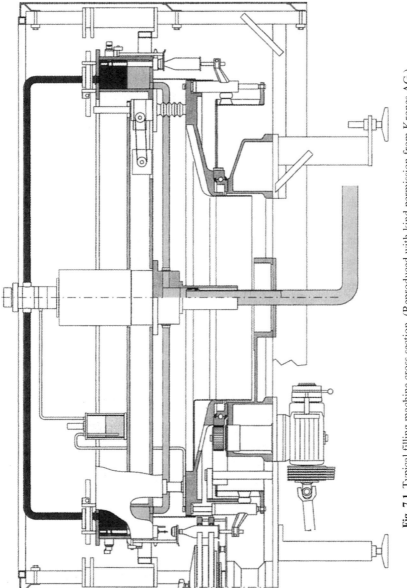

Fig. 7.1 Typical filling machine cross section. (Reproduced with kind permission from Krones AG.)

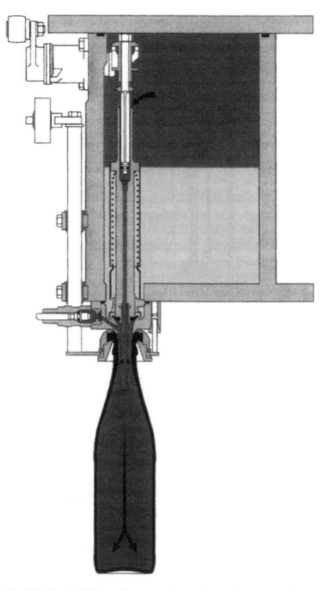

Fig. 7.2 Typical filling valve system for carbonated water products.
(Reproduced with kind permission from Krones AG.)

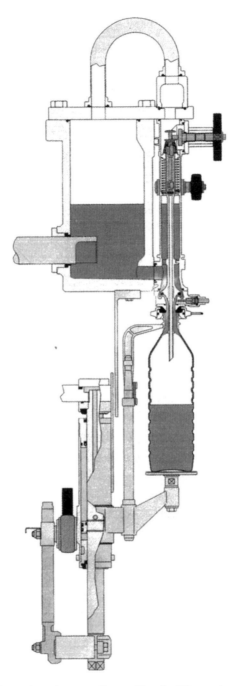

Fig. 7.3 Alternative valve mounting position for filling carbonated water, with the complete valve arrangement external to the product ring bowl. (Reproduced with kind permission from Krones AG.)

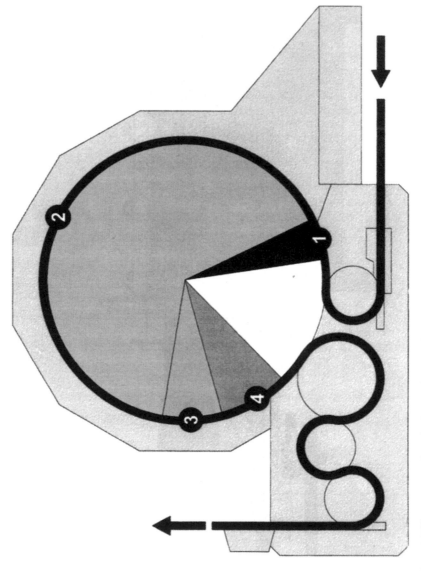

Fig. 7.4 Filling sequence for a basic machine handling carbonated product: (1) pressurising; (2) filling; (3) settling; (4) snifting. (Reproduced with kind permission from Krones AG.)

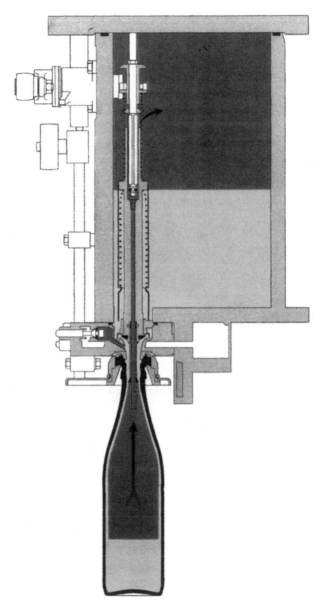

Fig. 7.5 Valve system incorporating evacuation channel.
(Reproduced with kind permission from Krones AG.)

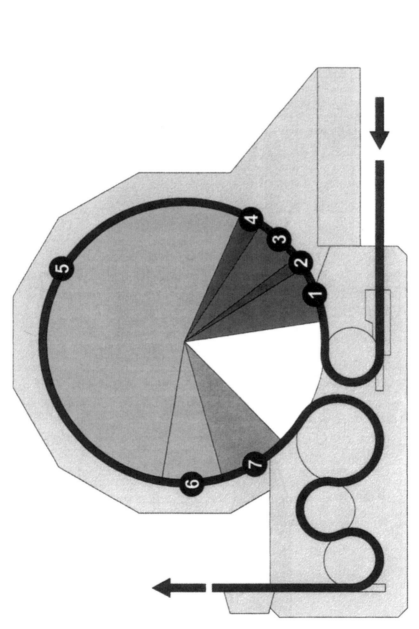

Fig. 7.6 Filling sequence incorporating an additional double vacuum/gas flushing sequence: (1) first evacuation; (2) CO_2 flushing; (3) second evacuation; (4) pressurising; (5) filling; (6) setting; (7) snifting. (Reproduced with kind permission from Krones AG.)

Figure 7.6 illustrates a filling sequence with an additional double vacuum/ gas-flushing sequence. This also has an alternative sequence.

Normal sequence	*Alternative sequence*
(1) First evacuation	
(2) CO_2 flushing	CO_2 flushing
(3) Second evacuation	
(4) Pressurising	Pressurising
(5) Filling	Filling
(6) Settling	Settling
(7) Snifting	Snifting

7.3 Electropneumatic valve system

The introduction of electropneumatic filling systems has provided the bottler with further process flexibility and ease of changeover, while maintaining filling quality. The basic arrangement of this new concept is similar to that of established mechanical cam-controlled machines. However, this system provides control of all filling process functions electropneumatically without the usual mechanical control cam fitted around the filler carousel. The electropneumatic system is thus independent of carousel rotation or rotational angle for arrangement and timing of process functions. Filler control valves are mounted above the ring bowl position. An operator terminal mounted at the control panel position provides an interface with the machine control system. Level control is carried out by conductivity probe, which provides the facility for fill level adjustment during machine operation. Changeovers for fill level or filler process times can be preprogrammed into the system to suit any predetermined bottle range.

Figure 7.7 shows a schematic arrangement of a typical electropneumatic control valve system. This arrangement is suitable for filling carbonated products with a process profile as illustrated in Fig. 7.4. The process is controlled by the predetermined parameters set within the programmable logic control (PLC).

Figure 7.8 illustrates a further enhancement of this system, introducing additional electropneumatic valves that can control extra process functions. In this instance, the addition of a vacuum-controlled evacuation process (suitable for glass bottles) or, alternatively, a preflushing process from a separate pure CO_2 supply can easily be accommodated. Evacuation/flushing discharge is via a separate return channel. This process is suitable for applications where quality of the final product is potentially sensitive. Independent of filling speed variations, the predetermined process function cycles are accurately maintained, guaranteeing constant filling quality.

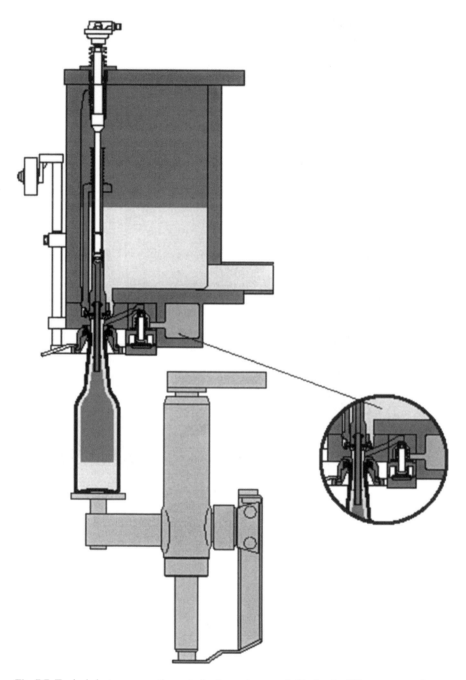

Fig. 7.7 Typical electropneumatic control valve system – suitable for the filling sequence shown in Fig. 7.4. (Reproduced with kind permission from Krones AG.)

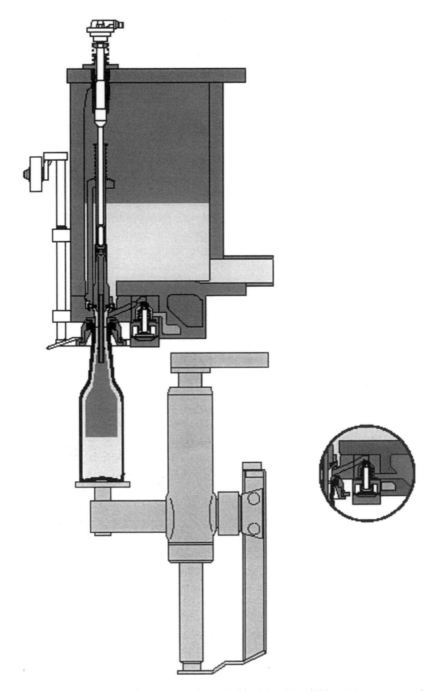

Fig. 7.8 An enhancement of the system shown in Fig. 7.7, using additional electropneumatic valves to control further process functions. (Reproduced with kind permission from Krones AG.)

7.4   Non-carbonated filling system

Figure 7.9 shows a schematic arrangement of an electropneumatic valve control system, which operates by gravity at atmospheric pressure. The filling valves are mounted on a rotating product ring bowl (as used for carbonated products), but in this case external to the main product carousel. The conductivity probe (fill level control) is shown here as part of the filling tube; when product contacts this probe, the filling valve closes. Filling speed is determined by the size (diameter) of the filling tube and the static liquid height in the filler product bowl (height above the filling valve). During filling, displaced air from the bottle is passed into the atmosphere and not into the filling system. Owing to non-contact of the bottles and filling valves, integrity of bottle quality is maintained and the system is ideal for the handling of all types of containers including lightweight, less rigid plastic bottles. Figure 7.10 shows an alternative arrangement for the product bowl, which is arranged centrally within the filler. The operating principle is the same as for the ring bowl design.

7.5   Bottle handling

The traditional method of bottle handling within the filling machine is by means of a scroll infeed, which carries the bottles from the infeed conveyor into a spaced 'pitch' suitable for transfer into the filler infeed starwheel. The starwheel system allows the bottles to enter the filling machine; bottles are correctly pitched in accordance with the valve location on the rotating filler carousel. They are transferred onto platforms and are then lifted into position at the filling valve (see Fig. 7.11). On discharge from the filling carousel, bottles are transferred from the lowered bottle platform into a similar starwheel system for transfer onto the discharge conveyor, or more usually into the monobloc capping carousel. The starwheel system is a well-proven method of handling bottles through the filling system and is particularly suited for glass and rigid PVC containers.

   With the advent of PET containers for carbonated products, alternative systems have been developed specifically for handling these lightweight containers. The most widely used is the 'neck support system' (see Fig. 7.12), which employs a bottle lifting and clamping device that uses the neck ring incorporated within the PET bottle design. This type of machine operates without lifting platforms and relies purely upon the neck ring support system for the necessary lifting and sealing into the filling valve position. The system relies on the integrity of the neck and neck ring support within the PET bottle design and, although PET bottles are prone to deformation and shrinkage, the neck support system positions them perfectly under the filling valves.

   With the increasing lightweighting of PET bottles, some machines retain the bottle lifting platform as well as the neck ring support system. The platforms are

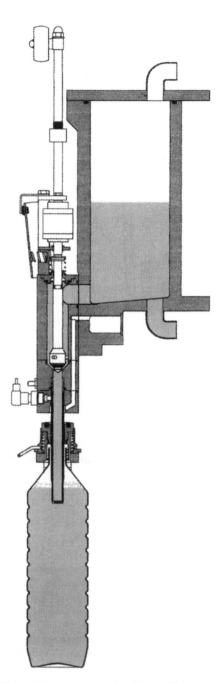

Fig. 7.9 Typical filling valve arrangement for filling still (noncarbonated) products. (Reproduced with kind permission from Krones AG.)

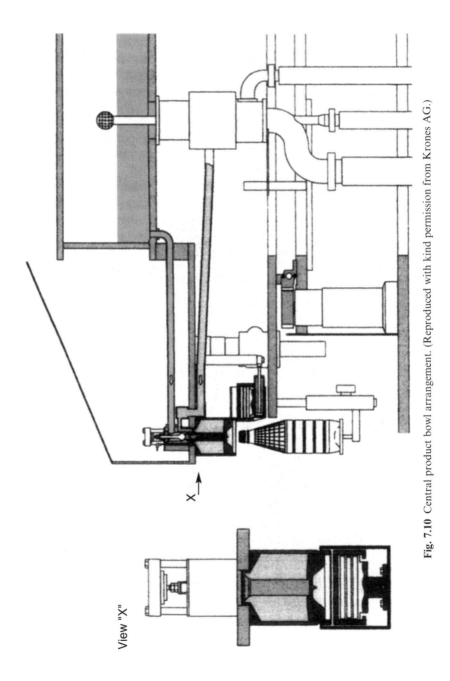

View "X"

X

Fig. 7.10 Central product bowl arrangement. (Reproduced with kind permission from Krones AG.)

Fig. 7.11 Typical starwheel handling system through a filling machine. (Reproduced with kind permission from Krones AG.)

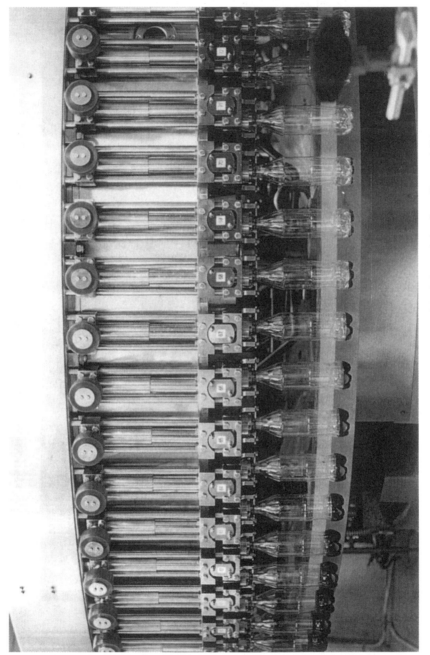

Fig. 7.12 A neck support system. (Reproduced with kind permission from Krones AG.)

then used purely to support the base of the bottle, not to lift it into the valve position. This provides some support to the bottle at the petaloid base when CO_2 pressure from the filling machine is introduced to the bottle. The platform takes the initial pressure shock load at the base of the bottle and helps to reduce the incidence of bottle bursting.

More recent developments have completely removed the need for the infeed starwheel system (which is dependent on bottle diameter) and replaced it with a 'total neck handling system'. This handles the bottle from the infeed scroll position through to the bottle discharge from the filling system without use of starwheels. For containers with one size of neck finish, changeover activity between bottle sizes is minimal.

7.6 Filler configuration

Traditional configurations utilise free-standing rinser/filler/capping machines with interconnecting conveyor system and controls. As line speeds have increased, monobloc systems have become more commonplace. The introduction of the filler/capper monobloc within the filling system has been followed by the development of a rinser/filler/capper monobloc system. Figure 7.13 shows a typical arrangement of filler/capper suitable for high-speed operation, and Fig. 7.14 shows a typical rinser/filler/capper monobloc system.

Some bottlers have introduced the superbloc concept utilising rinser/filler/capper/labeller in a single complete monobloc system. The introduction of the labeller into the bloc, while providing desirable advantages in terms of efficient utilisation of space and operating personnel, is dependent upon the type of labelling system required (see Fig. 7.15).

7.7 A cleaner environment for the filling process

As the demand for bottled water continues to increase, so does the need to increase production capacity. Filling efficiency becomes evermore important as the bottler tries to maximise the production hours in each working day, and this, in turn, can put a strain on the systems providing a clean and hygienic product.

Continually improving electronic technology applied to filling machines has further increased the level of automation, thus reducing the need for operator intervention for filling parameter changes during production. Also, the introduction of bottle neck handling systems for the transport of PET bottles within the filling system has created reductions in bottle size changeover times. Therefore, actual productive filling time has been increased.

A considerable amount of development work over recent years has been aimed at producing a cleaner environment for the filling process in order to

Fig. 7.13 Filler/capper monobloc. (Reproduced with kind permission from Krones AG.)

Fig. 7.14 Rinser/filler/capper monobloc. (Reproduced with kind permission from Krones AG.)

Fig. 7.15 Filler/capper/labeller. (Reproduced with kind permission from Krones AG.)

maximise package integrity and product shelf-life for sensitive products. Creation of optimum conditions in the following areas is necessary to provide a totally "clean" filling system:

(1) Clean filler design
(2) Clean and aseptic filling
(3) Clean environment
(4) Clean bottles and closures

7.7.1   Clean filler design

Internal cleaning (CIP) of the filling machine (described in Chapter 8) is now, in most cases, fully automated and can be carried out effectively and to a very high standard with minimum operator intervention. In order to provide an overall clean filling system, however, its external parts must be designed to be easily, quickly and effectively cleaned. Developments in electropneumatic technology have allowed the removal of external cams, levers and operating cylinders from the periphery of the filler bowl. This has led to the creation of smooth surfaces without crevices and protrusions, which can be difficult to clean.

Further changes in design have been made to other areas of the filling machine to incorporate sloping surfaces. This allows product and cleaning agents to run off the inclined surfaces to the floor, where they can readily be flushed away. Guarding systems with large doors, and crevice-free surfaces have also provided an even more cleanable surface. An example of a typical clean guarding system can be seen in Fig. 7.16.

By creating smooth surfaces in all parts of the filling area, automatic external CIP systems can be utilised to maximum effect with minimum operator intervention. Systems can be selected to provide spraying nozzles for the application of contact detergents and sanitisers for effective external cleaning. With the optimum choice of cleaning systems, both internal and external CIP can be carried out simultaneously, thus making more time available for production.

7.7.2.   Clean and aseptic filling

Filling machines described earlier in this chapter have incorporated the short tube filling valve. This valve controls filling to a level commensurate with the declared volume. The length of the tube projecting into the bottle normally fixes this level, or, in the case of electropneumatically controlled valves incorporating a level probe in the tube, it can be adjusted at the operator control desk. To allow this type of valve to function, the bottle must seal against it. During CIP of the filler, it is necessary to include this sealing in the cleaning circuit.

A recently introduced volumetric filling system uses an inductive flow meter on each individual filling valve to precisely fill the bottle with the correct volume of product. This filling system also has electropneumatic controls, allowing changes to filling parameters *without* operator intervention inside the machine.

Fig. 7.16 'Clean' guarding system on a filling machine.
(Reproduced with kind permission from Krones AG.)

With the smooth exterior design of the machine, optimum internal and external cleaning can be achieved.

Machines of this new generation can also be equipped to handle filling processes under sterile conditions. In particular, when handling sensitive still products, it can fill without the bottles touching the filling valves, which minimises the risk of contamination from bottle–valve contact. A filling machine thus designed, installed in a clean room, with bottle and closure disinfection systems, represent the optimum combination for the aseptic filling of sensitive products.

A PET bottle in position under a contact-free filling valve can be seen in Fig. 7.17.

The same filling valve with an automatic CIP cup being positioned can be seen in Fig. 7.18.

Fig. 7.17 Contact-free filling valve on flow-meter filler.
(Reproduced with kind permission from Krones AG.)

Fig. 7.18 Contact-free filling valve with automatic CIP cup.
(Reproduced with kind permission from Krones AG.)

The flow-meter filling system arranged for aseptic filling with contact-free filling valve can be seen in Fig. 7.19.

7.7.3. Clean environment
A clean environment around the filling process is just as important as a clean filler. Traditionally, the filling process has been separated from the rest of the filling line

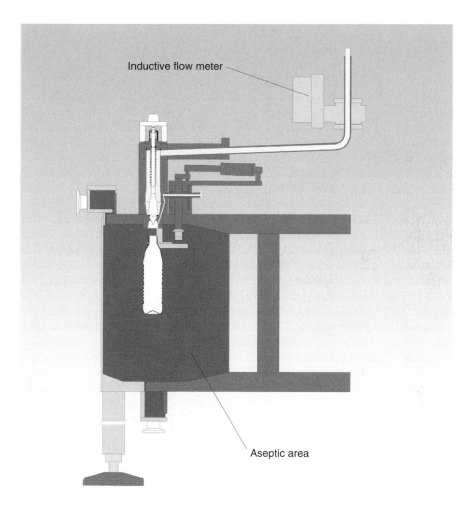

Inductive flow meter

Aseptic area

Fig. 7.19 VODM (volumetric filler with flowmeter) volumetric aseptic filler.
(Reproduced with kind permission from Krones AG.)

in a "clean" area or room. Developments in this area have introduced clean rooms, which, in the new style of clean guard systems, fully surround the filling process. The complete process through rinsing, filling and capping is housed in a clean room incorporating fans with microbial filters. The fans create a positive airflow inside the filling area, which is ventilated to atmosphere at a low level. By creating this environment around the filling process, the clean room size is minimised and is more effective. A filling machine clean room can be seen in Fig. 7.20.

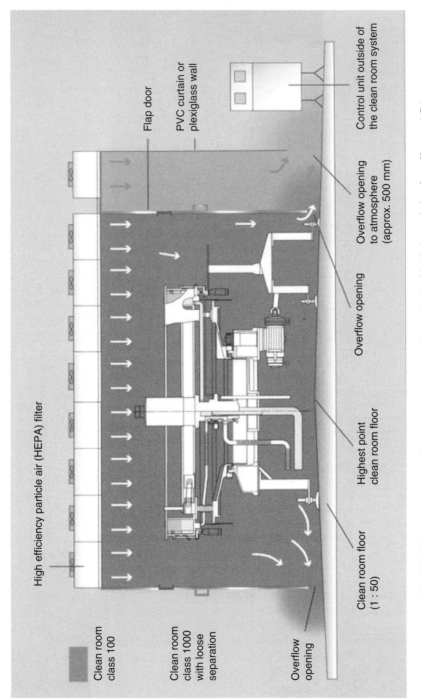

Fig. 7.20 Schematic diagram of a typical clean room. (Reproduced with kind permission from Krones AG.)

7.7.4.   Clean bottles and closures

In particularly sensitive areas where bottles and caps have to be stored prior to filling, system design has also been extended to cater for bottle and cap cleaning. Whilst, traditionally, PET bottles are rinsed internally before filling, systems are now available to internally and externally rinse bottles. In special cases, bottles can be sanitised and rinsed internally and externally.

Cap feed systems deliver caps to the capping machine from an external source of storage. In order to complete the cleaning of the total package, cap feed systems can be fitted with ultraviolet lamps. However, for more effective cleaning of caps, systems are available which transport the caps through a disinfecting bath, with complete submersion immediately before application to the bottle.

7.8   Bottle inspection systems

As line speeds have increased, the introduction of automatic bottle inspection machines (for glass bottle production) is a further necessary addition to the complete bloc configuration. Empty bottle inspection systems have been introduced to check the integrity of the glass bottle within the production line. These inspection systems have been developed to check for defects such as chipped neck finish (crown and screw thread) and other bottle imperfections at high production speeds.

These machines could also be introduced within the complete mechanical bloc configuration. Usually, the glass bottle inspection system is introduced within the filler operating area as a free-standing machine with an interconnecting conveyor between the inspection unit and the filler/rinser bloc. Figure 7.21 shows a typical superbloc configuration with an electronic empty bottle inspection system. In this case, the bottle inspector drive system is electronically interfaced with the filler drive system, i.e. the bottle inspector is electronically linked to the entire filler bloc. A rotary rinsing system is mechanically introduced into the filler unit, which is interfaced with a special starwheel arrangement that incorporates a metered dosing system. This in turn is connected to a rotary capping machine fitted with a bottle turning carousel (for product mixing) and also a complete labelling system. Figure 7.22 shows a complete filler bloc installation of this configuration.

7.9   Carbonation

The degree of carbonation of bottled water is expressed as the ratio of absorbed CO_2 to water in terms of volume. Thus, 4 litres of CO_2 absorbed into 1 litre of water represents a carbonation level of 4 volumes. The maximum amount of CO_2 that can be absorbed depends upon pressure and temperature: the higher the

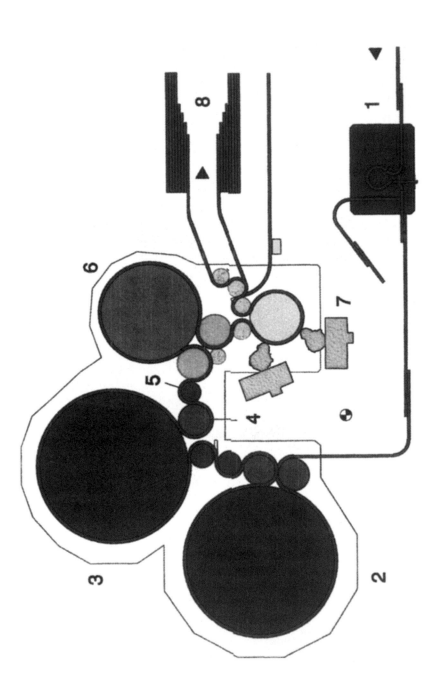

Fig. 7.21 A superbloc configuration: (1) empty bottle inspector; (2) rinsing; (3) filling; (4) flavour-dosing system; (5) capping; (6) bottle turner (mixer); (7) labelling; (8) multilane distribution. (Reproduced with kind permission from Krones AG.)

Fig. 7.22 A complete filler bloc installation of the configuration shown in Fig. 7.21. (Reproduced with kind permission from Krones AG.)

pressure and the lower the temperature, the more gas can be absorbed. Using Table 7.1, for example, at a pressure of 2 bars and a temperature of 5°C, just over 4 volumes of CO_2 gas can be absorbed.

Pressurised with CO_2, the product is injected into the saturation tank via a multistage rotary pump and check valve. A nozzle and injector are located in the carbonation pipe, with the nozzle accelerating the product, allowing CO_2 absorption to take place (Fig. 7.23). The drawn-off CO_2 is stabilised in accordance with the preset pressure and product temperature prior to product collection and settling in the lower parts of the saturation tank. An overpressure pump feeds the product to the filler.

Figure 7.24 shows a typical arrangement of a one-tank carbonator, suitable for outputs between 2000 and 45 000 l/h. Figure 7.25 shows a two-tank system, again suitable for the same outputs. This system incorporates a water de-aeration tank and provides stable product quality that will assist in the filling process. Incoming product is atomised by means of vacuum or pressure inside the de-aeration tank. A carbonator pump draws off the de-aerated water and passes it through an injector into the saturation tank where it is saturated with CO_2. Following a settling phase, an overpressure pump feeds the finished product to the filler. A pressure/temperature ratio controller determines and maintains the required CO_2 content.

7.10 Flavour dosing

Over recent years, further developments within the bottled water industry have led to the introduction of a number of marketing innovations such as flavour

Table 7.1 Volumetric chart for carbon dioxide absorption: volumes of CO_2 dissolved in one volume of water

Pressure (bar)	Temperature (°C)									
	1	2	3	4	5	6	7	8	9	10
0	1.68	1.60	1.53	1.48	1.43	1.39	1.33	1.30	1.24	1.20
0.2	2.00	1.92	1.84	1.78	1.72	1.66	1.60	1.54	1.48	1.44
0.4	2.33	2.22	2.15	2.08	2.00	1.94	1.87	1.80	1.74	1.68
0.6	2.63	2.53	2.44	2.37	2.30	2.22	2.15	2.08	2.00	1.93
0.8	2.95	2.83	2.72	2.63	2.55	2.48	2.40	2.23	2.25	2.18
1.0	3.25	3.10	3.00	2.90	2.81	2.72	2.63	2.55	2.45	2.40
1.2	3.50	3.46	3.25	3.12	3.04	2.95	2.86	2.78	2.68	2.59
1.4	3.80	3.65	3.50	3.40	3.30	3.20	3.10	3.00	2.90	2.82
1.6	4.11	3.95	3.80	3.68	3.58	3.46	3.36	3.25	3.15	3.05
1.8	4.40	4.25	4.06	3.95	3.83	3.71	3.60	3.49	3.38	3.28
2.0	4.70	4.50	4.35	4.20	4.09	3.95	3.85	3.70	3.60	3.48

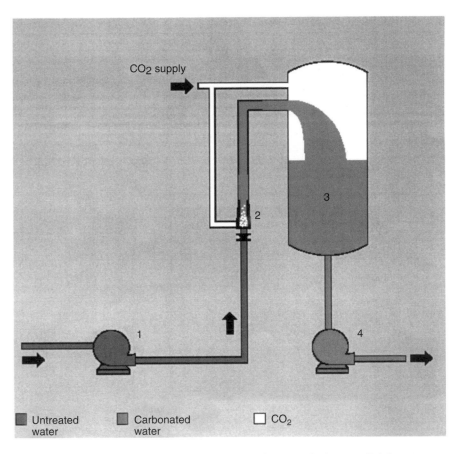

Fig. 7.23 Saturation tank and carbon dioxide injector: (1) feed pump; (2) injector; (3) carbonation tank; (4) dosing pump. (Reproduced with kind permission from Krones AG.)

dosing by the introduction of a very small metered volume of flavouring to the bottled product. One method of effecting this is by an additional feature within the expanding filler bloc configurations.

7.11 Cleaning-in-place

The principles behind cleaning and disinfection are elaborated in Chapter 8, but the way they are applied with regard to the filling equipment is worth examining here, since the regular cleaning of the complete filling system is essential in any bottle filling environment.

Fig. 7.24 One-tank carbonator system for outputs between 2000 and 45 000 l/h.
(Reproduced with kind permission from Krones AG.)

Fig. 7.25 Two-tank carbonator system incorporating product de-aeration, suitable for outputs between 2000 and 45 000 l/h. (Reproduced with kind permission from Krones AG.)

In the CIP process, CIP liquor flow is supplied, controlled and monitored by a separate free-standing supply system. Often this is part of an overall CIP process supplied to all areas of the bottling hall. Control of the CIP process within the filler provides a necessary link with this external CIP supply. The valves and controls are provided for the predetermined CIP programme to suit the cleaning requirements of a particular filling system. The filling machine normally incorporates the necessary valves and interconnecting pipe-work to ensure the cleaning of all its internal components. These valves and cleaning routes can be controlled by either manual or automated operation.

7.11.1   Manual CIP system

A manual CIP system applied to a filling machine is entirely operator-dependent but incorporates the most important features to carry out the CIP functions. However, such a system offers a difficult operator environment as all operations – opening/closing of the correct valves and the valve sequence to suit the CIP programme – totally depend upon the operator's intervention. Compared with an automated system, the scope of manual functions available within the CIP process is very limited and is subject to operator error.

The CIP process on the filler provides for the removal of any contamination within the filling machine. CIP is carried out within a closed circuit, enabling cleaning of the various product channels and filling valves. Different agents can be used for the cleaning of pipes and machines. Hot water and caustic, acid and sanitising solutions are commonly used. The exact combination of cleaning agents depends upon the particular operating conditions of a filling system. CIP follows a predetermined programme containing basic steps and comprises combinations of cleaning with hot water, cleaning with caustic, cleaning with acid, cleaning with caustic and acid, cold sterilisation with disinfectant, and cold water rinse. Many cleaning variations can be achieved by a combination of standard programmes.

Within the filling system, the following concentrations are typical: caustic up to 2%; acid up to 1%; disinfectant up to 0.1%. Duration of the cleaning process depends on the concentration and temperature of the cleaning agent as well as a knowledge of the kind of contamination to be dealt with. Table 7.2 shows a list of typical cleaning agents and treatment times applied within a modern filling machine.

Manual control of the CIP process is typically used when the filling system is to be cleaned by an open flow arrangement (not pressurised). CIP fluid fed to the filler is pumped through the product channels and discharges directly to waste. To achieve this, the valves are maintained in the open flow position by the CIP cam rail that is brought into use.

The typical functions, which in this case are totally operator-dependent, are controlled either by manual or automated operation setting of all valves according to programme (there is no safety device or check-back signals giving

Table 7.2 Typical cleaning agents and disinfectants used for a modern filling machine

Cleaning agent/ Disinfectant	Maximum Cl⁻ content in make-up water (ppm)	Maximum concentration (%)	Maximum soaking period (min)	Maximum temperature (°C)
Alkaline media based on **NaOH** (sodium hydroxide, caustic soda)	—	2	30	80
Acidic cleaning agents based on H_3PO_4 (phosphoric acid)	100	1	20	30
Acidic cleaning agents based on HNO_3 (nitric acid)	100	1	20	30
Disinfectants based on **PES** (peracetic acid)	100	0.1	20	20
Acidic disinfectants based on **halo acids**	100	1	20	20
Combined cleaning agents and disinfectants based on **NaOH** and **NaOCl** (sodium hypochloride) (pH value: ≥ 11)	80	1	15	40
Disinfectants based on **NaOCl** (sodium hypochloride) (pH value: ≥ 10)	80	—	15	20
Hot water sterilisation	—	—	45	95
Steam sterilisation (pressure 0.5 bar, steam consumption: 2.5 kg/filling valve)	—	—	45	110

confirmation of correct valve settings); pump starts by manual operation of push-buttons (ther is no confirmation of full CIP programme completion time-scales). Most modern high-speed filler installations demand a more sophisticated and reliable system.

7.11.2 Automated CIP systems

The fully automated programme built into the filler control system provides a guaranteed completion of any predetermined CIP programme. The process is completed without any operator intervention. By the use of programmable logic control (PLC), all preset criteria for the CIP process are monitored by this automated system. This ensures the completion of each pre-programmed sequence within the cleaning process. Several cleaning programmes can be stored for various applications. A typical automated cleaning process within a filler unit includes integration of each cleaning flow programme, confirmation that each function within the preset programme has been completed, and the timed opening and closing of each of the valves within the filler that control flow to each of the product routes.

The CIP process requires flooded and pressurised ring bowl (CIP supply from external source at 3 bar) and a high volume flow of CIP liquor through the system (usually in the region of 30 000 l/h). To provide a closed pressurised system for CIP flow, it is necessary to fit a special CIP cup to each filling valve to create a seal. This CIP cup provides the 'closed loop system' and the necessary flow path for liquid return to the CIP supply set by a special CIP return channel within the machine. Figure 7.26 shows a typical CIP cup installation. When the machine is to be placed into CIP mode, it is necessary to fit a CIP cup to each of the filling valves. This is normally done manually and can be time-consuming.

Within modern high-speed production units 'downtime' is becoming increasingly unacceptable, and the automation of the CIP cup feed system is a more recent development. The fully automated cup feed unit, shown in Fig. 7.27 provides the necessary automatic function of feeding CIP cups into the filler carousel while the machine is in slow rotation mode. Once CIP cups are manually or automatically positioned, the correct location of each CIP cup is verified by acknowledgement (by display at the filler control station); the machine is then ready for the CIP process.

Fig. 7.26 CIP cup installation. (Reproduced with kind permission from Krones AG.)

Fig. 7.27 Fully automated CIP cup feed unit. (Reproduced with kind permission from Krones AG.)

8 Cleaning and disinfection in the bottled water industry

Winnie Louie and David Reuschlein

8.1 Introduction

This chapter deals with the science and technology of cleaning and disinfection methods used within the bottled water industry. The choice of methods, combinations and frequencies will depend on equipment design, process layout, factory location and the quality of the incoming water. It will also be influenced by the normal operating pattern of the factory, whether it is running 24 hours a day, seven days a week or on an intermittent basis. No bottling plant can operate continuously without a well-considered and correctly implemented cleaning and disinfection regime. It will require the allocation of time to maintain product safety and quality standards. The amount of time will also depend on the nature of the facility and the technology available; for example, a dedicated water bottling line will require a significantly different program from a soft drink factory with lines that also produce bottled waters. In deciding the time required, methods and frequency of cleaning and disinfection, advice should be sought from the manufacturers of cleaning products and equipment suppliers.

The role of cleaning and disinfection in the bottled water industry is more than that of only cleaning the filler and filling room. In considering an effective program for a water bottling facility, it is important to take into account the whole operation, from the parking lot to the point of dispatch. The principal concern in this chapter, however, is to discuss the methods for sanitation of the primary product contact areas – pipe work, storage vessels and filling equipment.

Depending on which dictionary or other reference source is used, the terminology associated with sanitation can be ambiguous. Though cleaning and disinfection (referred to as cleaning and sanitizing in some countries) are two separate procedures, potential for confusion exists because in different parts of the world, the words used to describe the various products are also different. From the following list of definitions we can see that in the USA, for example, a sanitizer has only biocidal properties and is not recommended for the purpose of cleaning.

- Cleaning is the process that removes soil and prevents accumulation of residues, which may decompose to support the growth of disease or nuisance-causing organisms. It must be accomplished with water, mechanical action, and detergents.

- A cleaner (detergent) is a substance that breaks the bond between the soil and the surface being cleaned. Not only must it remove the soil, it must also hold it in suspension and allow it to be flushed away. It does not kill bacteria.
- Disinfection is the killing or inactivation of microorganisms, except for some spore forms. The efficacy of disinfection is affected by a number of factors, including the type and level of microbial contamination, the activity of the sanitizer and the contact time. Organic material and soil can block sanitizer contact and may inhibit activity; therefore, cleaning must precede all disinfection processes. There are three different levels of disinfection:

  (1) High-level disinfection refers to sterilization activities in which all microbial life, including spores and viruses, are destroyed. It is reserved for special applications such as disinfection of surgical equipment and medical devices.
  (2) Medium-level disinfection usually refers to elimination of microorganisms as well as the destruction of the more resistant types of viruses.
  (3) Low-level disinfection refers to the destruction of bacteria and is not effective against spores and viruses.

- A disinfectant is a chemical agent that is capable of destroying disease-causing bacteria or pathogens, but not spores and not all viruses. In a technical and legal sense, a disinfectant must be capable of reducing the level of pathogenic bacteria by 99.999% during a time frame of more than 5 but less than 10 minutes as tested by the Association of Analytical Communities (AOAC) method.

The main difference between a sanitizer and a disinfectant is that at a specified use dilution, the disinfectant must have a higher kill capability for pathogenic bacteria than that of a sanitizer.

- To sanitize means to reduce the number of microorganisms to a safe level.
- A sanitizer, according to the AOAC test method, should be capable of killing 99.999% (5 log reduction) of a specific bacterial test population (*Staphylococcus aureus* and *Escherichia coli*) within 30 seconds at 25°C (77°F). A sanitizer may or may not necessarily destroy pathogenic or disease-causing bacteria, as is a criterion for a disinfectant.
- Sanitation is the term used to describe the complete plant cleaning and disinfection program to ensure public health.

Although this chapter deals with cleaning and disinfection, the word 'disinfection' as used here refers to low-level disinfection, which in practice is achieved without the use of disinfectants as defined above, but rather through the use of cleaners and sanitizers. Today, with larger processing facilities and worldwide distribution, the importance of sanitation is greater than ever before, and a well-managed sanitation program must encompass the employee, the customer and the environment. An effective sanitation program is important for many reasons:

- To protect the company's reputation
- To reduce the potential for financial losses
- To remove and prevent bacterial buildup
- To reduce the chance of off flavors developing
- To maximize the shelf life of the product
- To ensure compliance with government regulations

However, the main reason that the bottled water industry needs well-managed sanitation programs is to ensure customer satisfaction and safety standards. A complete bottled water food safety program should include both cleaning and disinfection and the use of methods of microbiological testing as a means of monitoring the performance of the sanitation program.

8.1.1   Why clean?

In order to control cleanliness and minimize the spread of bacteria, it is important to know proper sanitation procedures, to determine application frequencies and to be vigilant in following the procedures. It is also useful to be able to understand and distinguish between microorganisms, such as pathogens (disease-causing organisms) and spoilage organisms. There are thousands of different kinds of bacteria, yeasts, molds and viruses that are categorized by their shapes and the way in which they grow. Bacterial cells exist in various shapes, but there are three basic forms: round or cocci, rod-shaped and spiral (Fig. 8.1).

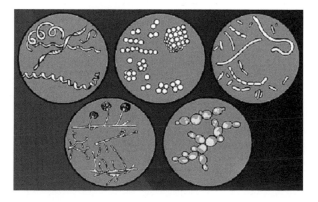

Fig. 8.1 Bacteria, molds and yeasts. Top row (from left to right): spiral bacteria, cocci bacteria and rod-shaped bacteria; bottom row: molds and yeasts.

Identification of different types of bacteria can provide some insight into their source and control. The major causes of food contamination are pathogenic microorganisms that live in soil, water, air, organic matter, on the bodies of animals and humans. Put simply, they are to be found everywhere.

Most bacteria do not have the ability to travel on their own, at least not very far. Those that can move independently – motile bacteria – use appendages called flagella. All bacteria have the ability to travel widely by 'hitching rides' on air, water, bottles, caps, people and anything else that goes from place to place.

Not only are bacteria plentiful and easily spread from surface to surface but they can also reproduce quickly. One bacterium becomes two, two become four, four become eight, and so on and on by a process of cell division called binary fission, which can occur as frequently as every 20–30 minutes. In a short time, one bacterium can produce millions of bacteria. This buildup of bacteria on surfaces is often referred to as biofilm (see Section 8.2.2).

Under ideal conditions, bacteria grow in phases and in a short time can get out of control. The first phase, called the lag phase, typically lasts 3–4 hours, during which binary fission occurs relatively slowly as bacteria adapt to their surroundings. After that, however, they reproduce faster and faster – this is called the logarithmic phase – and contamination becomes much more difficult to control. Eventually, they will reach a stationary phase where they can maintain a very high population, and finally, in the death phase, they begin to die due to lack of food, water and other nutrients.

A good food handling, processing and sanitation program will take advantage of the lag phrase, in which very little growth occurs. A quick, thorough response to control bacteria is therefore a critical factor in sanitation. The sooner bacteria can be destroyed, the greater the chances of eliminating contamination and biofilm buildup, thus reducing the potential for sickness and disease. Equipment that has already been sanitized but that might harbor bacteria surviving beyond the 3–4 hours lag phase should always be sanitized again before use.

8.2 Cleaners (detergents)

In the sanitation process, time, temperature, concentration and mechanical action are the four cleaning variables. Choosing the right systems is a necessary part of the process but they cannot work without the right cleaners and sanitizers. In most parts of the world, cleaners and sanitizers are two distinct categories and sometimes governed by different agencies. Selection of cleaning compounds, methods and frequency of use depends upon the following factors:

- The type and amount of 'soiling' on the surface
- Nature of the surface to be cleaned

- Physical nature of the cleaning compound
- Method of cleaning (foaming, cleaning-in-place (CIP), soaking, manual cleaning, etc.)
- Quality of water available
- Time available
- Temperature allowance

Heat breaks up fat and grease and assists in its removal, and an increase in temperature by 32.4°C (18°F) will double the activity of the chemical. However, excessive temperature can also cause cleaning problems. For example, temperature above the 'denaturization' point will increase the adhesion of protein to the surface. An effective cleaner must have the following properties:

- Rapid penetrating and wetting power
- Ability to control water hardness
- High detergent power to remove soil
- Suspending power to keep the removed soil from redepositing on the surface
- Easy rinsability
- Noncorrosiveness to surfaces being cleaned and to cleaning equipment

In practice, these functions are not performed independently but tend to occur all together. No simple chemical, such as alkali, acid, wetting agent, can supply all the properties, but by combining selected chemicals, cleaners can be prepared that are effective on given applications. Different cleaning compounds are required for different cleaning tasks; one group of cleaners that works satisfactorily in one plant may not be effective in another because of differences in composition of the water supply. However, chemicals should never be mixed at the point of use; any combination of chemicals must be performed by the manufacturers.

8.2.1   Cleaning chemistry

Bacteria are living organisms that have the same basic needs as man to sustain life and to multiply. They need food, moisture, sometimes oxygen or air, a place to live and time. Cleaning chemistry will remove one of those basics, namely food, and by scheduling cleaning, the time required for multiplication of organisms is addressed.

There are no magic wands in the area of cleaning chemistry; but the selection of cleaning chemicals can either make the job a lot easier or turn it into a nightmare. Even a properly designed cleaning product, if handled incorrectly, can become a dangerous liability.

One tool used in selecting the proper cleaning chemistry is the pH scale, which gives some general information on how alkaline or acidic the cleaning product is. Different organic challenges require different pH levels in cleaning chemicals. In general, fats, oils, greases, proteins and carbohydrates require a cleaner with a pH of above 12, which is highly alkaline. Low-pH products are usually used for removing mineral deposits such as calcium carbonate, which is sometimes known as 'stone'. The most efficient method for removing this buildup is by using a phosphoric acid or a blend of acids such as phosphoric, nitric, sulfuric, etc.

Cleaners act in two ways: they either interact with soils on a physical basis by changing their solubility characteristics or they interact with soils chemically to form a modified substance with desirable solubility characteristics in water. Cleaner components are generally classified in the following manner:

(1) Surface active agents (surfactants): organic materials generally composed of two parts – (a) water loving (hydrophilic) and (b) water hating (hydrophobic) or oil loving (lipophilic). Consequently, they have one part of their structure that wants to dissolve in water and one part that is insoluble in water; they provide three types of action: wetting/penetration, emulsification and suspension.

(2) Builders: a category that includes (a) alkaline builders, (b) acid builders, (c) enzymes, (d) water conditioners and (e) oxidizing agents.

   (a) Alkaline builders are generally used to provide a source of alkalinegative ions in cleaners. They are a rich source or donor of electrons, or negative ions. These electrons congregate at the surfaces of many soils and in much the same way as emulsification by surfactants; they disrupt the structure, swell the soil and break it free. The highly negatively charged particles are repulsed from each other and dispersed in the cleaning solution. Strong alkalis such as sodium hydroxide are used in heavy-duty alkaline cleaners for bottle washing or various CIP applications. Highly alkaline materials at high temperatures react with fats and oil to form soaps which are soluble in water.

   (b) Acid builders include phosphoric acid, nitric acid and sulfuric acid. In the food and water industries, phosphoric acid is commonly used to remove and help prevent mineral stone on processing equipment.

   (c) Enzymes are specialized protein catalysts or molecules, which speed up a chemical reaction. An enzyme reacts with a specific organic substance – either a protein, fat or carbohydrate. Enzymes are very specific in their action; they will only interact with the particular substance they are designed to work on. They are generally not used in the bottled water industry, as protein is not found in our normal process.

   (d) Water conditioners can be an important component in treating impurities in the water source. Minerals such as calcium and magnesium salts in the water may react with ingredients in the cleaning compound to form insoluble salts. These salts then form a film that builds up on equipment. To prevent this, materials are added to alkaline cleaners to interact with the calcium and magnesium.

(e) Oxidizing agents are used as a cleaning booster in many alkaline detergents. Sodium hypochlorite is sometimes used as an aid in protein removal at concentrations in the range 50–100 ppm as available chlorine. When it is used in a high-pH environment, it loses its sanitizing properties but remains very effective in solubilizing and removing protein soils or protein films from equipment surfaces.

There is a great deal of flexibility in the selection of cleaners to match specific cleaning requirements and conditions. To make the right choice, it is necessary to select the detergent system that adequately conditions the water and neutralizes its effects on the cleaning system, provides wetting or contact with the soil, dissolves the soil and holds it in suspension so that it can be flushed away.

8.2.2 The five factors

Before finally selecting a cleaner, the interaction of five key factors needs to be understood.

(1) The nature of the soil. An understanding of the soil to be cleaned is essential in determining the right choice for cleaning. This is further complicated by other factors; the solubility varies depending on the soil's condition, the quantity of heat, the duration of its application, the age and moisture content of the deposit. Here are some examples of soils and methods of cleaning them.

(a) Light soil – in this case, an oxidizing agent such as sodium hypochlorite in the presence of an alkaline cleaner will be effective. This will hydrolyze, meaning that it attacks the large molecule and breaks it up into smaller, more easily dissolved particles. Typical chlorine concentrations should be somewhere between 50 and 100 ppm in the alkaline cleaner solution.

(b) Mineral salt – calcium and magnesium in their insoluble form are responsible for most mineral salt deposits. However, iron and manganese are also objectionable because of their intense color. These deposits not only create sanitation problems, but if they are allowed to accumulate, they may contribute to corrosion and poor heat transfer. Acid is the most economical material for removing mineral deposits. Inorganic acids such as phosphoric acid are preferred because of their effectiveness, low cost and generally noncorrosive nature to stainless steel food-processing equipment. Organic acids, such as citric acid, may have special applications but are not widely used because of their cost.

(c) Grease and oil – including those approved for use in the bottled water industry are not solubilized by either acids or alkalis. Surfactants allow the detergent solution to wet these soils so that they can be suspended in water or be flushed from the surface to be cleaned.

(2) The role of water. More than 99% of the cleaning solution is water and it is necessary to know about the specific attributes and impurities in the water being used, including the following:

(a) Water hardness – the most important chemical property of water because it has a direct effect on cleaning and disinfection. It is responsible for excessive detergent consumption; it also encourages scale deposits, so that undesirable films and precipitates can be left on equipment following improper cleaning procedures. Hardness forms scale or leaves film. When calcium and magnesium are present in water as bicarbonates, it is referred to as 'temporary' hardness. Both of these salts are quite soluble in water and consequently, can exist in water at very high concentrations. However, when this water is heated, the salts convert to calcium and magnesium carbonates, which are insoluble in water and will precipitate in the form of scale. 'Permanent' water hardness (which cannot be forcibly precipitated by heating) exists when calcium, magnesium or both are present as chloride or sulfate salts.

(b) Microorganisms – water can harbor significant numbers of microorganisms, which can exist in water for extended periods even if nutrient levels are low. Groundwater may contain significant amounts of organic matter that can provide the nutrient source, either in dissolved or dispersed form.

(c) The pH of water also varies considerably. The normal range is from 6.5 to 8.5 (with a pH of 7 being neutral), and a pH outside these limits is considered unusually alkaline or acid. It may be necessary to treat water in these extreme ranges in order to achieve effective results. Acid and alkaline cleaners are generally not affected by water pH; their acidity or alkalinity far outweighs the effect of the water itself. In sanitizer solutions, however, the pH of the water supply and the resulting pH of the sanitizer solution can greatly affect their effectiveness as antimicrobial agents.

(3) The surface or material to be cleaned. It is essential to consider the composition of the surface being cleaned whether it is stainless steel, aluminum, brass, copper, iron, tile or plastic. Different materials interact with soil and with the cleaner in different ways. In the bottled water industry, stainless steel is the best material for product contact surfaces, and 304/316 stainless is often preferred because it presents a smooth,

cleanable surface, as well as protecting the organoleptic integrity of the water. Rough, cracked, pitted surfaces are much harder to clean because of the difficulty of removing the soil from crevices or holes.
(4) The method of application. There are a number of different ways to apply cleaners to the area being cleaned and each presents a distinct level of exposure to the employee.
   (a) Hand or manual cleaning – because the employee has the greatest potential for physical contact, with manual or hand cleaning the pH of the solution must remain between 4 and 10.5.
   (b) Spray or high-pressure cleaning – because of misting and atomization, there is likely to be some exposure of the employee to cleaning products. Highly alkaline or acid products should not be used unless employees are provided with suitable personal protective equipment.
   (c) Cleaners applied as foam or gel – have less potential for employee contact.
   (d) Where mechanical cleaning or CIP is used, no direct employee contact would be expected. This provides the least risk to the employee since the solution is contained within a vessel or lines.
(5) Environmental concerns. All cleaning solutions and soils eventually become part of the waste stream and need to be properly treated prior to disposal or discharge. This effluent may be treated at a public or private treatment plant; in either case, there are certain restrictions on the quality or characteristics of that waste stream. Major considerations are pH, phosphorus, biological oxygen demand (BOD), fats, oils, greases, the volume of water discharged, dissolved solids, conductivity and the presence of heavy metals.

8.2.3 Types of cleaner (detergents)

There are four basic categories of cleaning chemistry: alkaline, acid, neutral pH and solvents. Prior to using any cleaner, it is necessary to consider the organic challenge and the technology of the equipment and surfaces being cleaned.

(1) Alkaline cleaners have a pH of 11–13.5.
   (a) Heavy-duty alkaline cleaners – usually caustic cleaners such as sodium or potassium hydroxide. Chelators are added to tie up minerals, and wetting agents, to allow free rinsing. Because of their caustic nature, they should not be used on soft metals and should have very little or no human contact.
   (b) Medium alkaline cleaners – in most cases these are excellent products to remove fats, oils and greases, and they are commonly applied by using foam.
   (c) Chlorinated alkaline cleaners – may be either heavy or medium alkaline. Hypochlorite is added to the alkali to peptize the proteins

for easier removal. Excellent on fats, oils, grease, proteins and carbohydrates, they are also used for CIP cleaning of pipes, tanks, etc.

(2) Acid cleaners are at the other end of the pH spectrum; these include:
  (a) Phosphoric acid – effective on most light mineral salts and relatively
      safe for hand scrubbing. It is often used at a concentration of 2–3%
      for cleaning.
  (b) Sulfamic acid – excellent for use in enclosed vessels because it is lower
      in pH than phosphoric acid. Used at a concentration of 1–3% for
      cleaning.
  (c) Acid blends – there are various products that combine acids such as
      phosphoric, nitric, sulfuric and sulfamic. These products are very
      effective on mineral buildup.

(3) Neutral pH cleaners are designed specifically for use where an acid or
    alkaline cleaner can do damage to a specialized piece of material or
    equipment, such as packaging systems, scales or other sensitive equip-
    ment.

(4) Solvents are used where light oil, light grease and other soft organics are
    deposited. These products can contain a foaming agent to aid in the
    application and cleaning. Unlike high alkaline cleaners that digest the
    organics, solvents break down the organics.

In general, Table 8.1 should assist in choosing cleaning chemicals.

8.3   Sanitizers

Cleaning removes soils, but after the cleaning operations, equipment surfaces
and the environment can still be contaminated with microorganisms. If these
microorganisms are not destroyed, the bottled water being produced may be
contaminated. Disinfection is therefore the most critical step in a sanitation
program. To ensure a high degree of sanitizer efficacy:

- Analyze and determine the best sanitizer for the application. The supplier
  should have a high degree of expertise in this area.
- Ensure that all cleaning chemical residues have been rinsed thoroughly
  before disinfection.
- Dilute all sanitizers according to label directions, and follow all labels and
  instructions to the letter.
- Confirm approval by regulatory/environmental authorities.
- Ensure that the sanitizer application methods provide coverage for all
  food contact surfaces.

Table 8.1 Properties of detergents

Acid detergents		Comparative ability									
	Ingredients	Mineral/Scale removal	Emulsification	Penetration	Suspension	Rinseability	Foam	Noncorrosive stainless steel	Noncorrosive soft metals	Nonirritating	Passivation
Mineral acids	Muriatic hydrochloric	A	C	B	C	C	D	D	C	D	DD
	Sulfuric	B	B	C	C	B	D	D	C	B	D
	Sulfamic	C	C	C	C	C	D	C	C	C	C
	Nitric	C	B	C	C	C	C	C	C	C	D
	Phosphoric	C	B	C	C	B	A	A	C	A	A
Organic acids	Citric	C	A	C	C	A	AA	A	C	A	A
	Hydroxyacetic	C	A	C	C	A	AAA	A	C	A	A
	Gluconic	C	C	C	C	C	B	C	C	A	A
	Wetting agents	C	C	C	C	C	AA	C	C	A	A

Alkaline detergents		Comparative ability									
	Ingredients	Saponification	Emulsification	Protein control	Penetration	Suspension	Water conditioning	Rinseability	Foam	Noncorrosive	Nonirritating
Basic alkalis	Caustic	A	C	B	C	C	D	D	C	D	DD
	Silicates	B	B	C	C	B	D	D	C	B	D
	Carbonates	C	C	C	C	C	D	C	C	C	C
	Trisodium phosphate	C	B	C	C	C	C	C	C	C	D
Complex phosphates	Tetrasodium pyrophosphate	C	B	C	C	B	A	A	C	A	A
	Sodium tripoly phosphate	C	A	C	C	A	AA	A	C	A	A
	Sodium polyphosphate	C	A	C	C	A	AAA	A	C	A	A
	Gluconates	C	C	C	C	C	B	C	C	A	A
Organic materials	EDTA	C	C	C	C	C	AA	C	C	A	A
	Phosphonates	C	C	C	C	C	AA	A	C	A	A
	Ploymers	C	B	C	C	A	A	B	C	A	A
	Wetting agents	C	AA	C	AA	A	C	AA	A	A	A
	Chlorine source	C	C	A	C	C	C	C	C	B	B

A, excellent; B, good; C, no ability; D, negative performance.

- Train staff in proper use and handling of sanitizers, and also in using the right sanitizer for the particular plant conditions.
- Use automatic dilution systems to ensure the correct dilution rate.

Sanitizers can be sprayed on or circulated through equipment. They can also be foamed on a surface or fogged (under very extreme and exceptional circumstances) into the air to help reduce airborne contamination. The key to effectiveness of any application method is intimate contact of the proper sanitizer concentration with the microbial cell. The sanitizer needs to thoroughly cover the surface for the proper recommended contact time.

Ideally, disinfection on equipment should be performed just prior to start-up. However, there are times when equipment is left idle before production restarts. In this case, it is recommended that the sanitizer be applied to the equipment immediately after cleaning in order to leave the surface with minimal microbial contamination. This will minimize any regrowth that could occur during downtime. Following extended downtime, sanitizer should be applied again, and the equipment finally rinsed with product or treated water at start-up.

8.3.1 Regulatory considerations

The sanitizer used must always comply with the regulations applicable in the geographical location. For example, sanitizers in the USA are regulated by the EPA and are of two types: (1) no-rinse food contact surface sanitizers and (2) nonfood contact surface sanitizers. The second are generally referred to as environmental sanitizers. FDA (Food and Drug Administration) is charged with approving anything used on food contact surfaces that could potentially contact food items. Ingredients used in sanitizers must comply with FDA requirements as safe and effective. The active ingredients for approved no rinse food contact sanitizer formulations and their usage concentrations are listed in the US Code of Federal Regulations, 21 CFR 178.1010.

In the USA, the official challenge test demands that they must reduce microbial activity of two standard test organisms (*Staphylococcus aureus* and *Escherichia coli*) from a designated microbial load by as much as 99.999% or 5 logs in 30 seconds at 25°C (77°F).

Regulations for usage instructions are very specific for each product and label instructions must be followed precisely. In fact the EPA requires a warning statement on the label that says: 'it is a violation of federal law to use this product in a manner inconsistent with its labeling.' Sanitizers are treated differently worldwide and it is important to consult local regulations to ensure that they are used accordingly.

Sanitizer solutions can only be prepared using potable water and concentrations must be accurate. If the concentration is below the recommended level, it may result in inadequate microbial control. A concentration that is too high is also undesirable. No rinse food contact sanitizers are considered indirect food additives, so concentrations above recommended levels are not allowed. However, in the bottled water industry, the use of no rinse sanitizers without a final rinse is ill-advised, as it can potentially leave residues that affect product quality and jeopardize its inherent properties, thus altering the nature of the product.

8.3.2   Types of sanitizers and their uses

(1) Chlorine products – these are the most commonly used. Typical chlorine sanitizers include various forms of which the most popular is sodium hypochlorite.

   (a) Advantages: Chlorine is effective against a wide variety of bacteria, fungi and viruses including bacteriophage. All chlorine products, regardless of their type (elemental chlorine, hypochlorites or organo-chlorines) form hypochlorous acid (HOCl) in solution. HOCl is the most germicidal species of chlorine. The amount of HOCl is dependent on the pH of the solution. As the pH is lowered, more HOCl is formed. However, as the pH is decreased below 4.0, increasing amounts of toxic and corrosive chlorine gas (Cl_2) are formed. Chlorine is much more stable at higher pH, but is less effective. Hard water salts do not affect chlorine unless they cause an upward drift in the pH of the use solution. Chlorine is effective at fairly low temperatures and is not as temperature-sensitive as other common sanitizers. It has the advantage of being relatively inexpensive and is often preferred because it does not foam.

   (b) Disadvantages: Residual chlorine at low levels imparts a taste and odor to the water and obviously alters its nature.

   There is potential for toxic gas formation – care must be taken as deadly Cl_2 will be formed if the pH drops below 4. Chlorine is also corrosive to many metals; inorganic forms are more corrosive than organic forms and may adversely affect plastics and rubber. A concern for those who work with chlorine is that it is irritating to the skin and mucous membranes.

   Chlorine is unstable and dissipates rapidly from solution. It loses activity rapidly in the presence of organic materials, light, air and metals. Liquid chlorine deteriorates during storage and stability decreases with increasing temperature. Because chlorine products degrade with age, solutions need to be prepared more frequently than some other sanitizers with concentration level tested and adjusted to obtain the required level of available chlorine.

   There is also controversy over its environmental impact because of the formation of potentially toxic organochlorine by-products. This concern is based on findings that chlorine reacts with naturally occurring organic materials, primary humic acid, naturally present in some waters, which results in the formation of suspected carcinogenic trihalomethane (THM) compounds.

(2) Iodophors – these are compounds that contain iodine dissolved in a surfactant carrier and an acid. The surfactant carrier provides a soluble, stable medium for the iodine and in the diluted form controls the release of iodine. On a ppm basis, iodophors are one of the most effective sanitizers available, and are especially effective against most yeasts and molds.

   (a) Advantages: Iodophors provide a weak acid rinse for mineral control, and are less irritating to the skin than chlorine. They offer less toxicity, and have a broader effective pH range than chlorine. Generally more effective at pH 2–5, iodophors offer acceptable disinfection efficacy at slightly alkaline pH, depending on the formulation and conditions.

   They are less corrosive than chlorine when used below 48.9°C (120°F), and the activity is not lost as rapidly as chlorine in the presence of organic matter. This is especially true at low pH. Concentrated iodophors also have a long shelf life, unlike chlorine. Working with iodophors offers a number of other advantages: Their concentration is easily determined by common field methods and the amber color is an obvious visual indicator of the presence of active iodine, which make this type of sanitizer good for hand disinfection.

   (b) Disadvantages: Iodophors can cause staining problems and discolor equipment. Depending on the formulation, iodophors may be more adversely affected by water hardness than chlorine and they have poor activity against bacteriophage.

   The efficacy of iodophors is adversely affected by low temperatures. This effect can be overcome by using higher concentrations or longer contact times. In addition, they cannot be used at temperatures above 48.9°C (120°F) or on hot equipment. At that temperature, iodine begins to vaporize and the vapor is very corrosive to equipment, including stainless steel. In addition, some people find the odor of iodophors offensive. Iodophors are more expensive than hypochlorite.

(3) Quaternary ammonium compounds (QACs) – these are sometimes referred to as quat sanitizers. They are often the right choice when the situation calls for an effective environmental sanitizer. The maximum concentration allowed by FDA for use in a no rinse sanitizer is 200 ppm of the active QAC.

   (a) Advantages: QACs at normal use concentrations are nontoxic, relatively odorless, colorless and noncorrosive. They are stable to heat and relatively stable in the presence of organic matter. Most QACs have a neutral pH, but are effective over a fairly broad pH range. In most cases, the maximum efficacy is exhibited in the alkaline pH range. However, research has indicated that the effect of pH may

vary with bacterial species, with gram negatives being more susceptible to quats in the acid pH range, pH 7 and below, and gram
positives in the alkaline pH range, pH 7 and above.

These compounds possess some detergency because of their surfactant activity. They are active against a wide variety of microorganisms including yeasts and molds.

(b) Disadvantages: QACs are generally considered less effective against
gram-negative bacteria than against gram-positive bacteria. This
drawback can be overcome by using higher concentrations, or by
providing longer contact time.

QACs are less effective against bacteriophage and because they are
cationic molecules, they are incompatible with soaps and anionic
detergents (most general cleaners are anionic). Therefore, surfaces
must be rinsed thoroughly between the cleaning and disinfection
steps to prevent inactivation of the sanitizer. They also have low
hard water tolerance. It can be improved by the addition of chelating
agents such as ethylenediaminetetra acetic acid (EDTA). QACs are
not as effective at low temperature as the oxidizing sanitizers chlorine
and peroxyacetic acid. When used in mechanical operations, they can
cause a foaming problem, which is why they are not recommended
for use as CIP sanitizers.

(4) Acid anionic sanitizers – these are fast-acting compounds.

(a) Advantages: Acid anionic sanitizers work well on yeasts and viruses.
Bacteria do not survive well in an acid environment and an acid
sanitizer can work best when the pH is below 3. Antimicrobial
activity is drastically reduced or stops when pH levels climb to
neutral. Acid anionic sanitizers not only rapidly kill bacteria but
they also provide a method to acid rinse equipment, which leaves
stainless steel bright and shiny. They have very good wetting properties and are usually noncorrosive, which means they can be left on
equipment overnight and do not stain. Hard water and organic
challenge do not have a major effect on the ability of acid anionic
sanitizers to kill microorganisms. They can be applied by CIP or
spray or can be foamed on if a foam additive is used.

(b) Disadvantages: Acid sanitizers can lose all their effectiveness in the
presence of any alkaline residuals or cationic surfactants. Thus, it is
important to thoroughly rinse all cleaning agents from surfaces
before application of the sanitizers.

(5) Peroxyacid compounds – hydrogen peroxide products represent the
newest class of sanitizers, although they have been used extensively in
Europe since the 1970's. Peroxyacetic acid is a strong, fast-acting sanitizer that (like chlorine-based sanitizers) works on the basis of oxidation.
FDA regulation allows peroxyacetic acid to be used as a no rinse

sanitizer at the dilution specified on the label, but it is advised that within the bottled water industry, it should be thoroughly rinsed from all equipment prior to production.

Peroxyacetic acid is one of the most effective sanitizers against biofilms, which are either composed of bacteria that have attached to surfaces and have excreted an extracellular polysaccharide or slime layer (Fig. 8.2). This slime layer protects the cells from adverse environmental conditions. In fact, research shows that bacteria within a biofilm are up to 1000 times more resistant to some sanitizers than those cells freely dispersed in solution.

(a) Advantages: The low foam characteristics of these compounds, like chlorine, make them suitable for CIP applications. They offer a broad range of temperature activity, even down to 4.4°C (40°F). As acid-type sanitizers, they combine the sanitizer and acid rinse in one step. They leave no residues and are generally noncorrosive to stainless steel and aluminum in normal surface application. They are relatively tolerant of organic soil, which probably accounts for their superior activity against bacteria harbored in biofilms.

Peroxyacetic acid sanitizers are generally formulated with phosphate-free compounds, which means they are environmentally friendly. They are readily biodegradable and break down into water, oxygen and acetic acid, and this lack of environmental impact is a major positive benefit. They provide a broad spectrum of bac-

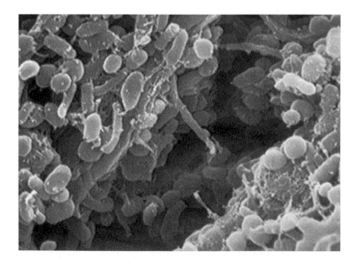

Fig. 8.2 Biofilm.

tericidal activity and can be used in a broader pH range than other acid-type sanitizers, with activity up to pH 7.5.

(b) Disadvantages: Peroxyacetic acid sanitizers lose their effectiveness in the presence of some metals and organic materials; e.g. when makeup water contains more than 0.2 ppm iron. They are corrosive to some metals, such as brass, copper, mild steel and galvanized steel. This corrosiveness is accelerated by the presence of high chlorides in the water (greater than 75 ppm). High temperatures will also accelerate the corrosion rate. Although use solutions are virtually odorless, full-strength peroxyacetic acid sanitizers have a strong, pungent smell. As with all chemicals, proper, safe-handling techniques must be followed.

(6) Hot water – another disinfection method providing a viable alternative to chemical sanitizer is one of the oldest and simplest: heat. It is critical that a proper time and temperature combination be used, i.e. 85°C (185°F) for 15 minutes.

(a) Advantages: Hot water has the advantages of being relatively inexpensive, easily available, and effective on a broad spectrum of microorganisms. It is noncorrosive, and provides excellent heat penetration into 'difficult to reach' areas such as behind gaskets, and in threads, pores and cracks.

(b) Disadvantages: It is comparatively slow compared to chemical sanitation, requiring a lengthy process involving heat, hold and cool down. It can lead to film formation or 'heat fixing' of any remaining soils, making future cleanup much more difficult. Hot water can shorten equipment life. Thermal expansion and contraction stresses equipment and can lead to premature failure. Equipment, including gasket materials, must be specially designed to withstand temperatures in excess of 82.2°C (180°F). Hot water in the system also creates potential condensation problems within the plant production environment.

Water heated in excess of 76.6°C (170°F) is hot enough to cause serious burns; as with chemical sanitizers, proper safe-handling procedures must be followed. Energy costs can be high to sustain high temperatures. There are also other costs, including that of the water, heating equipment and maintenance. In addition, water conditions are important, as high temperatures will increase the formation of hard water film and scale.

(7) Ozone – an allotrope of oxygen, ozone is a powerful and naturally unstable oxidizing gas. It is a more powerful oxidizer than chlorine with an excellent broad spectrum of germicidal activity. Because of its instability, it cannot be stored, but must be produced on-site where it is to be used. Like chlorine, ozone is affected by pH, temperature, organics and inorganics. It is most effective at a pH of 6–8.5. As the temperature

increases, the half-life of ozone is reduced. It is not tolerant of organic soil. There are safety issues to be followed for ozone, as it is a powerful irritant to the respiratory tract. There is a high capital cost associated with the use of ozone.

Disinfection is the most critical step in a sanitation program. Table 8.2 summarizes the advantages and disadvantages of sanitizers.

Generally, in selecting a sanitizer, it is important to consider the method of application, with spray, circulate and foam being the normal choices. A sanitizer can only reduce the number of bacteria; it must therefore be applied to as clean a surface as possible. Corrosion is one of the largest concerns, as it damages lines and filling equipment and acts as a harborage for microbes. All sanitizers should be used at the proper temperature within the recommend concentration guidelines.

8.3.3 Maximizing effectiveness

Whichever sanitizer is chosen, a number of additional factors contribute to the effectiveness of disinfection.

- Cleanliness of the surface – soil can chemically inactivate the sanitizer as well as physically protect the microbial cell from direct contact with the sanitizer. The surface must be cleaned and thoroughly rinsed, so that it is free of soil and residual detergent that can chemically inactivate the sanitizer.
- Intimate contact – in order for a sanitizer to be effective, it must come into contact with the cell wall of the organism. Harborages, such as pits, crevices and cracks, as well as soil residue can prevent this intimate contact.
- Suitable product temperature and concentration – chemical reactions are accelerated by a rise in temperature, so the efficacy increases with temperature and concentration. An exception to the rule is in the case of iodophors, which vaporize at temperatures above 49°C (120°F), so their use is somewhat limited.
- Contact time – the longer the contact time, the greater the kill.
- Proper pH is crucial – this is especially true with acid sanitizers and with chlorine since chlorine has greater activity as the pH is lowered.
- Composition of the water – this can make the sanitizer chemically inactive, or buffer the pH and diminish the sanitizer's effectiveness.
- Types of microorganism – not all sanitizers are equally effective against all microorganisms, or the various forms of the microorganisms. For example, cells in the spore state or in a biofilm are much more resistant than cells in the vegetative and freely suspended state.

Table 8.2 Advantages and disadvantages of sanitizer types

Chemical	Advantages	Disadvantages
Chlorine	Broad spectrum of activity Hard water tolerant Low-temperature efficacy Relatively inexpensive No residual activity Non–film forming	Potential for toxic gas formation Corrosive Irritation Unstable, short shelf life Potential for toxic by-products
Chlorine dioxide	Strong oxidizing chemical More tolerant of organic matter than chlorine Less corrosive to stainless steel Less pH sensitive	Safety Toxicity Sensitive to light and temperature Cost
Iodophors	Broad spectrum of activity Less irritating than chlorine Low toxicity Effective pH range Broader than chlorine Less corrosive than chlorine Stable – long shelf life Color of use solution provides visual control	Staining porous and plastic materials Poor activity against bacteriophage Poor low-temperature efficacy Corrosive at high temperatures (do not use above 48.9°C (120°F)) May produce excessive foam on CIP application More expensive than chlorine Odor may be offensive
Quaternary ammonium	Nontoxic, odorless, colorless Noncorrosive Temperature stable Relative stability in presence of organic soil Broad spectrum of activity Residual antimicrobial film Some detergency and soil penetrating ability Stable – long shelf life Mold and odor control	Incompatible with anionic wetting agents Low hard-water tolerance Limited low-temperature activity Excessive foaming in mechanical applications Antimicrobial activity may vary depending on formulation

(Continues)

Table 8.2 (*Continued*)

Chemical	Advantages	Disadvantages
Acid anionic sanitizers – older technology for CIP application	Stable – long shelf life Generally noncorrosive Nonstaining Low odor Not affected by hard water Removes and controls mineral films Good bacteriophage activity	Incompatible with anionic wetting agents Low hard-water tolerance Limited low-temperature activity Excessive foaming in mechanical applications Antimicrobial activity may vary depending on formulations
Carboxylic acid sanitizer	Low foaming CIP application Broad spectrum of bacterial activity Stable – good shelf life Not affected by hard-water salts Remove and control mineral films Nonstaining	Limited and varied activity against fungi pH sensitivity – optimum activity pH < 3.5 Inactivated by cationic surfactants Temperature sensitivity; use at >12.7°C (55°F) Corrosion potential and equipment compatibility issues
Peroxyacid compounds – newer formulations combine carboxylic acid for better mold and fungi control	Low foam Broad temperature range of activity Combine disinfection and acid rinse No residue Generally noncorrosive to stainless steel Relative tolerance to organic soil Phosphate-free Environmentally responsible Broad spectrum of bactericidal activity Active overbroad pH range up to pH 7.5	Metal ion sensitivity Corrosive to soft metals Odor of concentrate Varied activity against fungi

Method	Advantages	Disadvantages
Hot water – minimum of 85°C (185°F) for 15 min	Inexpensive Easily available Broad spectrum efficacy Noncorrosive Penetration	Slow Film formation Equipment damage Condensation formation Safety Cost
Ozone	Powerful oxidizing gas Broad spectrum germicidal activity	Unstable pH sensitive Temperature sensitive Safety issues Toxicity Cost

- Numbers of organisms present – a sanitizer is only capable of reducing the number of bacteria, which means the higher the initial number present, the higher the number of possible survivors. High numbers can overwhelm the sanitizer.

All these factors are interrelated and can normally be adjusted against each other. If the sanitizer can be prepared only in cold water, it may be necessary to increase the contact time or the concentration to obtain the effectiveness comparable to one at a higher temperature with a shorter contact time or lower concentration. Also, to ensure maximum effectiveness, sanitizer solutions should be freshly prepared for each use.

8.4   Types of cleaning and basics

Depending on the facility and the situation, there are several types of cleaning that make up a well-managed sanitation program, including dry cleaning, wet cleaning, manual and automatic, CIP, cleaning-out-of-place (COP) and hi-pressure.

In order for these to be effective, there are basic rules that employees and management must follow, generally referred to as the '6 × 4' process, comprising six steps necessary for cleaning and four variables that we control in the process. The six steps are preparation, pre-rinse, washing, post-rinse, inspection and disinfection. The four variables are time, temperature, concentration and mechanical action.

Many sanitation articles refer to the '4 × 4' process, often excluding preparation and inspection. However, good preparation prevents cross-contamination during cleaning and insures that the employee has necessary training and equipment to do the job. Inspection insures that the cleaning process is done properly with no short cuts. In modern line CIP applications, inspection may be facilitated by observing swing elbows or hookup stations. Emphasizing inspection as part of the process will gain valuable data that will insure control of the process and consequently clean lines and vessels.

8.4.1   Cleaning dynamics
Installing the proper sanitation equipment will accomplish two basic things: first, it will increase labor efficiency by reducing costs; second, it will provide a consistently effective sanitation program. Whether manual or mechanical soil removal is selected, the four variables, sometimes referred to, in automated cleaning, as the four Ts – time, temperature, titration and turbidity, apply to every cleaning operation. All four are equally important, and effective sanitation cannot take place without the use of all variables in one degree or another. There are two basic ways to remove soils: manually or by mechanical methods. Each is effective in particular situations and for particular types of equipment.

8.4.1.1   Manual method

The manual method is no doubt the best-known and oldest method of cleaning, using just a bucket and brush. Hand cleaning is good for small areas or small pieces of equipment that must be disassembled and manually scrubbed. With manual methods, one is restricted to small areas and mild detergents. If the person does not expend enough manual energy, poor cleaning will result. The detergent must be mild enough so that splashing will not cause skin burns. The contact time will vary with the size of the equipment but must be sufficient to ensure soil removal. Chemical usage can be high if an automatic dispensing system is not used, and manual cleaning is time-consuming with higher labor costs. In addition, cleaning aids like sponges or rags should never be used in a bottled water plant; they entrap soils, providing breeding conditions for microbes, are hard to clean and can contaminate surfaces. Color-coded equipment for different areas or tasks is recommended and all equipment should be maintained in clean conditions, fit for the purpose.

8.4.1.2   Mechanical methods

Mechanical methods offer advantages in controlling the temperature and concentration; however, it takes more engineering design to achieve proper mechanical action.

Depending on the concentration, temperature and mechanical action, it is possible to adjust the contact time to maximize the cleaning cycle. One can increase one or the other of these variables but they must all be in balance. However, the law of diminishing returns comes into play; just because good cleaning is achieved with a 0.5% solution @ 71°C (160°F) for 30 minutes contact time, does not necessarily mean that a change to 5% solution @ 71°C (160°F) for 3 minutes will achieve the same results. In most situations, minimal contact time at a given temperature and proper mechanical action are needed to get consistent results. In addition, it would take extra rinsing to remove the 5% solution. When optimizing cleaning, it is necessary to establish baseline data and monitor the results of the changes. Changes in cleaning parameters should be made only with the assistance of the chemical supplier and plant managment. The effects on the effluent system, water and sewer cost also come into play when determining the best cleaning parameters.

Production time may be the most costly part of the sanitation process. Generally, cleaning and disinfection takes place at the expense of operations, and often it is viewed as a necessary evil. This is not a suitable or sustainable approach, as cleaning and disinfection is an essential function of the process. It is up to the plant management to learn how to optimize cleaning time and performance. The ultimate cost, in addition to lost product and product recalls, may well be 'your brand', and in optimizing mechanical cleaning, investment must be weighed against the return. Several methods of cleaning need to be analyzed against the expense of time and utilities, and energy, water and sewer cost. Some examples of mechanical systems are discussed below.

Foam systems. Foam application systems are widely used and are very effective for a range of applications. Foam cleaners can be applied directly to the surface of the soiled equipment, where the foam remains for 3–5 minutes, providing extended contact time. A brush is then used to scrub down all surfaces and equipment properly rinsed with clean soft water to remove the cleaner and loosened soil. Foam in itself does not contribute to cleaning other than helping the chemistry to work more effectively. Care must be taken when using any high foaming product, and it is important to ensure that the product in question has a specified content of cleaning product, so that the factory is not just buying 'suds'. It is also important to read the Material Safety Data Sheet and to review the chemical breakdown of the cleaning product by comparing cleaning compound content and dilution rates. Both foam and thin film will allow the employee to clean more surface area faster than bucket and brush; however, they are only ways to dispense the cleaning chemical. The real-time savings come by combining them with a properly designed rinse system.

Foam tank. This is a 5-, 15- or 30-gallon tank, which must meet regulatory requirements as a pressure vessel. Water and an appropriate amount of cleaner are added to the tank, the lid is secured; with an air hose attached and air pressure is brought from 40 to 60 psi to provide thick foam.

Wall-mounted foamer. A wall-mounted unit works off a central system in which the chemical is brought up into the foamer from the 1 gallon container. Air and water are supplied to the unit and the chemical is automatically diluted. A wall-mounted unit can be configured to include cleaner and sanitizer dispensing systems with a rinse feature, enabling foam, rinse and sanitizer functions from a single unit.

Central system. In a central system, the chemicals and the pumping system are located in a central room. One pump brings prediluted cleaning product to foam or sanitation stations throughout the plant. Sanitizer is also pumped out to these stations through a separate line. Pump and line size are important, and the type of pump is also critical; it should be made of sanitary plastic tubing or stainless steel so it will have the highest resistance to chemical corrosion. This is an excellent method of cleaning and allows total control.

High pressure spray. A high pressure spray is not best suited for use inside a processing environment because a spray unit creates aerosols, meaning that pathogens can be carried on the airborne spray to other areas throughout the plant. The spray can also have the disadvantage that instead of ridding the processing area of organics it may only be blowing them from one side to another. Another concern with high pressure is possible damage to process equipment by blowing the grease out of bearings or forcing water into crevices and electrical outlets. Care and instruction need to be provided to personnel, as high pressure air can be dangerous.

A good sanitation program may also use COP and CIP systems, both of which allow the employee efficient use of time. COP washer tanks are designed for parts that are otherwise manually washed; the parts are placed in baskets that are loaded into the COP tank (Fig. 8.3). Each COP system uses a variety of different high-velocity water jets to clean the parts as they are submerged in a cleaning solution. This allows parts to be cleaned while the operator does additional tasks; higher temperature and concentrations are also possible than with manual methods. Ultrasonic baths can also be used for cleaning intricate, small parts.

The above methods can save time and increase sanitation efficiency; again, the only factors controlled are time, temperature, concentration and mechanical action. Nevertheless, a well-trained, well-motivated employee is the most important part of good sanitation. However, as bottling plants grow, high-volume production equipment and lines require stand-alone CIP (covered in Section 8.5).

8.4.2 Master sanitation schedule

For effective cleaning and sanitation, it is necessary to define a plan, known as a 'master sanitation schedule' or, as some call it, a 'task manager'. The master sanitation schedule can range from a simple form to a computer-driven database linked to scheduling, man-hour projections and Sanitation Standard Operating Procedures (SSOPs). In any case, it must cover the whole plant, from parking lot to roof maintenance to filling room.

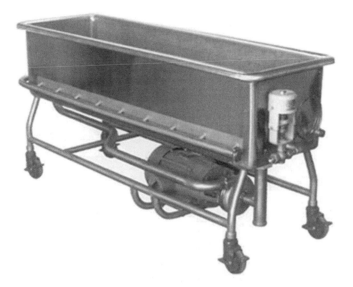

Fig. 8.3 A COP tank.

This tool ensures that the whole plant is cleaned and sanitized on a regular basis. To acquire information, data will be needed from many sources to determine the frequency for cleaning and disinfection. Equipment manufacturers will provide general guidelines; laboratory data such as microbiological swabs will also give guidance on appropriate frequencies. Employee feedback, sanitation inspections and audits will also be a useful source of monitoring to provide information. The key is that this is a living document, and by keeping the schedule updated it is possible have a cleaner and better running plant, reliably producing safe and wholesome product.

8.4.3   Sanitation Standard Operating Procedures

SSOPs should be in place for every task on the master sanitation schedule and should contain the following information:

- Equipment – name or reference
- Revision date – everyone needs to know it is current
- Supplies needed – chemicals, brushes, wrenches, etc.
- Safety equipment – personal protective equipment (PPE) needed for the job; lock out/tag out instructions if needed
- Cleaning procedure – this can be as detailed as needed on complex equipment and pictures can be attached; abstracts from the equipment manuals may be helpful
- The employee should be involved in describing the job and cleaning
- This document should be at the job site and updated if changes in procedures are made
- Sign off and audit by management and employee

An inspection and equipment release form should accompany the SSOP, enabling all parties to know that the job has been completed properly. The inspection should have an objective, nonbiased test if possible to ensure that satisfactory results have been achieved. Many inspections include quick testing methods to give employee feedback. The employee is the key to any sanitation program; he or she must be well trained on Good Sanitation Practices (GSPs) and must follow these and SSOPs zealously. GSPs involve employee hygiene, equipment and cleaning aids maintenance.

8.5   Cleaning-in-place

CIP is the process of bringing the cleaning solution to the equipment and piping. It can be manual or automated, and as plants grow, high-volume production equipment and lines require stand-alone CIP equipment. CIP for the bottled water industry is an integral part of the sanitation process. The same basic rules apply, meaning that time, temperature, concentration and mechanical action must be controlled. In addition, to ensure quality and repeatability, proper data acquisition and monitoring are necessary. The effectiveness of CIP is more than just pushing the buttons; it requires plant management involvement and must be properly maintained and monitored. CIP has to be on the same level as the plant's processing equipment. If all the proper components are not in place along with trained employees motivated to do their jobs, CIP will become 'circulate-in-place', not 'cleaning-in-place'.

CIP systems use fixed pipes, spray devices, valves, tanks, sensors and controls to provide closed-circuit cleaning and improve the efficiency and repeatability of the cleaning and disinfection process. CIP systems offer significant advantages over other cleaning methods including reduced labor, energy and water costs while providing better results due to the ability to use higher temperatures and concentrations. The automatic programming feature of most CIP systems provides a degree of repeatable performance not found in other methods. Additionally, since the processing equipment does not need to be taken apart and reassembled, the risk of recontamination is greatly reduced. An effective CIP system delivers some significant advantages:

- Reduced labor
- Energy control
- Water control
- Consistent results (repeatable with automated controls)
- Higher concentrations when needed
- Higher temperatures
- Safety – employees do not have to touch hot surfaces or come into contact with chemicals

To achieve desired results, several choices must be made to include location and type of system, hydraulics, spray devices and programming. CIP systems must be integrated into the process design.

In the 1920's, two trade associations and one professional society joined together to formulate uniform standards for dairy equipment. These standards became known as '3A standards'. Today equipment fabricators, equipment users (food processor) and regulatory officials work together to develop and set sanitary standards for food processing equipment of all types. Equipment that meets the design criteria is permitted to use the 3A symbol. In the global market, 3A works with the International Organization for Standardization (IOS), Technical Committee 199 and the International Dairy Federation (IDF) Group of Experts B3 for Sanitary Standards.

Once it is known that the equipment and affiliated piping are designed correctly, it is necessary to determine which type of CIP system is right for the operation, and the hydraulics and spray devices required.

Table 8.3 Typical cleaning flow

Pipe size		Velocity	Flow	Time (s) to fill
OD	ID	(ft/s)	(gpm)	10 gallon can
1.5	1.40	5	24	25
2.0	1.87	5	43	14
2.5	2.37	5	69	9
3.0	2.87	8	163	4
4.0	3.83	8	288	2

Most regulatory agencies require a minimum velocity of 1.5 m(5 ft)/second through all parts of the CIP system (Table 8.3). This velocity through a pipe creates turbulent (rather than laminar or straight line) flow (Fig. 8.4). In piping, if the diameter is smaller than 7.5 cm (3 in.), 1.5 m(5 ft)/s is acceptable for turbulent flow; for 7.5 cm(3 in.) and larger, 2.4 m(8 ft)/s is required. This provides the necessary scrubbing action inside the piping.

In vessels, spray devices achieve complete coverage and provide the hydraulics. The flow required depends upon the vessel shape and the means by which this is calculated is shown in Appendix 1.

If performed correctly, these calculations will give the flows required to clean both lines and storage vessels. CIP, however, is more than just a matter of putting water in. In order to achieve good hydraulics, proper solution return is needed to complete the loop. To obtain balance, the return should be slightly faster than the supply, otherwise water building up will cause 'puddling', inhibiting the cleaning process (Table 8.4). Other factors to consider when performing CIP on vessels are:

- Tank outlet valves – suitable size
- Return pumps – 12–18 in. below tank outlet
- Tanks pitched – $\frac{3}{4}$ in./ft

The selection of spray devices is also critical, as it is important to have complete coverage and flow rate compatible with the vessel. The spray device should be

LAMINAR FLOW　　　　**TURBULENT FLOW**

Fig. 8.4 Laminar and turbulent flow.

Table 8.4 Tank outlet valve flow

Pipe size		Gravity flow through
OD	ID	TOV (gpm)
1.5	1.40	40
2.0	1.87	75–80
2.5	2.37	115–120
3.0	2.87	190–200
4.0	3.83	250–275

self-draining, self-cleaning and made of sanitary materials (304/316 stainless steel). Fixed and rotating spray devices are both commonly used; the fixed types have no moving parts and are the most widespread in use today. They come in a variety of sizes and shapes (Fig. 8.5).

Spray devices must provide proper coverage at designed flow rates to make it possible to use large tanks and vessels that cannot otherwise be cleaned manually. Effective and repeatable CIP depends upon the proper selection of the spray device, based on the size and configuration of each tank and vessel. If the correct spray device is installed, cleaning will take less time, which saves water and chemicals and minimizes effluent charges. Fixed devices have advantages over rotating ones, in that they have no moving parts to break. All spray devices can clog with foreign materials such as gasket pieces, even if a strainer is in line. Checking for clogging should be a regular part of CIP preparation.

Fig. 8.5 Selection of fixed spray devices.

Having addressed the minimal CIP flow requirements, it is necessary to choose between manual and mechanical (automatic) systems. Manual CIP can only be relied upon for the simplest of circuits, generally referred to as pot-and-pump applications. Even then, there is likely to be little or no documentation to insure proper cleaning. With the many regulations covering the bottled water industry, it is essential to know that the proper time, temperature, concentration and mechanical action are achieved each and every time. This can only be done successfully with modern CIP systems designed for the plant's cleaning needs and that will be capable of repeatable cleaning performance. Automation is the only way to meet these standards, and the choice will be in determining the right kind of system and controls that best fit your plant.

It is crucial that all processing equipment, including storage tanks, piping, filter housings, UV units, contact tanks, fillers and tanker trucks be properly cleaned and sanitized to maintain high product quality and shelf life, and to prevent the spread of microorganisms.

Filling machinery in most cases will be integrated into the CIP system. Most fillers will require a pressure ring or CIP cup attachment for cleaning. It is important that filling equipment be designed for the CIP process with parts that are compatible with the temperature and chemical concentrations needed in CIP programs. Gaskets should also be the subject of a selection and maintenance program and be compatible with the water to be bottled, cleaning chemicals and methods used. When integrating fillers and other specialty equipment into the CIP system, it is important to consult the equipment manufacturer for recommended cleaning protocols, programming and flow requirements. The plant quality assurance department and chemical suppliers both need to be part of the CIP process to ensure chemical capability and efficacy.

8.5.1 Automated CIP

Consistent CIP results were made possible in the early 1950's, largely with the advent of the automated valve, which in the bottled water industry are usually air-operated, sanitary valves. They are used to control the direction of flow in the circuit or to block or shut off flow.

The simplest is the two-way (shut-off or blocking) valve, in which the actuator opens and closes to stop flow (Fig. 8.6a). This is used for simple tasks, and a typical application would be for a tank outlet valve. The three-way valve is used for directional flow control, but without stoppage of flow; this valve is commonly used and referred to as a divert valve (Fig. 8.6b). The flexibility provided by using these valves in various configurations gives the ability to move water and other liquids to any filler from any storage vessel in the plant.

Valve matrices that have been put in modern facilities require special programming and routing during cleaning. Each valve must be cleaned and sanitized

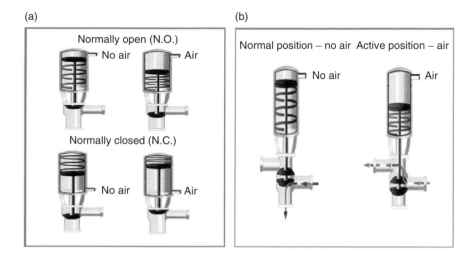

Fig. 8.6 Automated valves: (a) shut-off valve – blocking valve; (b) three-way valve.

on both sides, the valve seats, stems, o-rings and seals being cycled so that each valve is cleaned on all surfaces.

The plant also has the responsibility to prevent crosscontamination and deadlegs that could allow product to sit idle for hours at a time, jeopardizing both safety and quality. Several methods have been designed to prevent this, and some allow idle lines to be cleaned or purged while the rest of the plant is operating. The most common method is the block-and-bleed configuration, where flow is stopped and then diverted to the drain or the CIP return. Double block-and-bleed allows an atmospheric air break to insure that no product can be mixed. To eliminate extra valves, new sanitary valves have been designed for mixproof operation. With these, two different products can flow through one valve, allowing cleaning on one side and product on the other. The valve consists of two bodies, which are welded together. The seats for the upper and lower plugs are located between the bodies, two independent plug seals forming a leakage chamber between them. Any product leaking from one side or the other flows into the leakage chamber and is discharged through the leakage outlet. The cost is much greater but this valve allows more flexibility in cleaning and production.

8.5.2   Types of CIP systems

CIP systems are of four common types: (1) single-use, (2) reuse, (3) solution recovery and (4) multiuse. There are advantages and disadvantages to each, so it is important to know the process, equipment and environment in deciding which to use (Table 8.5).

Table 8.5 Advantages and Disadvantages comparison of types of CIP system

	Advantages	Disadvantages
Single use system	Versatile Multiple detergents/concentration Fresh water solutions Low volume wash water Multiple temperatures Less thermal shock	High water use High detergent use Longer delay to temperature
Three-tank reuse system	Easy to use Effective Low detergent use Short delay to temperature	Single detergent/concentration Limited wash temperature High sanitizer costs Heat shock
Solution recovery system	Versatile Reduced water use Some energy savings Better pre-rinse	Uses more detergent than reuse CIP system
Multiuse system	Multiuse/single use/reuse/solution recovery Circulate sanitizer/hot water Optimize programs Optimize water, energy, chemical use	Higher equipment costs

(1) Single-use system (Fig. 8.7): a fresh cleaning or sanitizer solution is prepared for each cleaning cycle and then discharged to the drain.

(2) Three-tank reuse CIP system (Fig. 8.8): separate tanks are used for fresh water and each cleaning solution. The same wash solution are used continually from CIP circuit to CIP circuit. The wash solution must be boosted for each use to maintain the specified concentration. The tank must be drained regularly to remove the accumulated soil and refilled with fresh water and fresh detergents. A fresh sanitizer solution is prepared for each cleaning cycle and discharged to drain.

(3) Solution recovery CIP system (Fig. 8.9): a recovery tank is used to recover the wash solution and post-rinse, which is then used for the second and third pre-rinses on the next CIP circuit. This type of system can be used in connection with a multitank (reuse) or a single-use CIP system.

(4) Multiuse CIP system (Fig. 8.10): through the use of three or four tanks and extra valves, CIP systems can be set up to operate either as a reuse or single-use with or without solution recovery. By using different programming techniques, selected programs can be run in any of the different modes.

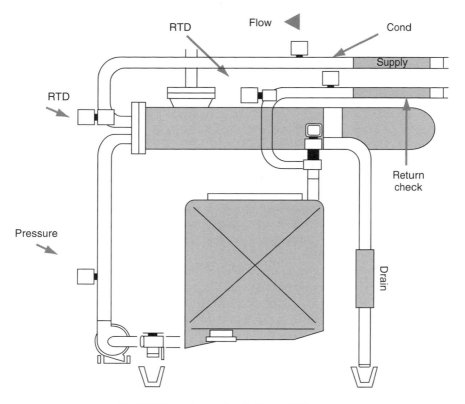

Fig. 8.7 Flow diagram for single-use CIP system.

8.5.3 CIP control and data acquisition

If the need is for one CIP system, multiuse is the best option. The controller will activate the pumps, valves, control the heating, flow control and safety interlocks. Controllers are available in several types, ranging from old-fashioned drums, cams and electronic sequencing devices to programmable logic controllers (PLCs) and dedicated PC microprocessors. A dedicated microprocessor has CIP logic programmed in and recognizes, for example, what valves to open or close to accomplish each step. The latter are today's choice; with them, it is possible to record performance data – time, temperature, concentration and mechanical action. Table A2 at the end of this chapter poses some questions of relevance to help in choosing a dedicated controller or PLC-based controller. The choice will depend on plant expertise available and/or local support for the controller requirements.

Fig. 8.8 Three-tank reuse CIP system.

Fig. 8.9 Solution recovery CIP system.

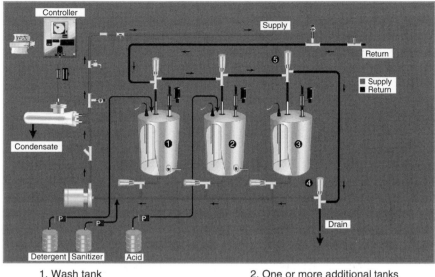

1. Wash tank   2. One or more additional tanks
3. Fresh water tank   4. Drain valve 3 way
5. Fresh water tank inlet valve 3 way

Fig. 8.10 Flow diagram for multiuse CIP system.

The wash program consists of clearly defined sequences or steps; e.g. a three-step program consists of (1) pre-rinse; (2) wash; and (3) post-rinse. This could be hot or cold depending on the utilities supplied to the CIP system.

8.5.4   CIP program and programming

During each CIP cycle, there is a subset of actions that must occur. In a rinse, the tank outlet valve (TOV) opens, the supply pump turns on, the routing valves open so water goes down the designated pipeline, the return valves open to drain, the fresh water valve opens, level controls in the rinse tank turn on, etc. Keeping track of this activity is difficult, so it is advisable to use a pin chart and a drawing with the CIP circuits highlighted to see solution path and identify valves that need to be sequenced. Pin charts generally contain: program, CIP function, CIP component (on/off), values (time, temperature), volume (water, cleaner, sanitizer) and flow rate. Table 8.6 is an example of a pin chart for a simple four-step line wash. It is a combination of the three steps above with a sanitizer step added.

Along with the pin chart, the programmer often provides a control description document, which provides a common language and the example below is typical of the steps one would expect in a program. In most plants, these

Table 8.6 Pin chart of a line program for a reuse CIP system

Function	Value per segement	Freshwater tank – outlet valve NC	Freshwater valve	Alkaline tank – outlet valve NC	Alkaline tank – inlet valve NC	CIP supply pump	Steam valve	Return pump	Drain valve NO	Detergent control system	Sanitizer pump	Valve sequence program
Rinse to drain	2 min	X	X			X						X
Pause	Restart	Manual preparation										
Rinse to drain	2 min	X	X			X						X
Alkaline wash	10 min			X	X	X	X		X	X		X
Rinse to drain	4 min	X	X			X						X
Sanitize	2 min	X	X			X					X	X
Shut down												

X = active/on

programs are more complex and valve sequencing will be involved. If there are any subroutes in the line circuits, each leg should run approximately 8–10 minutes at a concentration and temperature to insure proper cleaning. A sanitizer step must be directed down each leg for the proper contact time recommended for that sanitizer.

In tank programs, the rinse part of the program is 'pulsed' to ensure effective draining. Three pulses are recommended and this can be accomplished by shutting off the supply pumps while leaving on the tank return pump. This will evacuate all the water and remove any residual from the bottom of the tank.

The chemical suppliers may have standardized checks that both they and the plant operators can use. Chemical suppliers should train plant employees in testing so that the right detergent and sanitizer concentration is used in each circuit.

What data should be collected from CIP systems and how, is sometimes determined by the regulatory agencies under which the plant manufactures. In general, all plants will benefit from knowing the date, circuit ID or name, operator, time, temperature, flow rate, concentration and pressure. In addition, other components may be added to the system, such as a flowmeter, conductivity sensor and pressure gauge. Some of the data from theses devices can be recorded with a simple chart recorder, showing time, temperature and pressure; operators are required to fill in date and circuit information and sign at end of each shift. They can also record chemical concentrations on the chart, which should be reviewed daily by management and filed. By doing this the plant will have a history of the system and performance data on wash frequencies.

Today's modern PLC and dedicated processors can also perform data acquisition and the plant will want to look at the regulatory guidelines for electronic data collection to make sure it is in compliance; in the USA, 21 CFR part 11 is a good starting place. The list below is not all-inclusive, but gives a general idea of what to look for in electronic data acquisition and management.

- Direct wiring to PLC
- Automatic report generation upon conclusion of CIP
- Ability to have printed statement of what is being cleaned (e.g. Tank 1, Transfer line)
- Graphical representation of conductivity, supply temperature, return temperature and flow rate with some markers showing the initiation and conclusion of distinct steps
- Summary stating CIP start time, finish time, total cycle time, time since last CIP completed, and for caustic circulation, acid circulation, and final rinse: average flow rate, average conductivity, and average return temperature
- Ability to show any alarms (e.g. CIP aborted, steps skipped, exceed of maximum time for temperature, conductivity, cycle time)
- Historical data backup via a memory card or PLC

The CIP System in bottled water facilities today is a matter of routine. There are many choices, from the design of the transfer piping to the right system and controls. However, no matter what is chosen, it is only as good as the operator's ability to run and maintain it properly. The operators need to know the system and programs well enough to monitor the results. Performance checklists should include, but are not limited to, CIP performance checks for lines and CIP performance checks for vessels. Without well-trained, well-motivated employees, no sanitation program will ever be successful.

8.5.5   Hot CIP safety precautions
- Never operate the CIP system in a manner for which it is not designed. Always operate well within safe parameters. Always verify that all equipment to be cleaned can tolerate the temperature, concentration, pressure and flow rates generated by the units. Always be cautious to allow proper venting of atmospheric tanks by opening the manway during the CIP

process. Never allow microfilters and UV lights to be subject to water hammer or high pressures.

- The CIP system uses highly corrosive chemicals as a cleaning medium that can cause burns to the skin and eyes. Always wear personal protective equipment (chemical-resistant gloves, apron, goggles or face shield, and boots) when making and breaking flow panel connections or removing equipment for maintenance. This is especially important when opening and handling chemical drums and taking samples.
- Be careful when maneuvering chemical drums, portable return pumps and other heavy objects/equipment. Always get help or use proper equipment such as forklifts to prevent back strains and other forms of personal injury.
- Never bypass the safety feature of an electrical panel while the unit is in operation and obtain the help of a qualified electrician when troubleshooting electrical panels.
- Always keep guards on rotating equipment in place. Never operate any equipment without proper safety precautions in place. Never allow any CIP circuit to be operated knowing that there is damaged or defective equipment in line.
- Never operate the CIP system prior to proper calibration of all instrumentation. Never adjust or 'tweak' any equipment or instrumentation without completely reading and understanding the maintenance and operating manual first.
- Communicate to all personnel that a CIP cycle is running and minimize excess employee traffic in the area where CIP circuits and equipment are located.
- Remember that the CIP system can start and stop remote equipment automatically without warning. Never do maintenance or place yourself in a dangerous situation without following appropriate equipment electrical isolation procedures.
- Always completely walk the CIP circuit during the prerinse to identify potentially dangerous leaks prior to chemical addition and heatup.
- Never enter the tank of a CIP system or place any part of your body inside without following your plant's confined space entry procedures. Never open the manway of any CIP tank or tank to be cleaned while the unit is running.
- The CIP system may use high temperatures to aid in the cleaning process. Always be cautious of hot surfaces and never touch any CIP equipment unless the unit is shut down and allowed to cool. Always be extremely cautious of steam piping. Wear suitable clothing when working with hot CIP surfaces, circuit piping and equipment.

8.6   The Do's and Don'ts of cleaning and disinfection

THE DO'S

(1) Become familiar with products and their applications because misapplication can cause explosive chemical reactions or discharge dangerous vapors.

(2) Read the labels on all products and become familiar with their applications and properties, so that you will be aware of their compositions and potentially hazardous capabilities.

(3) Know what antidotes to use when someone becomes injured so you can assist in directing immediate first aid and minimize injury.

(4) Know the limitations and capabilities of products so you will understand the seriousness of injury that could occur when skin tissue is brought into contact with specific detergents.

(5) Know the maximum operating temperature of products so that while preparing cleaning procedures or observing cleaning operations you will know when temperatures exceed that maximum.

(6) Know which products are acidic or alkaline so that instant first aid action can be taken in case of injury.

(7) Always mix detergents in 'use' dilution – not their concentrates.

(8) Wear and instruct employees to wear necessary protective clothing such as goggles and/or facemasks when dispensing detergents.

(9) Wear footwear appropriate to the environment, which protects from moisture and provides nonskid soles.

(10) Always provide detailed written instructions when training personnel on the safe use of products to minimize misunderstanding and prevent unsafe action.

(11) Teach safety by example, as a picture is worth a thousand words.

(12) Use equipment that allows safe sampling and dispensing into test equipment when sampling solutions that usually are at elevated temperatures.

(13) Check solution temperatures with a thermometer, which gives maximum protection to its user from high solution temperatures.

(14) Know the conditions of application and chemical characteristics of sanitizers you are using to prevent injury to personnel and equipment.

(15) Know safety as applied to bulk handling programs so that you can observe an installation and know that safety is being practiced concerning eyewashes, showerheads and bulk tank labeling.

THE DON'TS

(1) Do not mix acid and chlorine products because this releases poisonous chlorine gas.

(2) Do not use a liquid chlorinated cleaner on an automated 'override' program because of the potential release of chlorine gas.

(3) Do not permit 'override' cleaning without instructions because of the possibility of damage to equipment or personal injury from chemical reaction.

(4) Do not add water to a pail of powdered caustic to dissolve the powder because of flashback, caused when water is added to powder. Rather, add caustic to a pail of cold water.

Never add water to concentrated chemicals.

(5) Do not use a lightweight plastic pail when dissolving caustic chemicals because heat generated by the action of water and caustic will soften the pail.

(6) Do not use hose stations where hot water is produced by mixing cold water and steam unless you have knowledge of the hot water generation process, because this can pose a real danger of blowing live steam.

(7) Do not try to remove caked powders from a shipping container because of hazard of product flying into the eye or making contact with the skin.

(8) Do not mix wetting agents and nitric acid products in concentrated form because the oxidizing reaction of acids and wetting agents can cause a violent chemical reaction and flashback.

(9) Do not charge detergent reservoirs that are at or above eye level because of the danger of splashing or spilling of product on the person.

(10) Do not mix detergents without knowledge of their compatibility because of the danger of reaction and flashback.

(11) Do not dispense products from shipping containers that are not labeled because of the possibility of misapplication of detergent, causing personal injury or damage to equipment.

(12) Do not store products in unlabeled containers because any person cleaning up may not know what product is in use, resulting in misapplication.

(13) Do not use a drum pump for dispensing two different products because the pump may not have been properly rinsed, resulting in the possible mixing of two incompatible products, causing a chemical reaction or flashback.

(14) Do not pressurize a shipping container for dispensing product, as no shipping container is designed to withstand more than atmospheric pressure.

(15) Do not dispense highly concentrated caustics or acids into open containers, but rather into a closed system to prevent personal injury from splashed products.

(16) Do not transport liquid products, especially highly corrosive types, in open containers because of splashing and spillage potential.

(17) Do not add detergents to hot water unless procedures are written with precautions because of the potential hazard of flashback.

(18) Do not start circulating hot water containing detergent without establishing circulation and checking for leaks, as leaks in such systems are dangerous.

(19) Do not add concentrated detergents directly to processing equipment. Rather, predissolve or have water in a vessel, because of the potential for chemical reaction with product contact surfaces of equipment.

(20) Do not enter any closed vessel immediately after cleaning, before venting or changing the air supply within. Use the 'buddy system' because of the potential presence of carbon monoxide gas.

(21) Do not use detergents at concentrations in excess of recommendations because of the potential for damage to equipment and personnel.

(22) Do not use cleaning solutions above recommended temperatures because of the potential of pumping problems and splashing of solution causing injury.

(23) Do not dispense a detergent from a shipping container or into a point of application without protective goggles and clothing.

(24) Do not perform a hazardous task without a planned route of exit to a safe area, in case unexpected things happen.

(25) Do not substitute detergent unless you have a thorough knowledge of the substitute product as equipment damage, cleaning failure or personal injury may occur.

(26) Do not wear pant legs on inside of boot. Rather wear them outside of boots to prevent detergent from going into an open top boot in case of spillage.

(27) Do not give verbal instructions on cleaning procedures because of the possibility of misunderstanding. Verbal instructions leave no document for future reference or review.

(28) Do not dispense used solutions at the end of a cleaning cycle to the floor where flooding will occur because floors become slippery when detergents are discharged to them and not properly rinsed.

(29) Do not swing discharge lines carelessly to the floor or drain when discharging spent solution because of the danger to other employees who may not be expecting your action.

(30) Do not manually clean equipment when it is operating.

Acknowledgements

Nicholas Dege – Nestlé Waters North America
Tables and graphics supplied by Ecolab Inc. Food and Beverage Division
"*Make the Right Choice*" Ecolab Training Series

Appendix 1

Calculations for establishing minimum flow rates for cleaning cylindrical vessels.

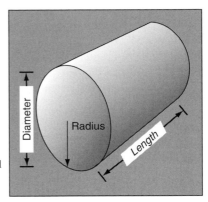

- Horizontal tanks or vessels: cleaning flow rate: 0.12–0.30 gpm/sq. ft. of surface area
- Cylindrical tanks: use the following formula to calculate surface area:

π (3.14) $\times$ r² (radius) = area of one end
π (3.14) $\times$ d (diameter) $\times$ l (length) = area of wall

For example, for a cylindrical tank 7 ft in diameter (3.5 ft radius) and 10 ft in length

Fig. A1.1 Cylindrical tank.

Ends: (3.14 $\times$ 3.5 $\times$ 3.5) $\times$ 2 = 38.5 $\times$ 2 = 77 sq. ft
Wall: 3.14 $\times$ 7 $\times$ 10 = 220 sq. ft.
Total surface: 77 + 220 = 297 sq. ft.
Recommended minimum flow rate:
297 $\times$ 0.12 gpm/sq.ft. = 36 gpm

- Vertical tanks or vessels: cleaning flow rate: 2.5–3.5 gpm/ft of linear circumference

π (3.14) $\times$ diameter = circumference

For example, for a silo that is 25 ft high and 8 ft in diameter

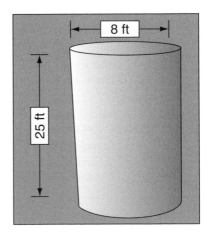

Circumference: 3.14 $\times$ 8 = 25.12 ft
Recommended minimum flow rate: 25.12 ft $\times$ 2.5 gpm/ft = 62.8 gpm

Fig. A1.2 Vertical tank.

Appendix 2

Table A2   Questions to ask when choosing between a dedicated controller and a PLC-based controller

---

What is the level of internal expertise/support available?

- A PLC requires an individual with expertise, programming software, and PLC communication hardware to a laptop or a PC
- Dedicated is easy for a nonprogrammer to support

Do you need the ability to fine-tune CIP cycles?

- A PLC can have CIP step variables made on the keypad; however, a remote PC with programming software is required for significant revisions
- Dedicated can have all CIP cycles fine-tuned from either a touch screen or from a remote PC

What is the control flexibility required?

- PLC controls the CIP unit and can control return pumps and line valve sequencing; standard controls do not include process control
- Dedicated controls the CIP unit and can control return pumps and line valve sequencing; it cannot control process operations

Are interlocks required between process and CIP?

- PLC can have start/run interlocks programmed and alarmed;. these must be specified for proper design and programming.
- Dedicated can have start/run interlocks programmed and alarmed; these must be specified for proper design and programming

Will the CIP controller need to interface to other controllers?

- PLC can interface via DH+ standard or RIO/Ethernet system selections
- Dedicated can interface via DH+ system selection

What is the communication protocol?
- PLC can communicate via DH+, RIO or Ethernet by system selection
- Dedicated can communicate via DH+ standard system selection; other communication options require electrical design review

Will CIP controller pulses process equipment?

- PLC controls process valve sequencing via communication with process controller, or custom programming per application requires electrical design review
- Dedicated can control process valve sequencing via discrete I/O or communication with process controller

---

Glossary of terms

antimicrobial agent a chemical agent that kills or suppresses the growth of microorganisms

biofilms composed of a collection of bacteria that have attached to surfaces and have excreted an extracellular polysaccharide, or a slime layer, which protects the cells from adverse environmental conditions

biological oxygen demand (BOD) a measure of the pollution present in water, obtained by measuring the amount of oxygen absorbed from water by the microorganisms present in it

chelate/chelators chemicals that are incorporated into the detergent formulation and that prevent scale buildup, i.e. the precipitation of calcium and magnesium salts onto the equipment surfaces

cleaning the process that removes soil and prevents accumulation of residues that may decompose to support growth of disease or nuisance-causing organisms; it must be accomplished with water, mechanical action and detergents

denaturization the process that changes the form of proteins, hardening them and making them less soluble; e.g. the way that heat acts on egg white, causing it to solidify

detergent see cleaner

disinfectant a chemical agent that is capable of destroying disease-causing bacteria or pathogens, but not spores and all viruses. In a technical and legal sense, a disinfectant must be capable of reducing the level of pathogenic bacteria by 99.999% during a time frame of more than 5 but less than 10 min as tested by the Association of Analytical Communities (AOAC) method. The main difference between a sanitizer and a disinfectant is that at a specified use dilution, the disinfectant must have a higher kill capability for pathogenic bacteria than a sanitizer.

disinfection the killing or inactivation of all microorganisms, except for some spore forms. The efficacy of disinfection is affected by a number of factors, including the type and level of microbial contamination, the activity of the disinfectant and the contact time. Organic material and soil can block disinfectant contact and may inhibit activity; therefore, cleaning must precede all disinfection.

emulsification a measure of a detergent's ability to break down fats and oils into smaller particles that are removed more easily during rinsing

Environmental Protection Agency (EPA) the US agency that it is responsible for the actual definition and regulated uses of both sanitizers and disinfectants

peptize mechanical action combined with a surfactant, resulting in two immiscible liquids; peptizing breaks down the protein bond and stabilizes it in suspension

pH scale a logarithmic scale that is used to measure the acidity or alkalinity of a solution; the pH of pure water is 7 with lower values indicating acidity and higher values indicating alkalinity

pin chart a chart used in programming. Historically, electrical controllers used pins to turn on electrical switches and a grid was used to assign the steps in order of operation, so that a programmer knew where to put the pins in the sequence. The term pin chart has carried over as a programming term in CIP applications.

quaternary ammonium compounds (QAC) sometimes referred to as quat sanitizers

sanitation the term used to describe the complete plant cleaning and sanitizing program protecting public health

sanitize to reduce the number of microorganisms to a safe level

sanitizer the AOAC test method requires that a sanitizer is capable of killing 99.999% (5 log reduction) of a specific bacterial test population (*Staphylococcus aureus* and *Escherichia coli*) within 30 s at 25°C (77°F). A sanitizer may or may not necessarily destroy pathogenic or disease-causing bacteria, as is a criterion for a disinfectant

sterilant an agent that destroys or eliminates all forms of life, including all forms of vegetative or actively growing bacteria, bacteria spores, fungi and viruses

9 Quality management

Dorothy Senior

9.1 Defining quality

In order to address quality management, it is important first of all to consider what is meant by quality. Defining quality may involve many routes but, for the purpose of producing goods, it is generally accepted to mean 'meeting specifications'. Specifications for a product may be defined by legislation, by the manufacturer of the goods or by the demands of the consumer. Bottled waters are no exception to this and, rather than being influenced by only one of these factors, the specification is usually a result of all the three. Importantly for bottled waters as is the case with all foodstuffs, this specification will include the essential requirement for food safety.

9.2 Quality policy

Commitment to a quality policy is necessary from the highest level within a company. Ideally, it is to this level that a quality manager will have a direct reporting line, providing confidence to senior executives through the management of a range of technical areas relevant to a water bottling company. Through this function, a quality management system can be developed, involving people working together towards a common goal. A comprehensive quality management system enables a company to control its processes of production and to influence those of its suppliers and distributors, thus ensuring that their product reaches the consumer fit for purpose and meeting specifications.

There are several formally recognised quality systems; however, they all generally specify what will be done and record the fact that it is done.

As well as quality, other issues that reputable companies address are those relating to health and safety and to the environment. Although not necessarily a requirement of legislation, it is beneficial to document the management system. This serves several purposes: it defines procedures and forms useful training material for employees; it provides confidence to suppliers, distributors and customers; and it demonstrates 'due diligence' to enforcement authorities.

9.3 Hazard Analysis Critical Control Point

In the UK and in most developed markets, bottled waters are classed as 'food' and are thus required by legislation not only to meet the quality expectations of their customers and consumers but also to do everything reasonable to protect the health of the consumer. As an essential part of this strategy, bottled water businesses are required by law to undertake a risk assessment of their processes and products.

As a system of food safety assurance, based on prevention of food safety problems, Hazard Analysis Critical Control Point (HACCP) provides a structural approach to the control of hazards. It is internationally accepted as the most effective way to control food safety. Essentially three types of hazards are considered: microbiological, physical and chemical. Although the system may also be applied to quality aspects, clear distinction needs to be made between what are considered to be food safety concerns and those related to meeting product specifications. Such decisions will contribute to the terms of reference. *Codex Alimentarius Food Hygiene Basic Texts* describe a seven-point HACCP, the principles of which are:

(1) Conduct a hazard analysis.
(2) Determine the Critical Control Points (CCPs).
(3) Establish the critical limit(s).
(4) Establish a system to monitor control of the CCP.
(5) Establish the corrective action to be taken when monitoring indicates that a particular CCP is not under control.
(6) Establish procedures for verification to confirm that the HACCP system is working effectively.
(7) Establish documentation concerning all procedures and records appropriate to these principles and their application.

HACCP can be applied not only to the company's existing product range but also to build product safety into new products/packaging formats. The system should also be reviewed in the events of installation of new equipment, technical developments and personnel changes. Though the principles of HACCP can be demonstrated through a generic plan, it is important to apply them to separate, specific operations. CCPs identified for a particular product, running down a particular line, may be different for another line or packaging format. HACCP involves people in the business, giving them responsibility for food safety. Senior management commitment to the system is fundamental.

Having defined the scope of the plan, it is recommended that a multidisciplinary team, with appropriate skills and/or knowledge of the process, prepare the HACCP plan. Operators involved in the production process may be required to undertake controls at CCPs. It is important that at least some members of the team are trained in the principles and application of HACCP to provide guidance for the implementation of the system. It is helpful to put together a product description, which will include:

- Type of water;
- Whether it is still or sparkling;
- Whether it is receiving treatment, and, if so, what kind of treatment;
- Type of packaging format being used;
- Which line it will run down;
- Prescribed shelf-life;
- Storage and distribution conditions;
- Intended use of the product.

A confirmed flow diagram for the process, identifying each stage, will form the foundation for listing all potential hazards. Based on the judgement of the team and use of a decision tree, CCPs are identified, i.e. steps at which control is applied to prevent or eliminate food safety hazards, or to reduce them to acceptable levels.

Ideally, acceptable levels will be measurable in some way and these will be the basis upon which monitoring of the CCPs is established. Should a CCP deviate from its determined limits, the team must prescribe corrective action to be taken to restore the CCP to its safe control.

To determine whether the HACCP is effective, it is important to audit and test it by reviewing the system and its records and confirming that CCPs are in control. The frequency of this verification and validation, which may be carried out internally or by a third party, may be at regular, defined intervals, or unannounced. It is essential to document HACCP procedures and to establish suitable record keeping. These will form the basis upon which verification and validation are performed.

Periodically a review of the HACCP system is required. The frequency of this may be determined based on risk factors and on any changes in the operation of the business.

The above description gives a skeletal overview of establishing a HACCP system and as well as the above Codex publication, Campden & Chorleywood Food Research Association (CCFRA) publication – *HACCP: A Practical Guide* – can also be recommended.

Some of the benefits of HACCP are:

- It moves towards a preventative quality assurance (QA) approach;
- It is complementary to other quality management systems;
- It covers all aspects from incoming materials to final product use;

- It focuses technical resources into critical parts of the process;
- Its application leads to reduced product loss;
- International authorities promote HACCP for ensuring food safety;
- It facilitates international trade;
- Its implementation can support a defence of 'due diligence';
- It complies with legal requirements.

Two important elements to stress for the success of a HACCP plan are management commitment (alluded to above) and a prerequisite programme. Prerequisites are the site-wide issues, which need to be managed well in order to underpin the product process and food safety; they vary from one operation to another but may typically include:

- Product and packaging specifications;
- Approved suppliers;
- Operating procedures;
- Process control;
- Quality assurance (QA);
- Quarantine arrangements;
- Positive release system;
- Shelf-life testing;
- Handling of customer complaints;
- Traceability;
- Calibration;
- Waste management;
- Record keeping;
- Auditing;
- Pest control;
- Compressed air;
- Lighting;
- Lubricants;
- Maintenance schedules;
- Cleaning-in-place (CIP) and other cleaning schedules;
- Chemical storage;
- COSHH or local equivalent (COSHH is an acronym for the UK legal requirements on 'control of substances hazardous to health');
- Training – induction, food safety and hygiene, health and safety, refresher.

This list is not exhaustive but gives some pointers for areas to be included (see also Chapter 6). A quality management system, covering a HACCP plan and a prerequisite programme, can be endorsed as well as audited by a third party, giving confidence to the company, customers and enforcement authorities. Figure 9.1 outlines a summary of the HACCP route.

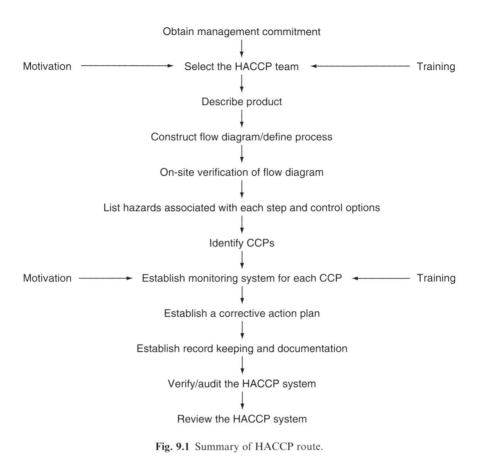

Obtain management commitment

Motivation ──────────────► Select the HACCP team ◄────────── Training

Describe product

Construct flow diagram/define process

On-site verification of flow diagram

List hazards associated with each step and control options

Identify CCPs

Motivation ──────────► Establish monitoring system for each CCP ◄────────── Training

Establish a corrective action plan

Establish record keeping and documentation

Verify/audit the HACCP system

Review the HACCP system

Fig. 9.1 Summary of HACCP route.

9.4   Process control

Process control is that element of quality management to which operators within the operation contribute by monitoring and recording performance and measurements for the part of the operation for which they are trained and responsible. This may well include the monitoring of CCPs. Process control can be broadly split into two: packaging materials in process and product water in process.

9.4.1   Packaging materials in process
First, there will be some assessment of delivered packaging materials to ensure that the consignment meets expectations and is accompanied by a certificate of conformance from the supplier. The water bottling company will decide what form such an assessment will take based on confidence in the supplier, on past history and on the particular packaging material characteristics.

9.4.1.1 Bottle handling

It is at point of use on the bottling line that the packaging is really put to test. Operatives depalletising bottles, for example, are in a position to make judgement on their fitness for purpose.

There are various possible ways in which bottles might become damaged during their progress along the filling line, and careful operatives will be alert to this occurring. Typically, for plastic bottles, damage may occur while going into or coming out of various machines, e.g. fillers, cappers, labellers. In particular, the worm feed can be a culprit if not set appropriately.

Glass bottles need particular care during filling operations. Conveyors need to be designed and set up to minimise bottle-to-bottle contact and the risk of impact damage.

During the bottling of carbonated water, occasional bottles can burst in the filling process. This may be due to flaws in the bottle, which fails through the internal pressure from the carbon dioxide gas. A strict procedure must be implemented on these occasions that minimises the risk of foreign body contamination from broken glass reaching other containers.

Where blow-moulding of bottles is part of the on-site operation, procedures for monitoring and recording standards will be undertaken by operatives in this department.

9.4.1.2 Closure application

Closure application is monitored in various ways:

(1) Torque test. This measures the tightness of application against the specification for that closure, as set by the closure manufacturer. This can be measured using a spring torque meter.

(2) Visual examination can also be carried out to ensure integrity of the tamper-evident seal, making sure no bridges are broken and, in the case of aluminium rolled-on-pilfer-proof (ROPP) closures, that the thread is formed correctly with no splits. The tuck-under of metal closures should also be examined to ensure that this has been completely rolled under. One additional check applicable to PET bottles is to ensure that the underside of the neck support rings on the bottles is not being cut by the 'antirotation' grippers on the capping machine. If not controlled, this can give rise to sharp burrs on the neck support rings that can actually injure consumers attempting to open the bottle.

(3) Manual removal of closures to simulate consumer use, carried out routinely, will provide confidence in satisfactory cap application.

(4) In the case of carbonated waters, the seal can be subjected to internal pressure to test its security with the container. This test is carried out within a pressure vessel.

9.4.1.3   Label application

Labels should be applicable to the product being bottled. They should be applied squarely and firmly without excess adhesive. An approved adhesive compatible with both bottle and label paper should be used, and examination of the gum pattern from the labelling machine will enhance confidence in the efficiency of label application.

Care needs to be taken to ensure that applied labels are not damaged along conveyor lines or in the process of packing bottles into trays and cartons. Particular care should also be taken that damage does not occur to labels as the goods are palletised.

9.4.1.4   Coding

As water is bottled, it is identified with a batch code and often the time of bottling. A Best-Before-End (BBE) or expiration date is also given. Such coding may be applied to closures, bottles or labels on individual units and may also be repeated on multipacks, outer cartons and pallets.

Process control can ensure that these codings are legible and relevant for the date of bottling. It is particularly important to ensure that coding is correct at the start of a shift and to monitor it throughout its duration.

The coding assists in traceability of product and packaging materials, in the event of any query, as well as informing the consumer of its durability.

9.4.1.5   Packing, wrapping and stacking

Operators can ensure that cartons and trays are correctly assembled and that glue application is satisfactory.

Shrink-wrap is applied to protect and secure a selected number of bottles in a pack. A firm pack should be achieved that is not too tight, and there should be no holes in the shrink-wrap material that could later lead to disruption of the pack.

Stacking of pallets is done for convenience for the distribution of goods. It should be done neatly with pack coding visible on all four sides of the pallet.

Stretch-wrap is applied to stacked pallets to secure packs in readiness for distribution. This needs to be firm enough to maintain integrity of the stack without pulling in at the corners.

9.4.2   Product water in process

Looking now at the process – the water itself – there are a number of checks that can also be made as part of process control.

9.4.2.1   pH

Each source of water has its typical pH. Testing the pH is a means of monitoring the integrity of the water. Carbonation of water reduces the pH to about 4.5. If a bottling line has been running on carbonated water and is changed over to still water, it can take quite a lot of flushing through to remove this acidity and to establish the water's natural pH.

9.4.2.2   Conductivity
Each source of water has a typical conductivity, which reflects the level of dissolved salts in the water. Where more than one source of water is bottled, conductivity readings will ensure that the correct source is being bottled.

9.4.2.3   Chlorine
If a chlorine-based sanitiser (a combined detergent and disinfectant) is used for cleaning and sterilising bottling equipment, it is important to carry out regular checks, especially at the start of shifts, to ensure that no traces of chlorine are present in the product water.

Where other cleaning or disinfecting agents are used, these should be the focus of monitoring.

9.4.2.4   Organoleptic evaluation
Nosing, tasting and visually examining the water regularly, especially at the start of shifts, will provide confidence in the organoleptic integrity of the water.

9.4.2.5   Filling volumes and levels
Filling volumes and levels are checked to ensure that, on average, the contents are as declared on the label and that levels are consistent.

Bottles may be overfilled provided that levels are consistent and that appropriate vacuity is maintained for any expansion of product through temperature changes.

Plastic bottles may be measured by weight to check the volume content. In the case of glass bottles that are declared measuring container bottles, templates may be used over fitted closures to check volume content. Both these tests are nondestructive.

9.4.2.6   Carbonation
Where the product is being carbonated, it is recommended to carry out regular checks throughout production to ensure that carbon dioxide volume is consistent and to the level prescribed for the product.

9.4.2.7   Reference samples
It is good practice to keep bottles from each day's production as reference samples for the duration of the product's shelf-life. These bottles should be representative of the production run, i.e. labelled, capped and appropriately coded.

Reference bottles can be useful in the event of any enquiry relating to a particular batch of product.

9.5   Quality assurance

Some areas of quality management are more suitably addressed by appropriately
trained and skilled technicians.

9.5.1   Microbiological assessment

Water is a vulnerable product and highly susceptible to microbiological contamination. There is no visible evidence of this and therefore routine microbiological
assessment is essential to ensure and confirm that hygiene standards are achieved
and legislative criteria are met. It is recommended to have laboratory facilities on
site to undertake microbiological monitoring.

It is suggested that sampling points for water in process be at any location
where something changes. For example, the first point would be at the source
itself. Following the route of the water, the next sampling point would probably
be on the pipeline that transports the water from source to the bottling plant. If
the water then flows into tanks, these would be the next area to sample. From
here, the water is likely to be filtered or treated and samples will be taken before
and after this stage. Finally, product water in bottle would be sampled. By
testing at all stages along the flow of water in process, any nonconformance
can be traced more easily to its point of origin.

Sampling frequencies will depend on the output from the bottling plant,
historical microbiological records and the confidence level in a sustainable
satisfactory status. To demonstrate due diligence, daily samples are likely to
be taken at points on the route from source to filtration or treatment and
several samples, including the start-up of shifts and end of run, for the bottling
process.

Sampling points at source, on pipelines and tanks, may be of varying types.
They may simply be fitted with small taps to allow water to be procured for
testing. On this basis, information on status relates only to the time of sampling.
There are also valves which are designed to provide representative samples
throughout a period of time.

9.5.1.1   Microbiological analyses

Total viable count (TVC) or heterotrophic count. The TVC is a nonselective test
that provides the means to culture single organisms of mixed species into visible
colonies of bacteria and thus facilitates counting them. These colonies grow on a
sterilised nutrient jelly-type medium (agar) that has been mixed in its warm,
molten state with a measured volume of sample water in a sterile Petri dish.
Through defined periods of time at specific incubation temperatures, living
(viable) bacteria present will multiply and develop into colonies. For the purpose
of this test, it is assumed that one colony develops from each separate organism
present in the sample. The resultant colonies are counted to give the number of
bacteria in the measured volume of water at the time of testing. Usually two Petri

dishes are prepared for each sample, one of which is incubated at 37°C for 24 h and the other at 22°C for 72 h.

Selective examination. Other microbiological analyses are carried out to examine the water for the presence or absence of specific organisms or groups of organisms within a specified sample size. Membrane filtration techniques are usually employed, as quite large sample sizes (up to 250 ml) are involved. The tests are performed using selective media that provide optimum conditions for the growth of a particular species while inhibiting the growth of others.

The particular organisms, or groups of bacteria specified are often termed 'indicator organisms' since, if they are found to be present, they can indicate that some form of pollution or contamination has taken place. Typically, coliforms and particularly *Escherichia coli* are tested for, and also faecal streptococci, *Pseudomonas aeruginosa* and sulphite-reducing anaerobes.

Bottled water should be held within the control of the producer until completion of all microbiological analyses, i.e. three days. Once satisfactory results from these tests have been obtained, the bottled product can then be positively released.

Microbiological criteria for bottled waters are specified by legislation. In Europe, this is determined by the Directives for water: 80/777/EEC for Natural Mineral Water (NMW), amended by 96/70/EC to include Spring Waters (SWs), and 98/83/EC for the quality of all other drinking water which includes other bottled drinking waters (table water) (see Table 9.1). Where positive results are obtained on selective examination, confirmatory testing may be advisable to determine more precisely the identity of allochthonous species.

In addition to the specific microbial parameters described above, there is a requirement for groundwaters, bottled without treatment, to demonstrate an absence of *Cryptosporidia* and other protozoa. It is inadvisable to undertake

Table 9.1 Microbiological criteria for bottled waters in Europe

Analyses	Maximum concentration			
	NMW/SW		Table water	
TVC at 37°C for 24 h[a]	20	cfu/ml[b]	20	cfu/ml
TVC at 22°C for 72 h[a]	100	cfu/ml	100	cfu/ml
Total coliforms	0/250	ml	0/100	ml
E. coli	0/250	ml	0/100	ml
Faecal streptococci	0/250	ml	0/100	ml
P. aeruginosa	0/250	ml		
Sulphite-reducing anaerobes	0/50	ml	0/20	ml

[a] To be tested within 12 h of bottling.
[b] cfu, colony-forming units.

analyses for these within the bottling premises. Samples for this testing can be supplied on a regular basis to government or other suitably equipped and qualified laboratories.

9.5.1.2 Plant, equipment and packaging materials

In addition to the microbiological assessment of product water, it is also good practice to monitor the standards of hygiene in filling rooms, and of bottling equipment and primary packaging materials.

Environmental monitoring can be carried out by the use of settle plates – poured plates of sterile nutrient agar are exposed to the air for a controlled period of time in selected areas of the plant and on a regular basis. The Petri dishes are then recovered, closed and incubated. Resultant colonies are counted. Automatic samplers for both microbes and particles are also available.

Swabbing is carried out on, for example, filler and rinser nozzles. After swabbing of equipment, the swab is streaked onto poured plates, which are then incubated. Here test results are recorded according to the degree of growth (low, medium or high) rather than individual colonies.

Microbiological assessment of bottles and caps is carried out using a known volume of sterile rinse. In the case of bottles, the rinse is poured into a test bottle, shaken around the bottle and then plated out in known aliquots with nutrient agar. After incubation the colonies are counted. The result is multiplied, according to the proportion of rinse cultured, to give a number of bacteria per test bottle. Similar bottles can be assessed before and after production line bottle rinsing to indicate efficacy of this process.

For caps, a known number of caps is placed into a sterile container with a known volume of sterile rinse. These are shaken and the recovered rinse is plated out as above and the resultant colonies are multiplied to give a number of bacteria per known number of caps.

The bottling company will set its own acceptability standards for these tests.

9.5.2 Assessment during shelf-life

As well as monitoring bottled water on the day of production, it is also beneficial to assess the product throughout its prescribed shelf-life and perhaps also beyond it. Through this a knowledge is built up of how primary packaging performs, what the microbial activity is and whether the organoleptic integrity is maintained up to the point of sale.

Different types of bottle and different sizes may give differing results over time, particularly in retention or loss of carbon dioxide in sparkling products.

9.5.3 New product development

QA technicians will play an active role in any new product development (NPD) – researching and evaluating any proposed new packaging formats or components.

Assessments will also be made of any deviations from the norm as a result of any process or equipment changes or developments, and any modifications to distribution and storage of product.

9.5.4   Sensory evaluation

There is clearly more to water than knowing the levels of its constituent composition or its lack of certain elements. It is the way in which all the components and factors involved work together, rather than merely their presence or absence, which brings about the whole quality of the water. Consumers buy and drink bottled water because they enjoy its taste and perceive it to be of good quality. Alert consumers learn more about water quality from their sense impressions than from a compositional report. The human capability for sensory evaluation is exemplary and very difficult for a machine to emulate and, for this reason, taste panel input is an important aspect of QA. Sensory science is rapidly gaining industry acceptance as a QA and NPD tool.

Sensory skills vary greatly from person to person, but it is possible through training and practice to improve the ability of individual tasters to identify particular taste and odour characteristics. An individual's interpretation may vary from day to day, or even from hour to hour, which is why taste panels need to involve several people (recommended 4 minimum). Our taste buds are on the surface of the tongue and around the mouth and throat.

Sensory evaluation is a subtle blend of odour and flavour/taste. The environment in which sensory evaluation takes place should be separate from other activities, where noise and odours can be minimised. Tasters should be able to sit down and be isolated from each other to enable them to concentrate. There should be no discussion on the tasting until all sensory evaluations are complete.

Presentation of the samples should be consistent. Water should always be at ambient temperature for tasting, and all samples at the same temperature. The containers may be clear, odour-free plastic – used once and discarded – or they may be glass. If the latter, care must be taken in cleaning glass containers to remove all/any detergents used and to make sure they are odour-free for use. Glassware should be dedicated to sensory evaluation and be kept scrupulously clean.

It is necessary for at least one person to have training in setting up taste panels and in addition, some rudimentary training given to those selected for the panel.

It is best to keep tests simple: e.g. the direct comparison between two samples; triangle tests where three samples are given – two the same and one different. Panellists are asked to select the odd one and to describe the difference; duo-trio tests where three samples are given – one is identified as a reference sample (R) and the other two are coded, one of which is the same as R. Panellists are asked to taste R, then the others and identify which of the others is different from R. There are also preference tests where panellists choose their preferred sample from two or more samples.

Sensory evaluation plays its part in shelf-life monitoring, comparing the same product just bottled with samples during shelf-life. It can assist in NPD, especially when new packaging components are being selected, or in relation to volume/packaging ratio. It is also a useful way to compare different brands of water.

9.5.5   Auditing

As well as auditing hygiene standards within the company's bottling facility, QA personnel will be involved in auditing suppliers of primary packaging materials and warehouses where finished goods are stored within the distribution system.

The British Retail Consortium (BRC) and the Institute of Packaging (IoP) have produced a *Technical Standard and Protocol for Companies Manufacturing and Supplying Food Packaging Materials for Retailer Branded Products* in 2001. This provides a basis upon which to audit suppliers of packaging materials, and the suppliers can be asked to be accredited to this standard. Through this process, 'due diligence' is passed back up the chain.

If product is contract-packed off-site, it is recommended that this is also audited. Last, but by no means least, it is good to audit product at point-of-sale in the store to ensure that it arrives there in pristine condition.

A documented code of practice for suppliers and for warehouses will provide an understanding of the standards expected.

In addition to the above auditing, a water bottling company may be subject to, or may choose to have, third-party audits (see Chapter 11). These would be undertaken by an independent body or authority, often as a requirement of a customer.

9.5.6   Calibration

Where pieces of test equipment are used to verify the meeting of specifications, it is important that these are calibrated on a regular basis, either internally using appropriate standards or by external bodies. Examples of such equipment include balances, conductivity meters, pH meters, pressure gauges, secure seal testers, thermometers and torque meters.

9.5.7   Accreditation

It may be advisable to consider gaining certification/registration through an accredited body for the QA laboratory. This will provide confidence for the bottling company and also credibility to existing and prospective customers as well as enforcement authorities.

9.6   Independent or government laboratories

Bottled waters are required to have extensive chemical analysis covering at least 50 parameters. Some of this analysis is very challenging as many parameters are at levels close to or below detection limits. In some cases, the analysis is undertaken to demonstrate absence of a parameter as, for example, to show freedom from pollution from pesticides, herbicides and fertilisers.

For some elements there is a maximum admissible concentration (MAC) as, for example, with toxic substances. Sophisticated techniques are needed.

This extensive and exacting analysis is required regularly, though not daily or weekly, and it is unlikely, therefore, with possibly the exception of a very large producer, that bottled water companies will have laboratory facilities or staff to undertake it. Independent or government laboratories are generally used for this purpose. Table 9.2 shows the usual parameters included in official analysis for bottled waters.

9.7 Recognition of source

The hydrogeology influencing a source of water is highly complex and specialist knowledge is needed. An appointed hydrogeological consultant would define the catchment area for the source and advise on all matters relating to the identification and protection of the source and its exploitation.

In Europe, there is a recognition process for a new source of natural mineral water, which may take about two years. During that time, many samples of the

Table 9.2 Parameters in the official analysis for bottled waters

Parameter	Expressed as
Physical and chemical characteristics	
Dry residue at 180°C	mg/l
Dry residue at 260°C	mg/l
Electrical conductivity (at given temperature)	μS/cm
Hydrogen ion concentration	pH
Radioactivity	
Total alpha activity	Bq/l[a]
Total beta activity	Bq/l
Toxic substances	
Antimony	Sb μg/l
Arsenic	As μg/l
Cadmium	Cd μg/l
Cyanide	Cn μg/l
Chromium	Cr μg/l
Lead	Pb μg/l
Mercury	Hg μg/l
Nickel	Ni μg/l
Selenium	Se μg/l

(*Continues*)

Table 9.2 (*Continued*)

Cations	
Aluminium	Al mg/l
Ammonium	NH_4 mg/l
Calcium	Ca mg/l
Magnesium	Mg mg/l
Potassium	K mg/l
Sodium	Na mg/l
Anions	
Borate	BO_3 mg/l
Carbonate	CO_3 mg/l
Chloride	Cl mg/l
Fluoride	F mg/l
Hydrogen carbonate	HCO_3 mg/l
Nitrate	NO_3 mg/l
Nitrite	NO_2 mg/l
Phosphate	P_2O_5 mg/l
Silicate	SiO_2 mg/l
Sulphate	SO_4 mg/l
Sulphide	S mg/l
Non-ionised compounds	
Total organic carbon	C mg/l
Total carbon dioxide	CO_2 mg/l
Trace elements	
Barium	Ba μg/l
Bromine	Br μg/l
Cobalt	Co μg/l
Copper	Cu μg/l
Iodine	I μg/l
Iron	Fe μg/l
Lithium	Li μg/l
Manganese	Mn μg/l
Molybdenum	Mo μg/l
Strontium	Sr μg/l
Zinc	Zn μg/l
Freedom from pollution	
Pesticides	μg/l
Herbicides	μg/l
Fertilisers	μg/l
VOCs	μg/l
Microbiological	
Parasites	
TVC at 37°C in 24 h	cfu/ml
TVC at 22°C in 72 h	cfu/ml
Total coliforms	
E. coli	
Faecal streptococci	
P. aeruginosa	
Sulphite-reducing anaerobes	

[a] Bq = becquerel.

water will be submitted to an independent or government chemist for analysis. The intention is to establish from this that the water is stable in its composition, allowing for slight seasonal fluctuations, that it demonstrates freedom from pollution and that it is safe to drink without treatment.

Once enforcement authorities are satisfied that the water meets legislative requirements, notice of recognition is published in the *European Journal* and in equivalent publications of the member states. Following recognition, full analysis is still required on a regular basis but not as frequently as during the recognition process.

9.8   Industry networking

The bottled water industry is young and dynamic and its associated legislation is continually developing. Membership of trade associations can provide a forum through which technical developments and other advances can be discussed with others within the industry. This also facilitates lobbying opportunities, where appropriate.

Bibliography

Codex Alimentarius Food Hygiene Basic Texts, Joint FAO/WHO Food Standards Programme.
Guide to Good Bottled Water Standards (2002) 2nd edn. British Soft Drinks Association.
HACCP: A Practical Guide, Campden & Chorleywood Food Research Association (CCFRA).
Technical Standard and Protocol for Companies Manufacturing and Supplying Food Packaging Materials for Retailer Branded Products (2001) British Retail Consortium (BRC); The Institute of Packaging (IoP).

10 Bottled water coolers

Michael Barnett

10.1 Introduction

The bottled water cooler industry is closely related to the bottled water industry as a whole. In recent years, the water cooler has been seen as the environmentally friendly approach to dispensing bottled water. By utilising the large 18.91 (US 5 gallon) reusable plastic bottle, as opposed to the small 0.5–2.0 l one-time-use bottles, substantial savings can be made, and, once beyond their service life, the larger plastic bottles are recycled into other nonfood-related products.

The compactness and elegant design of the modern water cooler and its availability on a rental basis, coupled with the guaranteed quality of bottled water, provided an ideal recipe for the growth of the industry at a time of ever-increasing concern about the quality of municipal water supplies for drinking purposes. This has been strongly assisted by the dramatic growth of the market for water in smaller bottles resulting from interest in healthy lifestyles, supported by strong brand advertising and promotion by the large national and international bottled water companies competing for market share.

10.2 World market

10.2.1 Europe

While bottled water in small containers can be traced back many hundreds of years, the bottled water cooler market as we know it today did not start until the mid-1980s when the American bottled water cooler concept was first imported into the UK and offered as the latest in modern office equipment to a booming commercial sector.

Despite the recession of the early 1990s, the water cooler market in the UK has grown steadily at 15–20% p.a. With some 530 000 water coolers installed by the end of 2003, it now leads the European market and is anticipated to grow steadily to 850 000 by 2007 (Zenith International). The residential market for water coolers is not yet developed and accounts for less than 2% of the total market.

This growth was not paralleled in Western Europe owing to restrictive legislation, particularly in France, Italy, Spain and Germany, and was referred to as the '2-Litre Rule', which was revoked in the mid-1990s. As a result, the development of the water cooler market in Europe was restrained for some eight years, when compared to the UK. In other countries, such as Ireland, Denmark, Sweden, Norway, Holland and Belgium, the water cooler industry commenced in the early 1990s and enjoyed an equal success as in the UK. Water cooler popularity in 2003 is shown in Table 10.1.

European water cooler placements at the end of 2003, excluding the UK, are estimated at 1130 000 water coolers and as a whole, the European market is estimated to grow to over 3 000 000 by the end of 2007 (Zenith International). This recent growth outstrips the market for bottled water in small containers. It is interesting and significant to note that both of the traditional European bottled water companies, Nestlé and Danone, established in supplying water in small bottles, only entered the bottled water cooler market as late as 2000 and then, primarily by the acquisition of existing water cooler companies.

10.2.2 Middle East

The Middle East, with its arid lands and the restricted availability of water, has long been a natural market for water in small bottles, but it has mainly been a client market for imported brands, usually from Europe. Some local bottlers of purified water and both natural mineral water and spring water do exist;

Table 10.1 Cooler units per 1000 population in 2003

Country	Units
Ireland	12
UK	9
Norway	7
Belgium	7
Netherlands	6
Portugal	6
Switzerland	5
France	4
Greece	4
Denmark	3
Sweden	3
Italy	3
Spain	3
Finland	2
Austria	1
Germany	1

Source: Zenith International (2003).

however, the generally low-income level of the population has meant that it was a luxury product and imported brands were generally perceived to be of better quality and prestige. The water cooler market, however, utilising local water sources (often purified) is a more recent phenomenon and can be traced back to the late 1980s.

Israel, with over 200 000 bottled water coolers, is the second largest bottled water cooler market in the Middle East, after Saudi Arabia. Saudi Arabia and the Gulf states have been desalinating and purifying drinking water for many years to support their fast economic and industrialisation programmes and growing populations. The supply of drinking water has been a matter of survival, rather than a desire for contaminant-free, good drinking water, as was the driving force for the industry in America and Europe.

An aspect of these markets, except for that of Israel, is that water coolers are not normally part of a rental programme as they are in America and Europe. The traditional bargaining and purchase of equipment in the souks and also more modern stores prevails and bottling companies are simply involved in providing the bottled water product. This is particularly evident where the water cooler is intended for residential or small business purposes. The lending of water coolers to large commercial users in return for a bottled water supply contract is also a common practice.

Egypt, Jordan, Lebanon and Syria are relatively recent emergent water cooler markets, and, despite Israel's meteoric growth in the late 1980s, Saudi Arabia currently remains the largest bottled water cooler market in the Middle East with Egypt promising to be the most significant future growth market in the region.

10.2.3 Asia

Asia traditionally boasts of some of the largest bottled water cooler companies in the world, with vast demand for good-quality, contaminant-free drinking water. Fast economic growth and related industrialisation have led to the contamination of many natural water sources, and hence purified water is favoured for bottling. Unlike in Europe, water companies and breweries supply bottled water in small containers as well as in the 18.9 l reusable bottle; they also often provide water cooler services. The vast demand for water coolers also meant that the market attracted some companies that do not pay sufficient attention to water quality and hygiene, and so care must be taken to ensure that only water from the most reputable companies is consumed.

Although Japan has a very large and long-established market for bottled water in small bottles, the bottled water cooler market commenced only in the late 1980s with natural spring water as the main source of supply. The market has not enjoyed the expected growth, apparently owing to public concern about bacteria entering the open reservoir of the water cooler and the size of a traditional water cooler relative to the small size of the average urban residence.

Until the early 1990s, the South Korean government had banned the sale of 18.9 l bottles intended for use with water coolers to Korean citizens on the grounds that the local water quality was of a good standard. The bottled water cooler market that evolved, however, was focused on the demand of the small foreign resident population, mainly US military personnel and diplomats. Since the rescindment of the law, many companies have entered the water cooler market and have been successful in establishing a thriving demand for both bottled water and locally manufactured water cooler appliances.

In the years since liberalisation, the Chinese market for water coolers has seen a slower than expected growth rate, mainly due to poor distribution logistics in the main conurbations and poor profitability due to very low pricing. It is estimated that around 500 000 water coolers are in service, with some ten major bottling companies supplying demand, which is in the main for distilled water. The unusually high demand for bottled water generally and distilled water in particular reflects both the public's distrust of drinking water other than purified water and water shortages in the municipal water system due to frequent drought conditions.

The neighbouring market of Hong Kong is a mature bottled water cooler market with an estimated 230 000 water coolers, with one bottler playing a leading role for over 20 years in both the small bottle and bottled water cooler market. As in China, purified water is the most popular product and there appears to be distrust by the population of local spring water or the municipal water supply.

Taiwan is traditionally a similar market to China, but has shown much slower growth. There does not appear to be the usual acceptance of the rental philosophy with respect to the water cooler, and as such, water cooler purchase and the supply of bottled water are not usually handled by the same company. This has probably been assisted by the preponderance of local water cooler manufacturers who sell bottled water coolers through retail distribution to the public. It is estimated that there are 60 000 bottled water coolers in place, supplied by four bottlers.

Philippines, Singapore and Malaysia have developed slowly since the early 1990s, but aggressive cut-price competition has resulted in very low margins that are insufficient to offer a quality delivery service to the consumer. In the Philippines, water coolers are frequently lent to customers free of charge. These economic constraints have led to private vending machine operators installing what are locally called 'filling stations', supplying water to consumers who bring their own bottles, thus becoming the major source of drinking water. A lack of water quality standards and control has exacerbated the situation.

Indonesia and Thailand are very large and well-established bottled water markets with over 1000 companies offering bottled water in all sizes of bottles. As many as 2000 secondary suppliers are said to supply bottled water on a seasonal basis. Several large bottlers dominate the markets and distribution patterns, offering both water in small bottles and bottled water coolers. Here too, aggressive competition has driven prices to levels that result in questionable quality and hygiene standards for many of the smaller bottlers. Poor road conditions outside of the larger cities have also militated against the growth of the water cooler market.

10.2.4   Australia and New Zealand

The Australian water cooler market is a mature market that began in the mid-1980s and flourished in the 1990s. There are over 165 000 water cooler placements, with one large company dominating the market with over 60% share. Buoyant growth has encouraged local water cooler production, which supplies the greater part of the demand. Spring water is the major source of water supplied to this market, with only a very small supply of purified water. New Zealand has a very small market for water coolers focused around the capital Auckland. It is estimated that some 20 000 water coolers are in service.

10.2.5   Central and South America

Markets here have been established since the 1950s and have adopted the water cooler concept for both commercial and residential consumers as a substitute for the poor quality of municipal supply.

Mexico has a very large bottled water market, both in small bottles and in the large reusable 18.9 l water cooler bottles. There are some 20 bottlers supplying demand; however, the general low-income level of the population has dictated that in the majority of cases, bottled water from 18.9 l bottles is not dispensed from bottled water coolers but is dispensed using other equipment such as tilters and crocks (see Section 10.3) that do not cool the water.

Brazil boasts of the largest bottled water consumption per capita in South America, but water coolers are only a more recent growth market, as they are also in the rest of the subcontinent. It is estimated that 150 000–200 000 water coolers are in service.

10.2.6   North America

North America is the place of origin of the bottled water cooler, and its development can be traced back over 100 years. As would be expected, it is now a stable market in an advanced state of maturity with an estimated seven million water coolers – 40% in commercial and 60% in residential locations. Here, too, bottled water companies supply water in small containers as well as the 18.9 l bottle, and brand names and market share play a significant role. Despite this, with the vastness of the continent, there are about 300 bottlers, encompassing small businesses trading on a local basis and multinational companies.

The Canadian market has been consolidated over the past three years from many small bottlers to just three major suppliers, only one of which offers national coverage. It is estimated that there are 30 000 water coolers in service, equally divided between commercial and residential consumers.

10.3   Equipment development

10.3.1   Dispensers

It is not known precisely when in history earthenware vessels for the storage of drinking water were first fitted with a simple on/off valve to dispense water. It would have been much later in the development of man and most probably in the dry and hot equatorial regions of the world, where drinking water is very much a precious resource, that the storage of drinking water in earthenware vessels became commonplace. Some forms of non-cooling water dispenser are illustrated in Fig. 10.1 – from the ceramic crock to the late nineteenth-century icer and the simple tilter and stand.

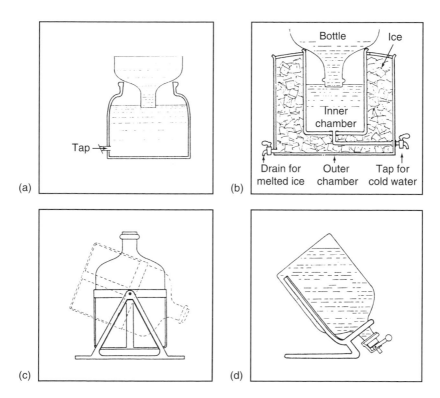

Fig. 10.1 Early and simple forms of water dispenser: (a) ceramic crock; (b) icer; (c) tilter; (d) stand.

As Fig. 10.2 illustrates, while drinking water dispensers have undergone many changes in the past 150 years, the most active period of their development has been in the past 20 years and it is envisaged that this will continue over the next decade. The more recent developments of bottled water coolers did not occur in isolation, but paralleled the development of its related service industry, i.e. the delivery of bottled drinking water to home and office.

Modern health and safety considerations for the short-term storage and dispensing of bottled water, and in particular microbiological considerations, have had significant influence on the development of water coolers in the 1990s. One area of particular focus has been the interface between the water bottle and the water cooler. Traditionally, the neck of the bottle was immersed in the open-topped cooling reservoir of the water cooler and the water level maintained until some water was drawn off from the dispensing tap. At that time, as the water level in the reservoir fell and the neck of the bottle became exposed above the water level, water flowed out of the bottle and into the reservoir and was replaced in the bottle by air which flowed in through the gap between the water level in the water cooler reservoir and the bottle. When the water level in the reservoir rose to cover the bottleneck once more, the flow of water out of the bottle would stop.

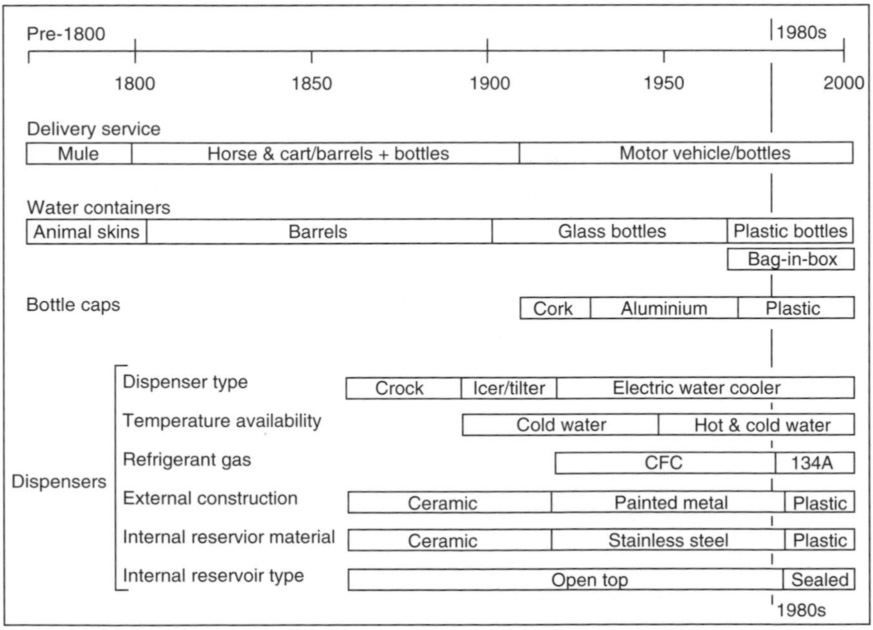

Pre-1800 | 1980s
1800   1850   1900   1950   2000

Delivery service
| Mule | Horse & cart/barrels + bottles | Motor vehicle/bottles |

Water containers
| Animal skins | Barrels | Glass bottles | Plastic bottles |
| | | | Bag-in-box |

Bottle caps
| Cork | Aluminium | Plastic |

Dispensers
Dispenser type: | Crock | Icer/tilter | Electric water cooler |
Temperature availability: | Cold water | Hot & cold water |
Refrigerant gas: | CFC | 134A |
External construction: | Ceramic | Painted metal | Plastic |
Internal reservoir material: | Ceramic | Stainless steel | Plastic |
Internal reservoir type: | Open top | Sealed |

1980s

Fig. 10.2 Development of bottled water coolers.

US Patent 1 142 210, filed by Walter Wagner of Chicago, Illinois, USA on 29 February 1912, but not developed and put into production until nearly 80 years later (see Fig. 10.3), addressed many of the microbiological contamination and usage issues related to the open reservoir system which were:

- External surface of the bottleneck immersed in drinking water in the internal reservoir
- Airborne contamination of the drinking water as air is drawn into the reservoir and bottle
- Open and exposed reservoir at the time of bottle changing
- Water spillage when inverting the bottle over the cooler
- Contamination of the open-topped empty bottle after use

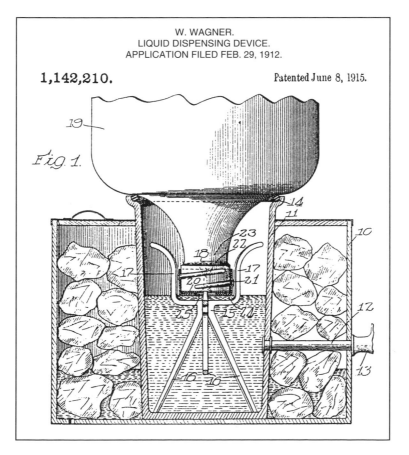

Fig. 10.3 US Patent 1 142 210, filed on 29 February 1912.

The Watersafe bayonet-and-valve (BV) closed-reservoir system was first introduced by Elkay Manufacturing Co., USA, for its bottled water coolers in 1990.
This innovation interpreted and commercialised the original patent using
modern materials and technology (Fig. 10.4).

The BV device requires that the bottle cap, which traditionally acted simply
as a closure to contain the water in the bottle during storage and transit, and was
removed prior to loading onto the water cooler, now play a vital role in the
dispensing operation. To this end, it has been redesigned as an integral component of the BV system so that it is not removed from the bottle prior to loading it
onto the cooler. The valve mechanism built into the BV cap is opened by the
bayonet component on the water cooler, and water flows from the bottle
through apertures in the bayonet and into the now sealed cooling reservoir.
When the empty bottle is lifted and removed from the water cooler, the valve
in the bottle cap closes, thereby preventing any contamination from entering the
bottle on its way back to the bottling plant for refilling.

In addition to eliminating external airborne contamination by sealing the
cooling reservoir and introducing the bayonet, the BV device incorporates a
submicron (0.5 μ) air filter to remove airborne spores and bacteria from the
incoming air supply that bubbles into and replaces the water in the bottle at the
time of dispensing water from the cooler.

Focus on the microbiological integrity of bottled water, initiated by introduction of the BV system, continued by reviewing the remaining water contact
surfaces up to and including dispensing, namely the cooling reservoir, piping/
tubing and dispensing taps.

Since the days of the first electric water cooler in the 1920s, the internal
cooling reservoir has been made of stainless steel. This material has many
benefits in terms of its corrosion resistance, good heat transfer from the refriger-

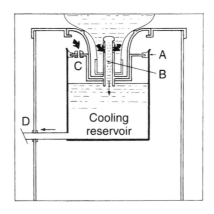

Fig. 10.4 Typical bayonet-and-valve device fitted to a bottled water cooler. A, seal; B, bayonet;
C, air filter; D, tap.

ant coils surrounding it and good surface finish that denies bacteria (biofilm) an anchorage facility and imparts no taste taint to the water. However, with the evolution of synthetic plastic materials as a spin-off from space technology, it was only a matter of time before the dominance of stainless steel as the only material for the cooling reservoir was challenged by plastics.

The challenge came in the mid-1990s with the introduction of a revolutionary new concept from a new entrant to water cooler manufacturing – Ebac – a British company and an established market leader in the manufacture of dehumidifiers. The 1994 Ebac water cooler (Fig. 10.5a) was both revolutionary and evolutionary, since it set out to address the requirements of modern-day bottled water cooling and dispensing by incorporating features from information gained through extensive research of bottled water cooler distributors. Some ten years on, the Sipwell water cooler embodies some of the innovations of the Ebac Water cooler, whilst contributing its own innovation in terms of both design and function, and demonstrates the evolution and influence of contemporary appliance design (Fig. 10.5b).

The most innovative aspect of the 1994 Ebac water cooler was its disposable Watertrail that includes all water contact components from the water cooler to

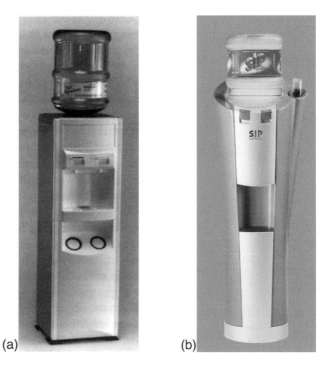

Fig. 10.5 (a) Ebac water cooler, c.1994; (b) The Sipwell contemporary water cooler, c. 2003.

bottle interface to the dispensing taps. The Watertrail commences with a BV system incorporating an air filter assembly connected by flexible silicone rubber tubes to a plastic bag that acts as a bladder and replaces the traditional rigid stainless steel cooling reservoir. In this manner, the water is in close proximity to the surrounding refrigerant coils and is only separated from them by a thin layer of plastic. From the plastic bag exit tubes that route the water to the dispensing taps and on some dispensers to the water heater. The complete Watertrail is a relatively low-cost disposable item, which is easily and quickly removed from the water cooler and replaced with a new one, the old one being disposed of. Sanitation of the water cooler is not required after installation of a replacement Watertrail, since each one is assembled and packed in a sterile clean-room environment.

Further innovations incorporated into the design of the 1994 Ebac water cooler were built-in dispensing taps, external all-plastic panels offering easy exchange for multicolour options, rear wheels for portability, built-in cup dispensers and covered drip tray incorporating a water level indicator.

The innovative disposable Watertrail and the focus on water cooler hygiene triggered the development by several water cooler manufacturers of a plastic, rigid, removable and reusable water cooler reservoir that replaced the traditional stainless steel reservoir. The objective is that, instead of having to sanitise the water cooler reservoir within the water cooler, it could be removed from the water cooler and exchanged for one previously sanitised: the one removed could be taken away to be cleaned and sanitised ready for reuse. A further benefit claimed for the removable plastic reservoir is that, in most cases, its removal from the water cooler also removes the dispensing taps attached to it, thereby replacing all the water contact surfaces in one operation, as in the Watertrail.

An alternative design for cooling the water in the reservoir, the Oasis concept, utilises a cooling metal probe, inserted into the centre of a rigid plastic cooling reservoir, which cools the water radially outwards. The refrigerant coils are embedded into the probe, as opposed to surrounding the external surfaces of the reservoir, and are said to be substantially more efficient in abstracting heat from the water and reducing losses to the surroundings.

Taps for dispensing water from the very early earthenware crocks have not changed much in engineering terms. These taps utilised the alignment of two apertures to obtain flow through the valve. Some taps in use today in other consumer appliances still use this principle.

As materials technology evolved and rubber seals were invented, metal taps became the norm, usually made of chrome-plated brass. Today, with the evolution of plastics and synthetic seals, the modern water cooler tap uses the principle of a spring-loaded plunger to operate a plastic diaphragm that seals an aperture opened by lifting the diaphragm away from the aperture's face.

The 1994 Ebac water cooler introduced the concept since adopted by other modern water coolers of squeezing a plastic tube with a spring-loaded roller to the point where flow is stopped. To facilitate flow, the roller is lifted clear of the tube by the use of levers.

While focusing on water cooler hygiene, research into external means of water cooler contamination has shown that the dispensing tap is a vulnerable component. Owing to frequent hand contact at the time of dispensing, there is a possibility for contamination of both the tap body and outlet nozzle. To overcome this, some manufacturers have introduced remote activation of the taps using both mechanical and electrical actuators or levers, and the dispensing nozzles are protected from hand contact with covers. Some even surround the tap nozzle area with beams of ultraviolet light.

The first water cooler providing hot water, as well as chilled water, appeared in the late 1940s, offering instant piping hot water at a temperature in the range 83–87°C, adequate for instant coffee, soups, etc. The heating took place in a built-in brass or stainless steel tank fitted with an electric heating coil. A requirement of the European consumer, particularly from the tea-drinking English, was for the achievement of a higher water temperature.

By the mid-1990s, while electric heating coils remained relatively unchanged, temperature control of water was achieved by electronic solid-state technology, attaining water temperatures as high as 97°C with reliability and safety. To complement these higher temperatures, dispensing taps were modified to incorporate tamperproof handles to ensure that only purposeful activation of the tap could provide hot water, avoiding accidental dispensing and scalding of the user.

Like the cooling reservoir, the traditional metal hot tank also underwent modernisation, with its material changed from metal to engineering plastic that withstands boiling water without deformation, as well as preventing the build-up of scale with the use of hard water.

With more sophistication and demands of the consumer, the technology employed by manufacturers also has advanced and the water cooler has not been an exception. Modern water coolers that dispense chilled carbonated water as well as chilled still water are also available. Essentially, these are the same as a standard water cooler, but additionally include a carbonation chamber and a CO_2 gas cylinder. Electronic solid-state circuitry controls the process and the normal chilled still water is carbonated, one cup at a time, on demand. The simple push of a button operating an electric solenoid valve is all that is required to obtain a refreshing cup of chilled sparkling water.

Water coolers have evolved to dispense water in various specialised containers made of glass and subsequently of plastic in a multitude of shapes and sizes. One such development, in the early 1970s, was the adaptation of the standard water cooler to dispense water in a container known as bag-in-box (see Section 10.3.2.3). Although water coolers required only a minor adaptation to enable them to dispense water from this container, they did not gain wide acceptance because the packaging system often introduced a taste taint to the water.

In some European countries, a regulation prohibited the sale of natural mineral water in bottles larger than 2 l. To overcome this barrier to the use of water coolers, an innovative Italian water cooler manufacturer produced an external water reservoir, which replaced the traditional bottle. This external reservoir was located on the water cooler where the bottle would normally be placed, and combined a housing and an adaptor for the location of six bottles of 2 l each. These bottles emptied into the upper reservoir, which in turn filled the internal cooling reservoir, as would a standard 18.9 l bottle.

Thermoelectric water coolers offer several benefits over traditional water coolers. They have no compressor or refrigerant gas and operate using a 12 V solid-state thermoelectric module (TEM) utilising the Peltier effect, which was discovered in 1834. This effect occurs whenever an electric current passes through a circuit of two dissimilar conductors, and, depending on the current direction, the junction of the two conductors will either absorb or release heat. The compactness of TEMs offers scope for original design and it is anticipated that advances in technology will increase the output and efficiency of these modules in the future, allowing them to compete effectively with the presently dominant refrigerant gas compressor technology. In the meantime, this technology is primarily employed in small-capacity water coolers aimed at the residential market.

Water cooler manufacturers are under pressure to reduce water cooler size for both office and residential consumers; however, restrictions imposed by contemporary compressor technology limit this for the present. Traditionally, water coolers have been floor-standing appliances, but since the mid-1970s, counter-top versions have been available. In essence, these are cut-down versions of the floor-standing models, where the space between the compressor and the internal cooling reservoir has been substantially reduced.

TEM offers the best opportunity for size reduction and it is anticipated that, once their operating parameters have been improved, these coolers will dominate the compact cooler market.

Water cooling has also found application in the traditional coffee brewing industry for multifunction appliances dispensing freshly brewed coffee as well as chilled and hot water from a bottled water source. They are increasing in popularity in America by virtue of their compactness. These appliances are not intended for large-volume water dispensing and utilise TEMs as the cooling source for the water.

10.3.2   Bottles

10.3.2.1   Wood and glass
Over the 100-plus years since the home delivery service of drinking water for reasons other than therapeutic, containers used have themselves undergone a metamorphosis. In America, the mid-1800s saw the horse-drawn wagon with several large 50 gallon wooden casks, bound together with rope, delivering water

to homes on a regular basis. This practice continued until the development of large glass bottles in the 1890s. These bottles were of thick-walled, clear glass, round in cross section and more akin to the traditional small glass bottles used in the beverage industry than the now standard 18.9 l water cooler bottle. It was not until the 1920s that the unique 18.9 l glass bottle intended exclusively for use on a bottled water cooler was introduced. This bottle closely resembled the modern plastic bottle. The most significant difference was its weight, which averaged 6 kg when empty and about 25 kg when filled with water.

10.3.2.2 Plastic containers

For many years and through most of the twentieth century, the glass bottle prevailed until the evolution of thermoplastics in the years following World War II. First, PVC evolved as the multipurpose thermoplastic, in both coloured and clear forms. At that time, it was considered to be the ideal mass production material, and in the bottled water industry, it replaced, in many cases, the small glass bottle for packaging still water. The dark green glass bottles were retained for packaging carbonated waters.

The evolution of plastics appeared to offer the solution to many problems of the water cooler industry relating to both weight and safety. Despite this, it was not until the early 1970s, and after over 10 years of research, when the Reid Valve Co., USA introduced the first polycarbonate 18.9 l (US 5 gallon) plastic bottle, that the modern age of the water cooler bottle had arrived. It was the product the water cooler industry had been waiting for. It was light, an empty bottle weighing approximately 750 g (an eighth of the weight of a glass bottle), and was strong and durable. It also demonstrated excellent long-term chemical characteristics for the packaging of food, without the migration of undesirable chemicals from the material into the product. Polycarbonate is also an environmentally friendly material, fulfilling the three Rs of the modern packaging industry: reduce, reuse and recycle.

Typically, the service life of an 18.9 l polycarbonate bottle is about 40–60 round trips, from the bottling plant to the consumer and back. In that time, one bottle will have dispensed, via the water cooler, about 950 l of water; i.e. a reuse ratio of approximately 950 : 1.

Despite the ideal properties exhibited by polycarbonate as a material for the 18.9 l bottles, the 1980s saw the redevelopment of PVC bottles, particularly in emergent economies, such as Mexico, Argentina, Brazil and Asian countries. The main reason was related to cost, but although the initial capital outlay for the PVC bottle was indeed less, the long-term cost with respect to durability shows little or no savings owing to its shorter service life.

Also the increasing use of ozone for the disinfection of bottled water and its dissolved residual content in the water after bottling have a detrimental effect on PVC, embrittling, crazing and discolouring the material. Polycarbonate bottles appear to be immune to the effects of ozone.

The advent of polycarbonate, PVC and more recently, multiple-use poly-thene water cooler bottles has demanded new materials for their washing and sterilisation prior to refilling. The very hot caustic wash solutions that were used with glass bottles proved detrimental to the polycarbonate material and new, non-caustic, low-temperature (55–60° C) wash solutions were developed. Sanitis-ing solutions similar to those used on glass bottles for disinfecting prior to refilling are used for polycarbonate bottles, but their concentrations have been adjusted to suit the new plastic material. Automated handling equipment in the bottling plant and storage racks on the delivery trucks have been developed to prolong the service life of bottles by reducing the flexing of the bottle sides during transit.

The modern polycarbonate bottle has further developed over the past 30 years through advances in both manufacturing and materials technology. New blow- and injection-moulding techniques have reduced the wall thickness of the bottles and have consequently reduced their weight by some 10–20%, without reducing their strength. Advances in materials technology have improved the clarity of the material and its flow characteristics in the mould, allowing for innovative bottle design and decoration. One such benefit has been the availabil-ity of built-in handles. Durability of the material has also improved, increasing service life and, in parallel, the environmental benefits of reuse.

Health and safety concerns for delivery personnel, as well as convenience for the end consumer during bottle exchanges on the water cooler, have increased the demand for bottles with built-in handles. Similar considerations may also influence bottle size and therefore weight in future. Today, plastic reusable bottles are available in 22.7l, 18.9l, 17.5l and 12l sizes, and in a variety of shapes (Fig. 10.6).

10.3.2.3 Bag-in-box containers

Bottles are not the only means by which drinking water may be packaged for dispensing through a water cooler. In 1961, liquibox, a leading American com-

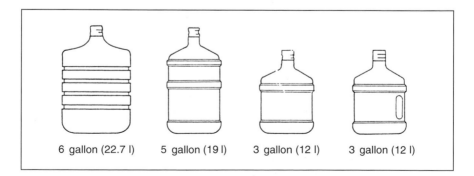

| 6 gallon (22.7 l) | 5 gallon (19 l) | 3 gallon (12 l) | 3 gallon (12 l) |

Fig. 10.6 Water cooler bottle shapes and capacities.

pany in the packaging of liquids, invented the bag-in-box concept for the milk industry. It was to fulfil the needs of being lightweight, stackable, nonreturnable (disposable) and aseptic. In simple terms, this is a plastic bag contained within a sealed carton, with the liquid contained being dispensed via a specialised built-in valve.

Unlike the bottle, this means of packaging does not require that air be introduced into the space above the water as water is dispensed from the cooler. The bag simply collapses as the water volume reduces. This feature provides for aseptic dispensing of the water and has created substantial interest from Japan, where the microbiological integrity of the water is a cultural as well as a medical concern.

In the 1970s, several American bottled water companies introduced the bag-in-box packaging for their water cooler products; however, after a short period these were discontinued. At the time, the main reason for their lack of popularity was related to a taste taint of the water, usually introduced by the packaging material. However, recent advances in materials technology are said to have overcome this problem. It is anticipated that with time this method of packaging will gain further market acceptability especially in the residential, hospital, school, specialised industry and military markets.

10.3.2.4 Caps

From the time when glass bottles were introduced for the supply of water for water coolers, the means of closing the bottles to guarantee the quality of the bottled product have improved substantially.

The plastic cap of the type introduced to coincide with the Watersafe BV system by Elkay Manufacturing Co in 1990, owes its origin to the same US Patent 1 142 210, filed on 29 February 1912. Despite taking nearly 80 years to reach the market place, this patent with its implications for both the cooler and the bottle cap has revolutionised the bottled water cooler industry. To summarise from the original patent, the inventor's intentions were to allow for a water bottle to be inverted over a water cooler without spillage and to avoid hand contact with the mouth of the bottle for hygienic reasons.

Recent improvements to the original invention for further safety and hygiene of the cap and bottle comprise a tamper-evident seal over the valve in the cap and the closing of the valve as the empty bottle is removed from the water cooler. Figure 10.7 shows the cap with valve. The loading of a new full bottle onto a water cooler entails removal of the tamper-evident seal on the cap to expose the valve below it. The bottle is then upturned over the bayonet on the water cooler top cover, which locates into the valve in the cap as shown in Fig. 10.8. As the bottle is lowered over the bayonet, the central section of the valve is lifted out of the cap and mounts onto the bayonet. Apertures in the bayonet direct the water into internal waterways that open into the water cooler's reservoir.

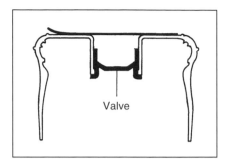

Fig. 10.7 Watersafe cap wih valve.

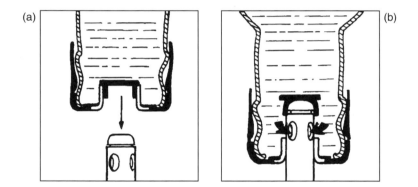

Fig. 10.8 Operation of bayonet-and-valve cap.

Several caps incorporating this, or similar valve design, are available today from a number of suppliers, each with a slight technical variation so as not to infringe each other's patents. There are also several novel inventions designed for insertion into the bottleneck, which have some of the nonspill benefits of the BV cap and which are intended for water coolers not fitted with the BV system, or for use with a crock.

10.4   Water categories for water coolers

Bottled water dispensed from water coolers may originate from many sources, but there are two major categories used (see also Chapter 3).

   (1) Natural Mineral Water and Spring Water. These are waters emanating from underground geological rock formations and are collected for bottling either from boreholes or emerging springs. Legislation in each

country often differentiates further between these two types of water and stipulates strict naming and labelling criteria based on natural source protection, total dissolved solids and the amount of processing the water may undergo prior to bottling.

(2) Purified water. This water may be from a groundwater source or from the municipal water supply and is produced by any one of the several methods of purification including reverse osmosis, distillation, de-ionisation and filtration. This water is also often treated by ultraviolet light or ozone for microbiological reasons and remineralised by the injection of soluble inorganic salts. Water processing is described in detail in Chapter 5.

10.5   The bottling process

The bottling of water in large reusable water bottles, such as the 18.9 l bottle, is different from bottling in small bottles, intended for single-trip use. The difference is not only related to the volume of water filled into the bottle, but more importantly, to the fact that the bottle is a reusable one that may be refilled over 50 times during its service life.

The polycarbonate bottle is now universally accepted as being the norm (see Section 10.3.2.2). On arrival of the empty bottles for refilling at a modern bottling plant, they will usually undergo some, or all, of the following processes (Fig. 10.9).

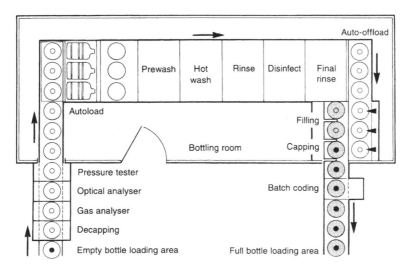

Fig. 10.9 Schematic diagram of a bottling plant.

(1) Visual inspection – manual determination of suitability for reuse: e.g. scratches, labels, etc.
(2) Decapping – removal of the cap, either manually or automatically
(3) Gas analysis – detecting the presence of organic contaminants in the bottle
(4) Optical analysis – detecting foreign bodies inside the bottle
(5) Pressure testing – detecting crack and pinhole leaks
(6) Autoloading – loading onto the wash conveyor
(7) Prewash – removal of surface grime and internal limescale deposits
(8) Hot wash – externally and inside of the bottle
(9) Rinse – externally and inside of the bottle
(10) Disinfection – sterilising from bacterial contamination
(11) Final rinse – product water rinse, externally and internally
(12) Auto-offloading – unloading onto the filling conveyor
(13) Filling – volumetric or timed filling of the bottle
(14) Capping – sealing of the bottle
(15) Coding – production batch number and best-by date: with ink jet/ stamping/paper label

Bottling equipment has undergone many stages of development as new technology evolved in the electronic and mechanical fields, often as a spin-off from other related industries. Traditionally, bottle washers and bottle fillers were separate items of equipment installed adjacent to each other in the bottling plant. As increased awareness of hygiene issues related to potential areas of contamination developed, the bottle filling operation was isolated and housed in its own 'clean-room' environment connected to the bottle washer by a conveyor belt. In time, the conveyor belt itself was seen as a potential site for contamination of the washed and sterilised empty bottles, and it was either enclosed in a tunnel or covered over to prevent foreign matter from falling into the bottles.

By 1990, equipment design had evolved to the point where one fully automated machine undertook all the operations from stages (6) to (15) previously detailed. Even the clean-room environment was built-in by creating a positive pressure area (PPA) for bottle filling. Today, this design of machinery prevails in most modern plants with up to 2000 bottles per hour operational capacity. Faster production rates usually still depend on the separate bottle washer and filler concept connected together by a covered PPA conveyor system.

An important aspect of the bottling of water in large reusable bottles for dispensing through water coolers is the frequent use in some markets of ozone gas for its disinfecting properties. With respect to bottling and the sterility of the final product, ozone's role is related to the final disinfection of the washed bottle in readiness to accept the product at the filling stage. Although the bottles have undergone many separate efficient washing and disinfection operations in their passage through the bottle washer, it is still possible for a single bacterium to have survived on the internal faces of the bottle. Once the bottle is filled with water, this one bacterium may multiply, thereby prejudicing the purity of the product. To overcome this, the water is ozonated just prior to bottling so that a residual of the ozone gas, in the range 0.2–0.4 mg/l, remains dissolved in the water at the time the bottle is capped. This residual ozone in the water is sufficient to sterilise the bottle's internal surfaces on direct prolonged contact. This method has been found to be so successful that sampling of the product water within 12 h of bottling does not yield any bacterial total viable count (TVC). A further application of the ozonated water is often made by spraying the inside faces of the bottle cap just prior to placing the cap on the bottle. The residual ozone in the water is sufficient to give a final sterilisation of the cap surfaces.

The maintenance of a sanitary condition for the water being bottled is paramount in the filling operation. To ensure that the washed bottles are not contaminated as they pass through the washer to the filling part of the equipment, they are kept in an enclosed area where air pressure is maintained at a higher level than that of the surrounding area (PPA), and where air has undergone filtration to remove submicron airborne bacteria, spores and other particles.

The chemical washing agents used in the initial stages of the bottling equipment must not be carried over in the empty bottles to the bottle filling area, neither should any be left in the bottle. For this reason, the bottles are loaded onto the wash equipment and travel through it in an inverted manner. The water jets in the wash and rinse sections spray upwards into the bottles and the water subsequently drains out easily. Externally, the bottles are washed and rinsed by jets mounted so that they spray the bottles from all directions, and with some equipment also by rotating brushes.

Periodically, bottles are manually removed from the bottling equipment at random, just prior to the filling stage, and a detergent and sanitiser 'carry over' test is performed to determine whether any residues are present. This verifies the efficacy of the final rinse stage of the equipment and ensures that an unadulterated product is being bottled.

Where once it was required to have different bottling equipment for different sizes of bottle, or many hours being spent in changeovers, modern equipment is extremely versatile and can normally fill all the standard bottle sizes and shapes. In the most advanced equipment, it is claimed that all that is required is the push of a button and the equipment adjusts to the new bottle dimensions, while the volumetric filler adjusts itself for the filling volume of the new bottle.

10.6 Handling, transportation and service

The bottled water cooler industry is a service industry throughout the majority of the Western world, since the product is delivered to the customer's premises, and in the commercial sector, the water cooler is maintained at the same time. This element of service is unique to water coolers, but is a necessary expense that can account for some 50% of the operating overheads of a bottled water cooler company. As such, the efficient handling and distribution of the filled bottles is a prerequisite to a well-run business.

As the large 18.9 l bottles come off the bottling conveyor, they are immediately transferred, either manually or by automated machinery, into bulk pallet containers or stillages holding between 16 and 40 bottles depending on the handling equipment and delivery vehicles in use. In this manner, they are then transported or stored in the bottling plant and warehouse. Where a central plant is bottling for several regionally located distribution depots, large trucks transport approximately 1000 bottles at a time to their destination, to be offloaded and stacked for distribution to the customers by smaller vans and trucks. The smaller vans delivering to customers, holding under 88 bottles, are usually fitted internally with fixed racks so that individual bottles are loaded manually, while larger-capacity trucks holding up to 256 bottles accept the preloaded bottle stillages directly and reduce the loading time required at the depot. The delivery vehicles are also often compartmentalised for the storage of water coolers and drinking cups that require delivery at the same time as the bottled water. The weight of the water cooler bottles can be as much as 24 kg for the 22.7 l bottles and, therefore, manual handling aids are required for the delivery person in the form of carrying handles and multi-bottle trolleys.

Once at the customer's premises, the bottles of water are normally stacked by the water cooler; however, large users may require delivery to a central storage area where the bottles may be stored in racks, or bottle stillages provided by the water cooler company. Building maintenance staff then attend to the water coolers and ensure that they always have an adequate supply of water. For the smaller user, there are bottle racks made of plastic or steel, holding from two to eight bottles to minimise the space required to hold sufficient bottles of water until the next scheduled visit.

The bottled water cooler industry differs from the bottled water industry that bottles water in small bottles in that the former is service-based whilst the latter is product-based. This difference is very significant to the organisational structure of the bottled water cooler company. Typically, water coolers are serviced with water supplies on a regular scheduled basis, which usually repeats every 14 days, but may be more or less frequent, dependent on water usage. The number of water coolers at each location may vary from just one in a small business or home to many hundreds in a modern high-rise office building. There are no hard and fast rules and a typical customer base will have a mixture of all types of customer, water cooler numbers and territories dictating that a suitable delivery pattern is established to meet all of the requirements.

Limiting factors on the daily bottle delivery capability of each delivery person will be dictated not only by these factors but also by the requirement to

service the water coolers at each customer location except in residential locations. This hygiene element involves the maintenance of the external surfaces of
the water cooler and dispensing taps using disinfecting sprays and wipes and
ensuring that drip trays are empty and free of undesirable matter.

As with all service industries, customer care is of paramount importance, and
while the regular delivery schedule satisfies the majority of the customer base,
there will arise situations where out-of-schedule water deliveries are required.
Similarly, water coolers will occasionally malfunction and it will be necessary to
exchange the appliance. Whatever the nature of the requirement, a speedy and
effective response is required from the service organisation. This dictates that in
addition to the regular delivery personnel, a support function of service personnel – both in-office and externally based – is required.

Bottled water delivery personnel are required to deliver a level of service in
excess of that normally expected of other delivery personnel, e.g. in parcel
delivery. This results from the intrusive nature of their job. It is not a doorstep
delivery service, but one in which they are required to enter the customer's
premises, remove the empty bottles, replace these with full ones, in some cases
supply cups, and maintain the hygiene of the water cooler.

10.7   Hygiene

The bottled water cooler industry's history owes its success to providing consumers with contaminant-free, good-tasting drinking water. While modern
source protection technology and state-of-the-art water technology can guarantee a safe product at the time of bottling, there are opportunities for the bottled
water in the bottle to become contaminated on its journey from the bottling
plant to the distribution depot and then at the customer's premises.

While under the control of the bottler and distributor, bottled water in small
bottles and in the large water cooler bottles requires the same careful handling
and storage. Both normally fall within the regulations that apply to all food
products. These regulations also extend to the delivery vehicles while the bottled
water is in transit on its way to the customer. Once at the customer's premises, its
storage should be treated with the same care, since external contamination of the
bottles may still occur prior to placement on the water cooler.

Unlike bottled water in small bottles which once opened is normally consumed within a very short period of time, the water cooler provides approximately 130–150 cup servings from each bottle. Depending on its situation, this
may take several hours, days or weeks. As such, control of the quality of
water in the bottle while it is on the water cooler and in the internal cooling
reservoir is of paramount importance. Similarly, the dispensing taps have a role
to play in the protection of the water at the time of dispensing. Some of the
technology employed by water cooler manufacturers to protect the water as it

passes through the water cooler – from the bottle to the cup – is detailed in Section 10.3.

Prior to installation of a new water cooler at the customer's premises, it must be flushed through and sanitised to ensure that any foreign matter is washed out and any bacteria that may have been introduced to the water contact surfaces during manufacture or subsequent storage are removed. Having been sanitised, the water cooler should be repackaged in such a manner that it is protected from any contamination during delivery and subsequent installation at the customer's premises.

Where water coolers utilise removable plastic reservoirs, or nonremovable stainless steel, it is considered good practice to sanitise these periodically with a proprietary disinfection solution. This eliminates the biofilm caused by natural harmless bacteria present in water and which normally establishes itself on contact surfaces within a short period of time. The same applies to the dispensing taps.

Similarly, particularly in warm and humid conditions, some unprocessed waters exhibit a tendency to turn green after prolonged periods of storage or exposure to bright sunlight. This 'greening' is caused by single-celled organisms known as chlorophytes, a simple form of algae. This is not thought to be of any significance to human health; however, their presence in water is unsightly and will also increase the organic content of the water, thus providing a food source for bacteria, which can impart a taste taint. The presence of chlorophytes in bottled water and water coolers is therefore undesirable, and measures must be taken to eliminate the cause through hygienic operations at the time of bottling and by good hygienic management of water coolers.

Many bottled water cooler companies offer their customers a water cooler sanitisation service on a quarterly or six-monthly basis at an additional cost over and above the normal water cooler rental price. This entails removal or sanitisation of the internal reservoir and taps, as well as the external cleaning of the water cooler with a bactericidal agent. In the UK, the Bottled Water Cooler Association mandates its members to undertake a quarterly sanitisation of rented water coolers. This practice has now spread to other European countries.

A considerable number of spring waters and natural mineral waters have a high mineral content, in particular, of calcium and magnesium bicarbonate. Such water is referred to as being 'hard', and when it is cooled these minerals may come out of solution and deposit on the walls of the cooling reservoir, creating potential sites for bacteria to colonise. Where the water cooler additionally has a hot water facility, this problem is more acute since the formation of the mineral deposit in the hot tank will not only reduce the efficiency of the heating coil but will reduce flow substantially. In time, all flow from the tap will be blocked. In such instances, the water coolers should be subject to regular descaling with a weak organic acid to ensure that the mineral deposits are removed.

10.8 Trade associations

The International Bottled Water Association (IBWA), now renamed International Bottled Water Association (United States), based in Alexandria, Virginia, USA, is a trade association founded in 1958 and has been a major influence on the development and regulation of the bottled water industry in the USA and worldwide. Since the historical evolution of the water cooler has been from that continent, this organisation has played a major role in the water cooler industry.

As well as the Environmental Protection Agency (EPA) and Food and Drug Administration (FDA) standards and regulations governing the production and packaging of bottled water in the USA, the IBWA has established additional standards for its members. A major instrument in the control of bottled water quality is the IBWA Model Bottled Water Regulations, which are not only applicable to its members but have also been used as the basis for bottled water regulations in much of the USA. This model code incorporates:

- Definitions/classifications of water by source
- Good manufacturing practices and operational requirements
- Source water monitoring
- Finished product monitoring
- Labelling requirements

It is also a condition of IBWA membership that bottlers be subject to an unannounced annual inspection programme by a third-party inspection organisation. The IBWA bottling plant inspection programme comprises over 65 items of compliance of which over 20% are related to the management of Critical Control Points (CCPs): these are items deemed essential to maintain production of a safe product. To obtain and retain membership, a minimum score as set by the IBWA is required, which must also exclude any CCP failures.

Up to the middle of the 1990s, IBWA operated outside of the USA through five international chapters: Europe, Middle East, Asia and Australia, Latin America, Canada, and each chapter disseminated information on behalf of the IBWA to its regional members. Today, these chapters have evolved into separate multinational regional trade associations, each focusing on its specific geographical area, but maintaining the criteria and standards originally set by the IBWA. These associations, though now independent, are represented and communicated through the International Council of Bottled Waters Associations (ICBWA).

The IBWA, European Bottled Water Cooler Association (EBWA) and Asia Bottled Water Association (ABWA), each holds an annual convention and trade show that attracts manufacturers and representatives from the bottled water industry worldwide and acts as a platform to display and view the latest state-of-the-art bottled water equipment and technology. Concurrently, educational seminars promote a better understanding of good manufacturing practices, product quality–related and legislative issues.

The international trade associations also play a vital role in liaising with regional government and both regional and international regulatory agencies to enhance understanding of bottled water and water coolers and some of the features that are unique to water dispensed through them.

Acknowledgements

I wish to acknowledge the invaluable assistance of Mr Donald Lovell, Elkay Manufacturing Co, Oakbrook, Illinois, USA, for historical and technical information; Mr Robert Hanby, Oasis Corp, Columbus, Ohio, USA; Mr Henry R. Hiddell, Hidell Eyster International, Hingham, Massachusetts, USA, for background to water cooler developments in the mid and far east; and Mr Richard Hall, Zenith International, Bath, England, for statistical information on the development of water coolers in Europe.

I would also like to acknowledge the Asia Bottled Water Association, ABWA (www.asiabwa.org), Australasian Bottled Water Institute Inc., ABWI (www.bottledwater.org), European Bottled Water Cooler Association, EBWA (www.ebwa.org), Canadian Bottled Water Association, CBWA (www.cbwa-bottledwater.org), Latin American Bottled Water Association, LABWA (www.labwa.org) and the International Council of Bottled Waters Associations, ICBWA (www.icbwa.org).

11 Third-party auditing of bottled water operations

Bob Tanner

11.1 Introduction

'It's only water', repeated the man who should have known better. The third-party audit report was not complimentary and had drawn attention to a number of items of nonconformance, but he could not comprehend their significance, or importance to his job – bottling drinking water. Fortunately, this dismissive attitude of a water bottling plant manager some ten years ago is not typical of the industry today as standards have improved significantly, and it is now very rare to find such uncommitted management. Third-party auditing companies must take some of the credit for that progress.

The responsibilities of water bottling plant managers have increased progressively in recent years, and ensuring product safety as well as quality is now a universally high priority. In addition, many water bottlers are moving to a 24 h, seven-day week, adding further challenge. Like other sectors in the international food and beverage industry, bottled water companies have not been immune from the crisis of confidence in food safety, shown in recent years by the public. Although there has never been a documented illness outbreak from bottled water, a number of well-publicised incidents of foodborne illness around the world, and other food safety concerns, are resulting in changes to strengthen food law and food law enforcement, foremost of which has been the introduction of Hazard Analysis and Critical Control Points (HACCP). In some parts of the world, particularly Western Europe, the law now requires all bottled water companies to have an HACCP-based food safety system in place, but whether mandatory or not, there is no doubt that HACCP is an invaluable food safety tool for all food and beverage companies.

The terrible events of 9/11 and the resultant 'war on terrorism' alerted us all to the dangers of the modern world and raised concerns about product safety and plant security. As elsewhere, water bottling plant operators were urged to review their own situation to ensure high levels of security and protection, particularly of raw water sources, water storage areas and water treatment and bottling plants. The World Health Organization (WHO) issued timely advice in this regard. WHO also published more general advice on water safety planning in the *WHO Drinking Water Guidelines*, 3rd edn, and, although mainly directed to the water supply industry, the advice has value for the bottled water sector.

The bottled water industry itself continues to change and evolve, as it responds to market forces and consumer expectations. The rapid and continuing growth of the Home and Office Delivery (HOD) sector – the so-called water cooler business – has been spectacular in Europe in the last decade and has resulted in numerous mergers and acquisitions. Some of the largest producers of bottled water have entered the HOD market bringing additional experience and expertise to that sector.

All these factors have broadened the role and value of the third-party auditing company and only a minority of bottled water companies today operates without the benefit of third-party oversight. With their independence, knowledge and expertise, third-party auditing companies can give confidence to bottled water plant managers and provide valuable and cost-effective risk management services. This fact has been recognised by many bottled water industry associations who require annual third-party audits as a condition of membership, from their appointed third-party auditing companies.

11.2   Conduct of audits

'Firm but fair' is the preferred method of conducting audits, and a proverbial arm around the shoulder of the plant manager, rather than a finger in the face, is likely to be more effective. To gain maximum advantage from the periodic visit of the independent auditor, the audit should be conducted in a spirit of openness and honesty, and, wherever possible, the auditor should be accompanied by a manager who not only knows the operation intimately and can therefore respond to the auditor's questions, but can also gain most from the auditor's expertise.

In order to provide greatest value, audits are normally unannounced so that the auditor sees things on a typical working day without special preparations having been made to enhance the audit results or to cover up aspects likely to be criticised. However, a good auditor should see through such cosmetic improvements, and, with the experience of many previous audits, will know the trouble spots, typical problems and areas likely to require particular attention. A good auditor should quickly establish credibility with the plant manager without which the whole process will lose value. Having found the problems the manager is already aware of, the auditor may go further and identify other matters requiring early attention, perhaps to pre-empt further problems.

The first or initial plant audit is normally made by prior arrangement to provide greater convenience to the company. This ensures that the right people are available and that adequate time has been allocated. Time must be allowed for a pre-audit meeting to provide an opportunity for the auditor to explain how the audit will be conducted and how the 'checklist' and audit report will be completed, and to answer questions. At the completion of the initial audit, further time must be taken to discuss the auditor's findings with assembled plant managers. There should be clear understanding of the reasons for the items cited by the auditor and of the required corrective action. This clarity is essential if repeat failures are to be avoided. Very rarely are items of noncompliance the result of a deliberate act or omission, but are mostly due to ignorance or oversight. Knowledge of the precise causes will always help to prevent recurrence. Although the audit is advisory, and not intended to replicate a regulatory inspection, issues should not be trivialised, and the auditor should make it clear what could happen during a regulatory enforcement visit.

The success of the audit will, to a large extent, be dependent on the competence of the auditor. The better auditing companies take particular care while recruiting auditors and will have training programmes in place to ensure that their auditors achieve and maintain the required standards. They will also have peer review procedures and will conduct regular 'check rides' in the interests of consistency and completeness in the provision of high-quality audits. Questions relating to auditor competency are legitimate concerns of companies receiving audits.

11.3   Setting the criteria for the audit

A bottled water audit is essentially an independent assessment of compliance, or conformance, with an agreed set of standards, regulations or other relevant criteria, and it is essential that all parties agree the criteria on which the audit is to be based well before the auditor arrives to begin the audit. There should be no surprises on audit day, and no confusion or doubts about precisely why the audit is taking place. To avoid any confusion, the audit criteria should be described in the contract for services, which should be agreed and signed between the third-party auditing company and the client whose premises are to be audited, well in advance of the audit. The criteria for audit can vary widely but the more common are described below:

- The company's own HACCP-based food safety system
- The requirements of the company's main customers
- The membership requirements of a trade association
- The legal requirements of the local food enforcement authority
- The legal requirements of the country or countries in which the client's bottled water products are sold
- A combination of two or more of the above

The most common type of audit for the bottled water industry is the bottling plant audit, which includes an audit of the raw water source and extraction arrangements, if existing, as well as the water bottling operation. The extent of water treatment processes to be audited depends on the type of water being bottled. For example, there will be practically no water treatment processes in a plant bottling Natural Mineral Water (NMW), as defined. In this case, focus would be on the company's implementation of a suitable code of practice.

A number of trade associations in Europe serving the HOD companies have adopted auditing programmes for those member companies that distribute bottled water as well as water vending machines – the so-called water coolers. Their 'distributor auditing' programmes involve third-party auditors in assessing all aspects of the service including cooler sanitation procedures, staff training, record keeping and recall programmes.

Many bottled water companies are registered under the ISO-9000 series of standards, verifying that the company's quality management system is in compliance with those standards. The third-party auditor may also be an ISO-9000 assessor, or lead assessor, and the audit may therefore include an ISO-9000 reassessment, in addition to the food safety–related criteria mentioned above. NSF International has a unique HACCP-9000 programme, which brings together the food safety–related criteria of HACCP with the quality management requirements of ISO-9000. Other combinations, or inclusions, are possible, such as ISO-14000, the Environmental Management System standard.

A competent third-party organisation can bundle related services together for the benefit of clients. This so-called one-stop shop has obvious advantages; however, there may be disadvantages to an integrated management system if each element is audited in its own right.

11.4   The bottling plant audit

Having agreed the criteria, whether or not the precise day of the audit is by prior arrangement, the audit will start with a brief on-site meeting at which the auditor explains the auditing plan. Certain documents will be asked for: e.g. the HACCP plan; a recall plan; authority to extract water, if applicable; sampling and testing reports, including laboratory results; cleaning and maintenance schedules; and staff training records. Some of these documents may be reviewed before the audit starts and others may be requested for review at the conclusion of the plant inspection. The process flow diagram in the HACCP plan will be particularly useful if the auditor is making a first visit to the plant. The auditor will lead the audit, rather than be led, and will expect to be accompanied by the plant manager, production manager or a responsible person familiar with the whole operation. (This may in fact be the quality manager.) The audit will then commence and will 'follow the flow' of water from source to final product dispatch.

11.4.1   Source
Having established the client's right to extract raw water, the auditor will want to visit the point of extraction, whether it is a well, spring or any other source of raw

water. Whilst most sources are underground, occasionally a surface water source is used. Unlike underground sources, surface waters, such as rivers or lakes, may be heavily polluted and the auditor will be looking at the methods of water purification in the bottling plant, as well as extraction arrangements at the source.

Some companies using underground water have determined the age of their water using carbon-dating techniques, sometimes with surprising results. Some underground waters have been found to be many years old. As a general rule, the longer the water has remained in the ground, filtering through the different geological strata, the purer it will be in microbiological terms. It will also have acquired a distinctive flavour due to the absorption of mineral salts during its time underground.

The auditor will determine whether the catchment area and the extraction point are free from sources of potential pollution. In particular, it will be the auditor's duty to establish, often on the basis of hydrogeological and hydrological information supplied by the client, that there is no risk of contamination from nearby habitation or industrial activities. Intensive agricultural practices (with their inevitable use of agrochemicals) can pose a risk, and roads and railways may also be a source of contamination if herbicides and pesticides are used on them. For underground sources, the point of emergence should be protected from obvious sources of pollution such as groundwater infiltration or flooding, and ideally should be housed in a small building to protect it from dust, insects, birds and animals. Access should be limited to a small number of authorised keyholders and the entrance should be equipped with alarm and, if in an isolated location, covered by closed-circuit television (CCTV) cameras. Where the 'source' of bottled water is the local piped supply of municipal or other private water, the auditor will ask to see an invoice as proof and will later review submitted test results.

11.4.2 Exterior of bottling plant

The areas around the bottling plant should be well maintained and free from uncontrolled vegetation and other sources of contamination. (Though this should not preclude the site from being well cared for with suitably controlled planting schemes.) Equipment and material stored around the building should not provide harbourage for rodents or other vermin and litter should not be allowed to accumulate. However, consideration needs to be given to storage of materials for recycling, awaiting collection. Ponding of water must be avoided. Wind-blown spores and moulds are less well-understood sources of possible contamination resulting from unkempt exteriors of bottling plants. In some locations, special measures will have to be in place to control the risk of dust or sand from blowing into the bottling plant. The experienced auditor will assess the likely risks from the plant's environment, even if not present at the time of the audit, and will note these in the audit report for the attention of plant management.

11.4.3   Plant construction and design

Plant construction, design and layout are important considerations when designing a bottling plant for the production of high-quality and safe bottled water products. It is not essential for the building to be brand new, purpose-built to achieve that objective, and even converted barns or cowsheds have (following extensive modification) been the origin of efficient bottled water premises.

The third-party auditor will quickly assess the suitability of the buildings for bottled water production and will have to regard the functionality of the various rooms, their possible adverse effects on the process and the integrity of the product. 'Is there enough space' is a primary consideration. All too often, particularly in the water cooler industry, a company starts small but grows quickly and expands its operation with the result that premises are overcrowded, space is limited and there is inadequate room to facilitate normal operations such as routine cleaning, servicing and pest control. The auditor assesses the construction and condition of the walls, floors, ceiling and roof, remembering that these are food rooms and must meet, at the very least, the requirements of food hygiene law: are all internal surfaces clean and in good repair, with smooth easily cleanable finishes? Anything less will be cited by the auditor. A review of the cleaning and maintenance schedule, and service records, will help to reveal the extent of compliance.

The bottling room, containing the filling machine, is the most critical area, perhaps for obvious reasons. This is where the product water is exposed to the air and where bottles are open and vulnerable to contamination from final rinse until capping. Is the bottling room, or specifically the filler, physically separated from other plant operations and storage areas? From domestic household activities, like washing machines for clothes? In smaller plants – particularly for the HOD market – bottle washing, sanitising, rinsing, filling and capping may all take place in one small machine, the so-called monobloc. The filling and capping area is contained in a cabinet, placed over a conveyor and equipped with an air pressure system. Filtered air is blown into the cabinet to maintain positive air pressure. The auditor will pay particular attention to this area and will check to determine whether the air filter is clean and in good condition; there is no surplus lubricant or other source of potential contamination; and that the equipment is clean, well maintained and working correctly.

The bottling room should be self-contained, with self-closing doors and any windows fixed shut at all times. It should have tight-fitting walls and ceiling, and where conveyors pass through walls, opening should only be large enough to allow product to pass through, and should be closable. (Though these may never be closed, if the plant is running continuously.)

High-quality lighting and ventilation systems are also important in a bottling plant. Lighting levels of at least 50 footcandles must be maintained in the filling room and other inspection points, toilets, etc., and a minimum of 20 footcandles maintained in other areas, including product storage. The auditor will carry a light meter to determine the adequacy of lighting levels and will make judgement about the adequacy of artificial lighting in plants likely to operate after dark, even during daylight auditing. Lighting units should be suitable for the purpose, kept clean and be easily cleanable. Light bulbs or tubes positioned over the filling area or over open bottles or other water-contact materials should be safety-type, shatterproof, or be otherwise protected from risk of breakage contamination.

Ventilation systems must be adequate for the purpose and effectively remove steam and prevent condensation build-up on surfaces. Maintaining a positive air pressure in the filling area will help to minimise the risk of dust and other debris. All air movement is thus away from the critical filling point. The bottle filling room should also be under positive air pressure, but less so than the immediate filling area. The auditor will conduct simple tests to determine the direction and adequacy of airflow and will make judgement about the condensation risk, whatever the timing of the audit. Ventilation systems must be maintained in a clean condition at all times and air filters must be regularly inspected and changed as necessary, and records kept of this action. Above all, it is vital to ensure that ventilation filters are appropriate for the function intended. For example, air being supplied to filling rooms and other open bottle areas must be appropriately controlled and in some cases (depending on the external environment) HEPA-filtered (high-efficiency particulate arrestor) to eliminate the risk of airborne contamination.

Sterile or ozonated air may be used to air-rinse bottles immediately prior to filling on larger bottling lines. Fixtures, ducts and pipework should be positioned so as to avoid dripping or condensation contaminating the product. Source and product water must be conveyed or stored in a sealed piping system, without leaks or other risk of contamination. Storage tanks must meet the same criteria. Care is needed to prevent tank implosion. Treatment for the air under atmospheric pressure can be provided to maintain the integrity of water. Tank filters, whether wet- or dry-type, must fully protect the stored water and records must be maintained showing regular inspection, conditions found and any remedial action taken. The use of wet-type filters in areas where freezing is a risk is obviously unwise and there have been incidents of serious tank implosions in freezing temperatures.

Bottle washers can be a source of serious contamination, if not functioning correctly. Not only will bottle washing, sanitising and rinsing be inadequate but the washer itself may contaminate bottles because, for example, detergent and sanitiser are not being changed frequently enough, according to suppliers' guidance. Placing heavily contaminated bottles – particularly 'green' bottles – in a bottle washing machine must be avoided.

11.4.4 Hygiene measures and controls

The auditor's next task is to look in-depth at all those other matters that may constitute poor, or even unacceptable, hygienic practice, i.e. that may pose an avoidable risk to the product. A distinction is made between the water intended to be bottled, i.e product water and water intended only for operational use in the plant, i.e. for cleaning, washing, flushing toilets, bottle washing, etc., which is called operations water. However, the criterion for operations water sourcing is exactly the same as for raw water intended as product water, i.e. from an approved, properly located, protected, operated and accessible source and is safe, of sanitary quality, conforming at all times with applicable laws and regulations. Operations water must therefore be of potable quality, i.e. drinking water quality. It is not uncommon to find the raw water source also being used for operations water, and this is acceptable provided it receives the same pre-treatment required to render it suitable for drinking. However, although only one pipe may be used to convey water from source to plant, it is very important that a nonreturn valve is fitted at the point where the operations water line leaves the source water line. This prevents the possibility of (polluted) operations water flowing back into the product line and eventually entering the bottle.

The auditor will make a careful inspection of water flows and plumbing and will study sampling and testing records to be certain that the system has been installed correctly. Later additions or alterations can sometimes present a hazard. Based on the type of water being produced, and the plant location, the auditor will be aware of the regulatory requirements for annual and periodic testing of source and product water – for chemical, microbiological and, perhaps, radiological parameters – and will be checking to ensure full compliance. The regulatory requirements of the intended country or countries for sale of bottled water products will be taken into full account in this determination.

'People make the difference' is an old adage, but it is not generally recognised that the single biggest risk to bottled water production is people. The plant will not run without people, no matter how automated it may be, and a high degree of training and supervision, personal cleanliness and conscientiousness are necessary to avoid problems. The fewer the people who have to be in close proximity to the product in process, the easier it will be, so access to some areas must be strictly limited. Personnel involved in cleaning, maintenance and repair activities must demonstrate the same level of care and attention while working in these critical areas, even when the plant or line is not in production.

Hand-washing facilities, comprising hand washbasins, with hot and cold water, soap and hand-sanitising solution, nail brush and hand-drying facilities, must be conveniently positioned to aid frequent use, particularly by workers in the filling room.

There must be washbasins adjacent to the toilets with clear signs reminding staff to wash their hands after use. Disposable paper towels with regular, suitable disposal, or hot air dryers should be supplied for hand drying. Cloth roller towels should be avoided. In addition to washbasins, wall-mounted dispensers, containing a suitable sanitising product may be located in the bottling area. Lockers and lunchrooms, where provided, should be physically separated from plant operations and must be clean at all times. Dustbins should be provided and equipped with foot-operated, close-fitting covers to avoid resoiling hands.

11.4.5 Plant operations

The auditor will ask to see written and recorded procedures for the cleaning and sanitising/disinfection of the facility, with particular emphasis on the filling area and the equipment therein. The auditor will also make a careful inspection to determine the state of cleanliness of the plant and equipment. He or she will look for evidence to make sure that what is written is actually done. All caps and seals should be properly stored in closed containers, in a clean and dry place. They should be examined before use, and handled, dispensed and used in a safe and sanitary manner. Unused caps should not be kept in the cap hopper or in the capping machine during periods of shutdown. Conveyor lines of open bottles should be covered to protect the bottles from risk of contamination. This is of particular importance in the case of glass bottles, as fragments of flying glass from the occasional, and inevitable, bursting bottle can travel surprising distances.

Ideally, the auditor will prefer to see bottling in process, as he or she must be assured that it takes place in an acceptable way. This can be accomplished even during shutdown by examining the bottling line in detail and checking the condition of the rinser (if used), the filling machine, the capper and the labeller. The equipment should be free from scale, rust or flaking paint, and there should be no leaking lubricant from motors suspended above the line or from other machinery. Only toxic materials essential for maintaining the sanitary condition of the plant and its equipment, or for use in the facility's laboratory, may be used or stored in the plant. All such items must be clearly labelled and controlled in use and storage.

The use of pesticides, while necessary to prevent access of vermin to the facility, must never pose a threat to the product. Similarly, electronic flytraps should be installed to prevent risk of flying insects falling into product containers.

11.4.6 Equipment and procedures

Equipment in bottling plants must be suitable for the intended use and be of such material that it can be thoroughly cleaned by normal cleaning methods. The design must preclude the risk of product adulteration by metal fragments, lubricants or contaminated water.

All product water contact surfaces must comply with the appropriate legal requirements. The auditor will ask for documentary evidence of this and will want to see the material suppliers' statements testifying to this compliance.

Source and product water must be conveyed or stored in a sealed piping system under pressure, free from leaks and without risk of contamination. Storage tanks must meet the same criteria except that storage under atmospheric pressure is acceptable. Tank filters, whether wet- or dry-type, must fully protect the stored water, and records must be maintained showing regular inspection and periodic replacement.

Where water-dispensing equipment, e.g. water coolers, is refurbished within the facility, servicing should take place outside the bottle filling area to avoid risk to the product. A purpose-made room should be provided and equipped to enable this important work to be done effectively and efficiently. New units, and serviced units ready for reuse, should be stored properly to prevent recontamination.

11.4.7   Process and controls

Water treatment methods must, of course, accomplish the intended purpose and this will form part of the auditor's enquiry. Detailed records should be kept of the dates on which the treatment equipment was inspected, of the conditions found and the action taken to attend to any problems. The performance and effectiveness of treatment equipment must be reassessed and recorded regularly. Treatment methods, and treatment equipment, must preclude risk of product contamination or adulteration.

Where returnable bottles are used, washed containers must be checked regularly for the possibility of caustic (or non-caustic) carryover from the washing machine, and the machine itself must be checked to ensure optimal performance. Here again, the importance of recording all actions cannot be overstressed. Records should be kept of the concentration of sanitising agents used in the washing machine, and of the time the agent was in contact with surfaces being sanitised. The volume and type of bottled water products produced, the date of production, the lot codes used and the distribution to depots, wholesalers and shippers should also be recorded.

The filling, capping and sealing processes must be monitored. Large modern plants today have electronic equipment to do this, while many plants still rely on the human eye. It is very important to verify if each container is filled to at least the volume of water stated on the label. Controls should be in place to ensure this, in addition to the adjustment on the filling machine. The threat of legal action for selling under-volume is as great as, if not greater than, the risk of a food hygiene offence and, in these cases, the 'due diligence' defence is not available. Each bottle or unit container must be identified by a production code, which may be 'open' (meaning that the consumer can read it and determine the actual date of production) or 'closed' (meaning that, though the consumer can clearly read the text, the date can only be deciphered by the bottler).

The auditor will also ask whether the company has a documented recall programme for use in the unlikely event of a product recall. Many companies have not made adequate provision for this 'insurance policy', yet the consequences of not having one in time of need can be disastrous. Finally, the auditor will check the company's sampling programme. Samples may be taken for internal quality control purposes only, or for legal compliance. Product water samples should be taken after treatment prior to bottling to ensure uniformity and effectiveness of treatment. The analytical methods used, as with all tests, must be approved by the government agency having jurisdiction. Representative bacteriological samples should be taken daily of each type of product, and chemical and physical samples should be analysed annually for each product water. This frequency does vary in some countries, as does the required frequency for radiological testing of product water. All sampling and testing records should be retained for at least two years.

11.4.8 Personnel

As previously mentioned, staff supervision is essential and the overall hygiene of the bottling plant should be under the supervision of one individual, whose responsibilities will include personnel supervision. Plant personnel must be adequately trained, and preferably qualified, in basic food hygiene so that they fully understand what they are responsible for and how to avoid risks. Training and 'qualifying' employees in the food industry also has an additional benefit in that it can raise the morale of the individual significantly. This may be the only 'qualification' they have received and the company's investment in their training and education will pay dividends.

The auditor will need to know what arrangements are in place for employees to notify the company when they suffer from a disease or illness that could be transmissible to the product. Naturally, it is important that the employee does not suffer financial consequences from this notification or he or she will keep quiet and work as normal, yet pose a risk to the product and the company.

Personnel practices will be assessed by the auditor, who will want to see that clean protective clothing is worn in the filling areas; that hair restraints are worn; that intricate jewellery, such as rings, brooches, studs, pins, clips and other adornments are not worn; that hand-washing practices are acceptable; and that a high degree of personal cleanliness is evident. Tobacco must not be used in any form and there should be no eating at the place of work.

11.5 Conclusion of audit and follow-up actions

Having completed the physical inspection of the bottling plant and source, and having reviewed appropriate records and other documents, the auditor will then spend time compiling the audit report, prior to the exit interview with appropriate plant management. At that interview, the auditor will explain the findings of the audit and describe any items of noncompliance with the agreed criteria on which the audit was based. These will require plant management to prepare their proposed 'corrective actions' to resolve noncompliances which should be submitted to the auditor within a reasonable time, typically 30 days after the audit, depending on their severity. (Some items may require immediate attention to avoid more serious consequences.)

The auditor may use a scoring system to help the client understand how the plant is performing against the agreed criteria. Typically, this will be a percentage score and each item on the auditor's checklist or audit report form will have a numerical value and points will be deducted based on severity or extent or noncompliance. Repeat items from previous audits may receive heavier penalty. Items of noncompliance with the company's own HACCP plan, particularly Critical Control Points (CCPs), will be of particular concern and the plant may be said to have failed the audit in such circumstances.

The auditor may not necessarily be a consultant but will be prepared to answer questions relating to the audit report or provide clarification if needed. On receipt of acceptable corrective action proposals, the auditor will confirm the acceptance and, typically, will plan to verify that corrective actions have been taken during the next audit. Audits are usually made at annual intervals, unless more frequent audits are required by the client, or if, in rare circumstances, the poor conditions found at the first audit warrant an earlier revisit.

Third-party audits can provide significant added value to bottled water companies. They assist with risk management responsibilities and help to provide 'due diligence' defence in times of need. Giving an audit score encourages the company to improve from year to year, but achieving a high score should not be the sole motivation. The assurance of a competent, experienced and internationally respected, independent, third-party auditing company that a bottling plant has achieved and is maintaining the highest standards possible must be the overriding objective.

12 Microbiology of natural mineral waters

Henri Leclerc and Milton S. da Costa

12.1 Introduction

The virtues of mineral waters have been expounded by poets and novelists from the times of ancient Greeks and Romans. Since then, numerous establishments have been built in Europe, where people could seek treatment for several diseases from the reputedly beneficial health properties of these waters. It was, therefore, natural that a business based on the bottling and transport of the mineral waters slowly began to take hold in Europe.

The mineral water market has increased dramatically since the 1970s because of the introduction of polyvinyl chloride (PVC) bottles for water. European regulations were instituted to control bottling and the marketing of Natural Mineral Water (Directives 80/777/EEC and 96/70/EC of the European Parliament and of the Council). These Directives were a compromise between the 'Southern European' who considered the beneficial action of the water to health and the 'Northern European' concept based on the mineral content and/or the presence of gas imparting a pleasant taste to the water.

Natural Mineral Water is defined as a microbiologically wholesome water, originating from an underground water table or deposit and emerging from a spring tapped at one or more natural, or borehole, sources. Natural Mineral Water can be clearly distinguished from ordinary drinking water by its nature, which is characterized by the mineral content, trace element components or other constituents; perhaps by certain beneficial health effects; and by the fact that it has not been treated, preserving the original qualities of the underground water, which has been protected from pollution.

These characteristics, which may make Natural Nineral Water beneficial to health, have to be studied extensively in the context of geological, hydrological, physical, chemical and physicochemical, microbiological, and if necessary, pharmacological, physiological and clinical characteristics. The composition, temperature and other essential characteristics of Natural Mineral Water must remain stable over time, within the limits of natural fluctuations, to show that the aquifer is stable and is not affected by surface alterations.

Natural Mineral Waters, like all subterranean waters, contain a natural bacterial flora. The presence of the normal flora of mineral water has given rise to a number of questions and debates about its effect on health, primarily because this flora is not yet very well characterized. However, Natural Mineral Waters cannot be subjected to any type of disinfection that modifies or eliminates their

biological components; therefore, they always contain the bacteria that are primarily a natural component of these waters.

In this review we discuss the results that have been published, in the last 20
years, in the area of mineral water microbiology and, as the reader will become
aware, show that there is a continuing search for new information on groundwater microbiology and microbial ecology.

12.2   Groundwater habitat

Before the 1970s, the study of life in groundwater habitats was relatively limited.
In the 1970s, however, it became increasingly obvious that certain waste disposal
practices were contaminating subsurface environments with effects on groundwater quality. This led to the current interest in the study of these environments.
There has also been an increasing interest in demonstrating that a variety of
shallow (Balkwill & Ghiorse, 1985; Madsen & Ghiorse, 1993) and deep environments (Balkwill, 1989; Frederikson & Phelps, 1997) contain substantial numbers
of viable microorganisms, and in using these microorganisms to degrade potential pollutants, i.e. in bioremediation. Subsurface microbiological research to
study microbial community structure, microbial activities and geochemical properties of groundwater environments has progressed with the development of
aseptic sampling techniques (Chapelle, 1993; Madsen & Ghiorse, 1993; Wilson,
1995; Fredrickson & Phelps, 1997).

The ecology of microorganisms is concerned with the relationships between
different species and between microorganisms and the environment. The basic
unit of ecology is the community or biocenosis. Both the biotic components – the
community – and the abiotic, physicochemical components make up the ecosystem. The major focus of ecological research is to determine how communities are
structured, how species in the community interact with each other and how
communities interact with their physical surroundings. When speaking of a
particular part of the community, i.e. microorganisms, the term environment
or habitat is often used to designate both the physicochemical and the biotic
components of the ecosystem.

12.2.1   Physical component
The terrestrial subsurface is an important component of the geological landscape
through which water flows in its cycle between the atmosphere, soil, lakes,
streams and oceans. The most fundamental distinction between subsurface
hydrological environments is the difference between the saturated and unsaturated zones. The unsaturated zone is usually divided into three components (Fig.
12.1): (1) the soil zone, with the A, B and C horizons, generally 1 or 2 m thick,
which contains living roots and supports plant growth; (2) the intermediate zone,

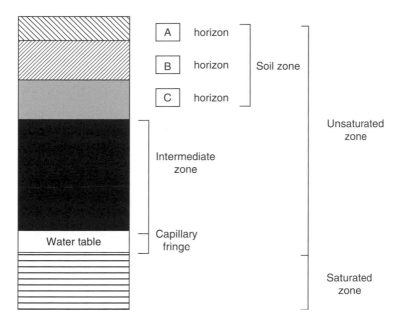

Fig. 12.1 The three components of the unsaturated zone.

which is the underlying material, consists of sediments or rocks that have not been exposed to extensive pedogenic (soil-forming) processes; and (3) the capillary fringe, which is the boundary between the unsaturated and the saturated zones. This boundary will fluctuate according to seasons, rainfall or pumping rates, and this will induce chemical oxidation–reduction reactions. The unsaturated zone is characterized by pore spaces that are incompletely filled with water. Any pore space that is not filled with water contains gas, and the capillary forces between water and sediment particles prevent water from flowing to wells. Thus, wells drilled into the unsaturated zone do not yield appreciable quantities of water.

For saturated environments, a rigorous distinction between local, intermediate and regional flow systems, related to the topography of recharge and discharge areas, has been long recognized by hydrologists. According to Toth (1963):

(1) A local system has its recharge area at a topographic high and its discharge area at a topographic low that are located adjacent to each other.
(2) An intermediate system is where recharge and discharge areas are separated by one or more topographic highs.
(3) A regional system is where the recharge area occupies the water divide and the discharge area occurs at the bottom of the basin.

There is an important difference between Toth's classification that depends only on the distribution of recharge and discharge into the hydrological system and the early empirical classification, related to depth as deep and shallow. Thus, Toth's classification can be more universally applied than the empirical classification (Chapelle, 1993). These three different hydrological settings are similar to the zonation that had been described empirically by Norvatov and Popov (1961) and that includes an upper zone of active flow strongly influenced by local precipitation events, a medium zone of deeper flow only moderately affected by local precipitation events and a lower zone of relatively stagnant water unaffected by local precipitation.

In a hydrogeological sense, groundwater refers to water that is easily extractable from saturated, highly permeable geological strata known as aquifers (Davis & De Wiest, 1966; Freeze & Cherry, 1979). However, to microorganisms living in microhabitats, all available forms of water may be important. Therefore, a broader definition of groundwater that includes capillary water, water vapour and water within aquifer formations was defined by Madsen and Ghiorse (1993). The groundwater habitat includes part of the unsaturated zone that may contain significant amounts of biologically available water. Also, unsaturated zones may be saturated transiently during recharge events and they may influence both the chemistry and microbiology of the saturated zone. Thus 'groundwater refers to all subsurface water found beneath the A and B soil horizons that is available to sustain and influence microbial life in the terrestrial subsurface' (Madsen & Ghiorse, 1993).

12.2.2   Chemical component

Within a given environment, such as groundwater environment, most microbial processes are consistent with oxidation–reduction reactions, which can be viewed as the microbial food chain. A compound that is an oxidized end product for one microbe may be a reduced substrate for another. By sequential coupling of microbial processes, virtually all the energy that is biologically available in a given substrate, or a group of substrates, will be extracted by the microbial population. The electron source, or donor, for the oxidation–reduction couple is organic carbon and is probably the most limiting substrate in groundwater systems. According to Ghiorse and Wilson (1988), allochthonous dissolved organic carbon (DOC), most of which is the recalcitrant fraction of organic substances, might govern subsurface microbial metabolism. This fraction arises from surface plant material (humic substances composed of polyphenolic subunits), delivered through hydrological recharge and groundwater flow.

The most commonly available terminal electron acceptors represent various degrees of oxidation–reduction potentials (Table 12.1). They are inorganic compounds that are relatively stable in water. Oxygen is the electron acceptor that provides the greatest energy yield; in the presence of oxygen, aerobic metabolism dominates. However, aquifers normally constitute a closed environment, and

Table 12.1 General characteristics of the electron acceptors most commonly found in groundwater

Electron acceptor	Reduced product	Level in groundwater (mmol/l)
O_2	H_2O	0–0.4
NO_3^-	N_2	0–20
Mn(IV)	Mn(II)	Low
Fe(III)	Fe(II)	Low
SO_4^{2-}	S^{2-}	0–15
CO_2	CH_4	0–4

oxygen availability is generally limiting. During aquifer recharge, oxygen rates are fixed via equilibrium with the atmosphere in the unsaturated zone or the capillary fringe. Any additional oxygen entering along a flow path within the aquifer can be considered as insignifiant (Smith, 1997). When oxygen is present, it is the primary electron acceptor, not only because it has the highest reduction potential but also because it inhibits anaerobic processes such as iron and sulfate reduction, and methanogenesis.

The carbon, oxygen and hydrogen cycles are driven either by solar energy through photosynthetic fixation of inorganic carbon or, in turn, by chemical energy with the oxidation of organic carbon to CO_2:

$$CO_2 + H_2O \xrightleftharpoons[\text{aerobic respiration}]{\text{photosynthesis}} O_2 + CH_2O$$

In groundwater systems removed from photosynthesis, the truncation of the cycle leads to depletion of oxygen and the accumulation of CO_2 and CO_2-derived carbonate species. In the case of local flow systems, characterized by relatively close interaction with the surface, most of the CO_2 generated by oxidation of DOC is readily exchanged with soil gases and, ultimately, liberated into the atmosphere. In contrast, the accumulation of CO_2 associated with intermediate or regional flow systems leads to the formation of effervescent mineral springs. The occurrence of CO_2-charged waters at the springs is thus an inevitable manifestation of carbon oxidation under anaerobic conditions. Many naturally carbonated mineral waters get their CO_2 from magma; this can be shown by isotopic analysis.

After oxygen, nitrate is the next most favorable electron acceptor in subsurface environments. Oxygen represses the formation of nitrate and nitrite reductases. If the enzymes have already been induced and the cells are subsequently exposed to aerobic conditions, oxygen will compete with nitrate for electrons delivered by the respiratory chain, and will inhibit the function of the nitrate-reducing system. Through the nitrogen cycle, nitrate produced by oxidation of ammonium (at or near the soil surface) can percolate vertically down to the groundwater system. If the aquifer is aerobic, nitrogen cycling truncates at this point and nitrate accumulates in groundwater. As the groundwater becomes anaerobic, as frequently occurs along the flow path, the nitrogen cycle continues through denitrification and nitrate does not accumulate in the aquifer.

Iron, manganese and sulfate are naturally abundant in many underground systems. Fe(III) occurs most frequently as insoluble oxides and hydroxides, and as coatings on mineral grains with various degrees of crystallinity and structure. Mn(IV), like Fe(III), may be present in large quantities in an aquifer, which may lead to water containing Mn(II) and Fe(II). Sulfate can be a minor groundwater constituent or an abundant reservoir from sulfate-bearing minerals. Finally, when all other electron acceptors have been depleted, CO_2 becomes the main terminal electron acceptor, being reduced to methane during methanogenesis. In this scenario, however, a considerable supply of degradable organic carbon must be available for methanogenesis to take place.

12.2.3 Biological component: source of microflora

The colonization of subsurface habitats has generated interest and speculation among geologists and microbiologists. Comparisons of microbial communities within vertical profiles extending from surface soil to subsurface sediments have shown that the bacteria found in different geological strata can be morphologically and physiologically distinct, but they do not reveal the developmental history of the subsurface microflora. Several hypotheses have been discussed by Madsen and Ghiorse (1993). It is feasible that the organisms and their descendants may have remained with the sediments ever since initial colonization during surface deposition. Such communities would be cut off from surface influences, perhaps for millions of years and they could have evolved unique phenotypes. Most recent studies have shown that the physiological and morphological characteristics of aerobic heterotrophic bacteria from different deep geological strata and different sites vary extensively with depth, probably in response to physical and chemical differences among geological formations. It can, therefore, be concluded that the microbial community at each depth in aquifer sediments is very diverse.

However, it is equally feasible that the stratified distribution of subsurface microorganisms could have been caused by vertical and horizontal colonization patterns of waterborne soil microorganisms via hydrological flow. Virtually all of the experimental evidence to explain the origin of subsurface microorganisms addresses their transport from surface environments. Laboratory studies suggest that cell surface properties affect transport of microorganisms through soil (Gannon et al., 1991a), and that mineralogy and/or solution chemistry influence attachment of microorganisms to aquifer material (Scholl et al., 1990; Gannon et al., 1991b). Factors such as motility, chemotaxis and metabolism can govern the penetration of bacterial cells into columns packed with sand (Reynolds et al.,

1989). Other factors such as sorption kinetics, association between cells and fine particles, and cell surface hydrophobicity, as governed by low-nutrient conditions, are major factors for the dispersal of subsurface bacteria (Lindqvist & Bengtsson, 1991). When examining samples of groundwater pumped from aquifers, a third possible source of microorganisms is direct colonization from the surface (Madsen & Ghiorse, 1993).

The drilling process, therefore, can effectively inoculate any aquifer system. For this reason, sediment samples could be preferred to well (borehole) water samples in performing subsurface microbiological investigations. Indeed, the major chemical and physical factors that govern bacterial abundances in soil, such as available organic carbon, nitrogen, phosphorus, sulfur, pH, temperature, light and biological factors, are strongly modified in the subsurface along the hydrological flow paths. Therefore, the distribution patterns of microorganisms, especially bacteria that colonize the groundwater and surface soils, are inevitably distinct.

12.2.4 Limits of microbiological studies

To deal with bacterial abundance and distribution in subsurface environments it is necessary to use specific methods for sampling sediments. The major difficulty in studying these environments is their relative inaccessibility. In most cases the groundwater samples are collected through pumping from the aquifer. Unfortunately, there are several potential problems with such a method. The first is that the drilling process can contaminate the environment under study, and it may be difficult to know whether the bacteria recovered are autochthonous or introduced by drilling. A second problem is the fact that a new environment is created by drilling, and the alteration of the physical conditions may significantly affect microbial processes in the vicinity of the well. A third problem with sampling water from wells is that most microorganisms in the subsurface tend to be sediment-bound. Thus, the species and the types of free-living microorganisms may be significantly different from those attached to sediment surfaces.

The question of attached or unattached microbial community was raised by microbiologists during early enquiries into the nature and the distribution of the subsurface microflora. It is generally accepted that the majority of bacteria in most ecosystems are attached to surfaces and are not suspended in the aqueous phase as are planktonic bacteria (Costerton & Colwell, 1979; Fletcher, 1979; Bar-Or, 1990; Savage & Fletcher, 1993; Wimpenny et al., 1993). Major differences between sediment-bound and unattached bacteria were documented by Kölbel-Boelke et al. (1988) and Hirsch and Rades-Rohkohl (1983, 1988, 1990), and this information suggests that unattached and attached groundwater communities are different. Nevertheless, they probably represent overlapping populations of a dynamic community, and it is likely that a constant exchange of cells exists between sessile and free communities. The most important advances in this respect were made by Kölbel-Boelke et al. (1988), who compared a water-bound bacterial community to a sediment-bound community from the same aquifer. There appeared to be important physiological differences in all of the communities, but there were also similarities. It is therefore assumed that water pumped directly from the aquifer through a borehole can conveniently be sampled for microbiological analysis.

It is important also to consider that the data documenting microbiological communities in the subsurface environments are closely dependent on the methods used for study. Microorganisms can only be observed directly by light or electron microscopy. Fluorochrome (acridine orange (AO) or 4',6-diamidino-2-phenylindole (DAPI)) analysis of sediments revealed that microorganisms were neither uniformly distributed in the sediment nor capable of dividing in situ. Transmission electron microscopy also showed that about two-thirds of the bacterial cells had gram-positive cell walls (Ghiorse & Balkwill, 1983; Wilson et al., 1983). This observation is a surprising result because culture methods indicate a preponderance of gram-negative cells.

Culture methods have been the mainstay of microbiology. In investigating the microbial ecology of the subsurface environments, they have demonstrated the phylogenetic diversity of microbial communities in sediments and ground-water samples (Kölbel-Boelke et al., 1988; Balkwill, 1989), and the preponderance of gram-negative bacteria in these environments (Balkwill & Ghiorse, 1985). The most significant disadvantage of culture methods is the selectivity introduced by the media used. Many of the bacteria present in groundwater are not able to grow under the culture conditions. Other bacteria grow so slowly or give only pinpoint colonies that they cease to grow upon repeated transfers into fresh media. The principal conclusion is that the information on diversity and community structure may be grossly biased. For example, caulobacters, as typical freshwater bacteria, should be regarded probably as second only to pseudomonads in breadth of distribution and numbers. *Caulobacter* spp. are oligotrophic, i.e. well adapted to conditions of nutrient limitation, and are probably widely distributed in shallow aquifer systems. Members of this genus have, in fact, been isolated from groundwater by Hirsch and Rades-Rohkohl (1983) using specific enrichment procedures, but many strains may not be recovered using these, and other, growth conditions.

The plate counts can dramatically underestimate the total number of bacteria present in samples taken from groundwater environments. In the late 1970s, several easily performed noncultural methods (Zimmerman et al., 1978; Kogure et al., 1979) showed that many of these nonculturable cells were indeed viable and able to metabolize actively, i.e. they are viable but nonculturable (VBNC). A variety of methods may be employed to determine the ratio of a cell population that are in a VBNC state. Most studies compare VBNC cells to total direct counts, as determined by fluorescence microscopy staining of nucleic acid with acridine orange (Daley & Hobbie, 1975; Hobbie et al., 1977) or DAPI (Porter & Feig, 1980; Hoff, 1988).

The two most commonly used methods to determine VBNCs are the direct viable count (DVC) originally reported by Kogure *et al.* (1979) and the method using *p*-iodonitrotetrazolium (INT) violet as electron acceptor. In the first procedure, the bacterial population under study is incubated for 6 h or more with small amounts of yeast extract and the deoxyribonucleic acid (DNA) synthesis inhibitor nalidixic acid. If viable, the cells are able to respond to the nutrient addition; they become significantly elongated but are unable to divide. With the second method, the reduction of the soluble INT by metabolizing cells leads to the formation of insoluble INT-formazan in the cell membrane. The newer redox dye 5-cyano-2, 3-ditolyl tetrazolium chloride (CTC) can be reduced to CTC-formazan, which fluoresces red and accumulates intracellularly.

Other techniques using fluorescently labeled monoclonal antibodies, or [^{3}H]thymidine-prelabeled cells have also been described (Oliver, 1993). Such methods showed that some bacteria, in response to certain environmental stresses, may lose their ability to grow on nonselective media, while remaining viable. They are of considerable interest to our understanding of microbial ecology and for detecting pathogens and indicator bacteria in subsurface environments.

To obtain a better understanding of different subsurface microorganisms, White *et al.* (1983) and Balkwill *et al.* (1988) have introduced biochemical marker techniques. Indeed, it is well known that different microorganisms exhibit significant biochemical differences in the fatty acids that constitute cell membranes and internal reserve granules. By direct extraction and characterization of these fatty acids from sediments, data concerning total biomass, cell membrane type, occurrence of eukaryotic cells and nutritional status of the microorganisms may be derived.

Another alternative to these methods is to evaluate the net effects of microbial processes rather than the species of organisms present. Aquifer materials placed into laboratory containers for measurement of microbial activity are referred to as microcosms. Microcosms have been used extensively for measuring CO_2 production (Chapelle *et al.*, 1988) or xenobiotic degradation (Wilson *et al.*, 1983). Another possibility is to use groundwater chemistry as an indicator of microbial processes by selecting a series of wells placed along the flow path and sampling their chemical parameters.

The goal of nucleic acid technology is the characterization of natural populations without the need to isolate or cultivate individual community members. In most characterizations based on these methods, there is a requirement for purification of nucleic acids from the microbial population because of contamination by humic materials, but the advantage of nucleic acids extraction is that bacterial populations are more numerous per unit volume than by using other methods. The advent of ribosomal ribonucleic acid (rRNA) sequence analysis has revolutionized bacterial phylogeny and might well revolutionize microbial ecology. When rRNA molecules are the target in hybridization studies, the potential sensitivity is greater because growing cells can contain 10^4 ribosomes per cell. The main approach in application of 16S rRNA gene technology is sequence analysis of 'clones' taken from the natural population. The advantage of the probing approach was shown by Giovannoni et al. (1988) using kingdom-specific probes targeted for conserved 16S rRNA sequences found in all living cells (universal probes).

By linking these probes with fluorescent dyes, phylogenetic diversity could be analyzed by epifluorescence microscopy and flow cytometry (Amann et al., 1990a, 1990b, 1995; Amann, 1995; Wallner et al., 1996). This new approach can reveal microbial community structures, and it is clear that the use of 16S rRNA gene technology may bring new information in environmental microbiology, especially in the case of microbial communities of Natural Mineral Waters (Wagner et al., 1993; Wallner et al., 1993, 1995). A limitation of the approach is that phenotypic information is not obtained. Moreover, it is not known whether the sequence was derived from living cells, dead cells or extracellular DNA.

12.2.5   Major microbiological groups

In contrast to surface freshwater, which may be influenced by suspended particles and their attached biota, most groundwater is interstitial, i.e. remaining within a matrix of minerals with variable chemistry, porosity and degree of saturation. Thus, in groundwater habitat, all life forms larger than microorganisms are excluded.

Madsen and Ghiorse (1993) have presented a generalized model for the relationship between geological stratigraphy and microbiological parameters (Fig. 12.2). In going across the A and B soil horizons into the C horizon, the bacterial abundance decreases in direct proportion with nutrient levels. The C soil horizon marks the beginning of the unsaturated subsurface zone with considerably fewer bacteria than in the B soil horizon. Below the C horizon, microbial abundance increases substantially at the water table and, just above, in the capillary fringe zone. It can be speculated that the interface zones between the unsaturated and saturated zones may be the sites of oxygen transport and nutrients recharge, especially in shallow unconfined aquifers (Madsen & Ghiorse, 1993). Thus, depth per se appears not to govern bacterial abundance and activity in the saturated zone. Rather, the abundance of bacteria appears to be related to hydrological, physical and geochemical properties of each stratum.

Within a given environment, the collective result of all microbial processes, most of which are oxidation–reduction reactions, is viewed as a microbial food chain. The process itself is an important indicator for determining the nature of the microbial community in a particular habitat. The food chain in aquifers is primarily heterotrophic, depending on dissolved organic materials. However, according to Madsen and Ghiorse (1993), aerobic chemolithotrophic food chains have not been described in the terrestrial subsurface. In the most simple cases, as in largely aerobic aquifers, the major bacterial groups are aerobic gram-negative

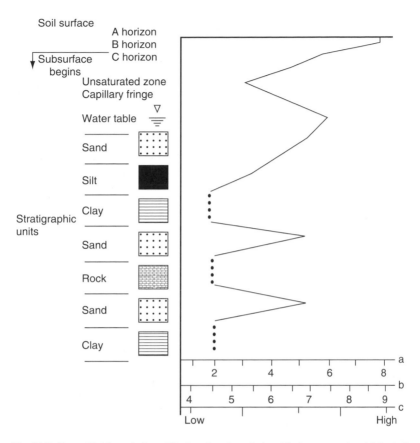

Fig. 12.2 Generalized vertical profile showing the relationship between microbiological parameters and stratigraphy. (Reproduced with permission from Madsen & Ghiorse 1993; published by Blackwell Scientific Publications.)

rods with cytochrome oxidase using oxygen as a terminal electron acceptor. Members of the genus *Pseudomonas* and related organisms appear to be particularly common in groundwater systems. Other groups such as *Caulobacter*, *Cytophaga*, *Flexibacter* and flavobacteria are also widely distributed. The most extensive characterization of bacterial distribution in a typical aerobic aquifer has been performed at the Savannah River Site in South Carolina (Balkwill, 1989; Frederikson *et al.*, 1989; Sinclair & Ghiorse, 1989). A major finding was that these sediments contained a significant number of protozoa ranging from below detection to as many as 10^3/gram of sediment. However, bacterial cell numbers were always higher, with total counts (AO) of 10^6–10^8/gram, while viable counts (colony-forming units (cfu)) of these samples showed large variations.

The three deepest units at the Savannah River Site function as intermediate flow systems but, unlike most intermediate aquifer systems, they are aerobic. Not suprisingly, the microbial community has an activity comparable to that of aerobic bacterial populations. In spite of this largely aerobic environment, anaerobic bacteria are also part of the microflora (Jones et al., 1989). The numbers of sulfate-reducing bacteria can attain up to 10^4 cfu/g, and there was also methane production in a number of zones.

Because an aquifer is water-saturated, the oxygen concentration of the bulk groundwater is a good indicator of its availability to free or attached bacteria. There is little likelihood that anaerobic microorganisms that are common in many soils are important in the groundwater environment. Only at very low oxygen concentrations, below approximately 30 μmol/l, is there a potential for nitrate competing with oxygen for available electrons. Thus, the presence of different nitrogen species in groundwater systems depends on the concentration of oxygen. Under aerobic conditions, nitrification may take place and nitrate accumulates. Under anaerobic conditions, nitrate will be converted to nitrite via denitrification, while nitrification is blocked. Denitrification is governed by organisms such as those belonging to the genera *Pseudomonas*, *Paracoccus*, *Thiobacillus* and *Bacillus*, among others.

It has long been recognized that microorganisms play an important role in iron cycling in groundwater systems. Fe(III) oxyhydroxides are reduced by several microorganisms, including the unclassified strain GS-15 and *Shewanella putrefaciens* by using Fe(III) as terminal electron acceptor (Chapelle, 1993). The oxidation of Fe(II) to Fe(III) at acidic pH can be mediated by representatives of the genera *Gallionella*, *Leptothrix* and *Thiobacillus*. When sulfate is an electron acceptor, it is reduced to sulfide, and many groups of anaerobic bacteria are capable of carrying out sulfate respiration. The presence of sulfate-reducing microorganisms in groundwater sediments can be easily observed. These may include species of sulfate-reducing genera such as *Desulfovibrio*, *Desulfotomaculum*, *Desulfobacter* and *Desulfonema*.

Intermediate flow systems generally correspond to confined aquifer systems with depths of less than 300 m. Microbial populations are largely dependent upon organic material present in sediments as a primary carbon source. These carbon constituents are often refractory, limiting the growth potential of microbes, and thus environmental conditions are typically oligotrophic. In these anaerobic aquifer systems, counts of total bacteria are in the order of 10^5–10^6 cells/g of sediments and those of viable bacteria are in the order of 10^4 cells/g (Chapelle, 1993).

Some major characteristics of microbial distribution in shallow and deep subsurface sites can be deduced from several observations.

(1) The population density of cyst-forming protozoa is low in shallow, pristine aquifer sediments and even lower in deep aquifer sediments;

their abundance is related to sediment texture, and the dominant populations are flagellates and amoebae.

(2) The abundance of actinomycetes and fungi in subsurface habitats appears modest and in the order of 10 cfu/g.

(3) Bacteria with simple life cycles are by far the most widely represented organisms in shallow or deep groundwater habitats. There are aerobic gram-negative bacteria of which *Pseudomonas* and related genera are particularly common; the total counts of bacteria range from 10^6 to 10^8/g, whereas plate counts show large variations.

(4) Some metabolic groups such as nitrate reducers, Fe(III)-reducers, sulfate reducers and methanogens can also be isolated from anaerobic aquifer systems.

12.2.6 *Nutrient limitations and starvation survival*

In shallow or deeper aquifers, the supply of readily utilizable carbon and energy sources may be extremely small. Only the most refractory organic material will survive the long and complex pathways through subsurface sediments. Thus, organic carbon is the most limiting nutrient in these aquifer systems. Low-nutrient environments, termed oligotrophic environments, primarily lack organic matter for the growth of heterotrophic bacteria.

Limitation or starvation for one or more nutrients is common in most bacteria in natural environments such as groundwater (Morita, 1997). Even bacteria such as the enterics that colonize the gut must be able to survive in another environment as they are transmitted from host to host. Thus, bacteria have evolved mechanisms to cope with a feast-fast mode of existence.

To confront nutrient limitation, bacteria may develop defence mechanisms to enhance their ability to survive periods of starvation. Some differentiating bacteria respond to starvation by marked alteration in their ultrastructure, producing spores or cysts. Spores are essentially dormant, waiting out lean periods to germinate as nutrients become available. Nondifferentiating bacteria respond more by an alteration of their physiology rather than developing resistant structural modifications. These organisms have an enhanced capacity for scavenging nutrients and possess many of the resistance characteristics of endospores (McCann *et al.*, 1992). The heterotrophic bacteria of groundwater habitats could have this type of survival strategy.

These cells respond to starvation by forming ultramicrocells that can pass through 0.2 μm membrane filters. The size difference between an exponentially growing cell and a starved cell is sometimes very large. For example, actively growing *Vibrio* cells may have a volume of 5.94 μm^3, while starved cells have a volume as small as 0.05 μm^3 (Moyer & Morita, 1989). This miniaturization is due to reductive cell division and a continuous size decrease by breakdown of internal macromolecules. These can be carbon storage polymers, such as glycogen or poly-béta-hydroxybutyrate, or constituent proteins and nucleotides. The energy needed for the stress response is derived from extensive degradation of internal macromolecules (Kjelleberg *et al.*, 1993).

Bacteria respond to specific nutrient limitation by two mechanisms: first, they produce transport systems with increased affinities for the nutrient most easily exploited; second, they express transport and metabolic systems for alternative nutrients. Thus, these bacteria may be able to escape starvation by more efficient scavenging of a preferred nutrient or by using another, relatively more abundant, source.

Evidence has been accumulating for years that bacteria subjected to nutrient starvation become more resistant to various environmental stresses. It is clear that the stress responses discussed earlier, involving enhanced scavenging capacity, are insufficient to ensure survival. This aspect has been studied rigorously by Matin and coworkers, who showed that, upon exposure to nutrient limitation, bacteria synthesized new proteins that increased their resistance to a number of stresses. This resistance failed to develop if synthesis of starvation proteins was prevented, and increased the longer the culture was allowed to synthesize the starvation proteins (Berg *et al.*, 1979, 1984; Reeve *et al.*, 1984a, 1984b; Matin & Harakeh, 1990). These additional proteins are often referred to as stress proteins. Different sets of genes are expressed and proteins are induced by different stresses. Well-characterized prokaryotic examples include the heat shock response (Neidhard *et al.*, 1984; Hightower, 1991), the SOS response (Walker, 1984), oxidative stress response (Morgan *et al.*, 1986), starvation response (Groat *et al.*, 1986; Matin *et al.*, 1989; Matin, 1990, 1991), response to anaerobiosis (Spector *et al.*, 1986) and responses to micropollutants (Blom *et al.*, 1992). The term stress proteins was introduced to recognize the more general nature of the response. It is most likely that some of them are related with the cell's catabolic activity/energy metabolism (Matin & Harakeh, 1990). A major unifying theme that emerged in the last decade is that stress proteins work together to regulate the response, to assemble/disassemble structures and to provide a molecular shuttle service for polypeptides by chaperoning (Hightower, 1991).

12.2.7   The viable but nonculturable state

Under certain conditions of metabolic stress such as starvation, bacterial cells may enter into a VBNC state. It has been realized for some time that plate counts can dramatically underestimate the total number of bacteria, determined by AO, present in samples taken from natural environments. In the late 1970s, several noncultural methods (Zimmerman *et al.*, 1978; Kogure *et al.*, 1979; Rodriguez *et al.*, 1992) were developed to determine cell viability. These studies led to the concept that some bacteria, in response to certain environmental stresses, may lose the ability to grow on media in which they are routinely cultured, while remaining viable. A bacterium in this VBNC state is defined by Oliver (1993) as 'a cell which can be demonstrated to be metabolically active, while being incapable of undergoing the sustained cellular division required for growth in or on a medium normally supporting growth of that cell'.

Several discoveries have shown that many bacterial cells respond to adverse environmental conditions, such as temperature, salinity (Xu et al., 1982) and nutrient deprivation, by entering the VBNC state (Baker et al., 1983). In this state, cells are reduced in size, become ovoid and cannot be grown by conventional bacteriological techniques (Oliver, 1993). Minicells or minibacteria have also been observed in some mineral waters (Oger et al., 1987). It has been speculated that the VBNC state represents an additional response to starvation displayed by bacteria for survival (Colwell et al., 1985; Roszack et al., 1987b), and many bacterial species, including pathogens, have been reported to enter this state under laboratory or field conditions (Oliver, 1993). This state could have important consequences with regard to ecology, epidemiology or pathogenesis, since potentially pathogenic VBNC cells could persist in the environment and regain growth capability and infectivity much later than vegetative cells.

The relationship between the starvation response and the VBNC response is complex, but it has been suggested that the VBNC state may be distinct from the starvation response for several motives (Walch & Colwell, 1994). A large number of environmental factors other than starvation, such as temperature, pH, salinity and osmotic pressure, may be involved in the induction of the VBNC state (Oliver, 1993). It is important to note that starved bacteria, after variable periods of time, respond rapidly to nutrients, while VBNC cells cannot grow on conventional bacteriological culture plates. A variety of methods have been described for determining the nonculturable state. The existence of a VBNC state, in response to natural environmental stress, has been observed more often than not with gram-negative bacteria representing members of the Enterobacteriaceae and Vibrionaceae including *Aeromonas* and some genera such as *Campylobacter*, *Helicobacter* and *Legionella* (Oliver, 1993). However, little is known about the VBNC state in most representative bacteria living in groundwater habitats.

12.3 Bottle habitat

Microbiological analysis of Natural Mineral Water at source has always revealed the presence of some bacteria that are capable of growth and that can form colonies in appropriate culture media. After bottling, the number of viable counts increases rapidly, attaining 10^4–10^5 cfu/ml within 3–7 days. Buttiaux and Boudier (1960) were the first to describe this phenomenon, which has been largely corroborated by others (Schmidt-Lorenz, 1974; Ducluzeau et al., 1976a, 1976b; Delabroise & Ducluzeau, 1974; De Felip et al., 1976; Schwaller & Schmidt-Lorenz, 1980, 1981a, 1981b, 1982; Warburton et al., 1986, 1992; Gonzalez et al., 1987; Oger et al., 1987; Bischofberger et al., 1990; Morais & da Costa, 1990; Ferreira et al., 1993; Warburton, 1993).

In Fig. 12.3, it can be seen that the colony counts of the water from five springs and from the mixed water derived from these springs are less than 1–4 cfu/ml, and in the storage tank and immediately after bottling they are, on average, only slightly higher. During storage at 20°C, bacterial populations increase in numbers, reaching a peak of more than 10^5 cfu/ml by the end of one week. During the next four weeks, the bacterial counts decrease slowly or remain fairly constant. At the end of the two years of storage, colony counts are still about 10^3 cfu/ml. The results of most studies (Yurdusef & Ducluzeau, 1985; Gonzalez et al., 1987; Morais & da Costa, 1990; Vachée et al., 1997) are broadly in agreement with the one reported by Bischofberger et al. (1990). Some of the results on colony counts from bottled still mineral waters are also described in Table 12.2.

These bacteria, capable of growing on simple organic compounds (principally carbohydrates, amino acids and peptides) found in culture media such as 'plate count agar' or R2A agar (Reasoner & Geldreich, 1979, 1985; Reasoner, 1990) are heterotrophic. They are also psychrotrophic because they can grow at temperatures as low as 5°C, and their maximum growth temperature is about 35°C (Mossel et al., 1995). Therefore, incubation at 20°C for three days has prevailed in the monitoring of plate counts of mineral waters. Furthermore, these bacteria do not have growth factor requirements such as vitamins, amino acids or nucleotides and are, therefore, prototrophic, in contrast to auxotrophic bacteria which require many of these growth factors.

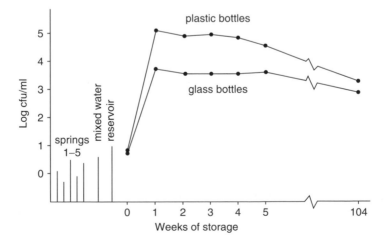

Fig. 12.3 Heterotrophic plate counts (HPC) of noncarbonated mineral water at five springs of a mineral water source, the mixed water from these five springs, storage reservoir, and plastic and glass bottles immediately after bottling and during storage for 104 weeks. After 14 days of incubation at 20°C cfu were determined by the membrane filtration method on 1 : 10 diluted plate count agar. (Reproduced with permission from Bischofberger et al., 1990).

Table 12.2 HPC at 22°C from bottled noncarbonated mineral waters sampled at retail outlets

Source	Number of samples examined	HPC (cfu/ml)					Reference
		$< 10^2$	$10^2\text{--}10^3$	$10^3\text{--}10^4$	$10^4\text{--}10^5$	$> 10^5$	
UK	44	18	11	18	36	16	Hunter *et al.* (1990)
France Portugal Belgium	23	26	4	34	34	—	Manaia *et al.* (1990)
France Italy Spain	50	2	8	56	30	4	Leclerc (1994); Vachée *et al.* (1997)

12.3.1 The bottle effect

Organic carbon should be expected to be the most limiting nutrient for growth and activity of bacteria. Many mineral waters contain 0.1–0.2 mg/l of dissolved organic carbon. By taking into account the level of organic carbon metabolism, it is theoretically possible to predict the number of culturable bacteria in a bottled mineral water. Carbon accounts for about 50% of the bacterial cell mass. The mass (dry weight) of bacterium is approximately 2×10^{-14} g. If, hypothetically, one admits the incorporation of the total available organic carbon into the bacterium, one will obtain for 0.1 mg/l of organic matter (1×10^{-4} g) a number of bacteria equal to $2[(1 \times 10^4)/(2 \times 10^{-14})] = 1.10^{10}/l$, i.e. 1×10^7/ml. These calculations indicate that other limiting factors may exist (e.g. phosphorus). Indeed, as reported by Van der Kooij *et al.* (1982a, 1982b; Van der Kooij, 1990), only a small portion of the organic carbon in mineral water is expected to be available as a source of carbon and energy to microorganisms.

Placing samples into containers terminates the exchange between cells, nutrients, and metabolites with the in situ surrounding environment. Compressed air is used at virtually all stages of the water bottling process. The microbiological quality of the process air must be of a very high standard. On the other hand, the complexed organic matter present in low concentration can be dramatically modified through bottling, under the influence of increasing temperature and oxygenation. A high increase in cell numbers due to the bottle effect was reported for the first time by Fred *et al.* (1924). ZoBell and Anderson (1936) described the bottle effect (originally named the volume effect) observing that both the number of bacteria and their metabolic activity were proportional to the surface area/volume ratio of the flask in which the seawater was stored. The greater the surface area in relation to the volume of water, the more rapidly growth of bacteria takes place; hence small containers provide considerably more surface area (Bischofberger *et al.*, 1990; Morita, 1997).

The explanation for this is that nutrients present in low concentration are absorbed and concentrated onto the surface and, thus, can be more available for the bacteria. The similar increase in bacterial numbers occurs when underground surface waters are placed in a container (Heukelekian & Heller, 1940). Flask surface adsorption of organic matter is the basis for the adhesion of bacteria to solid surfaces as demonstrated in both the aquatic environment and in the laboratory, and because of the increased concentration, the nutrients are more available (Morita, 1997). It is also possible that many of the more labile compounds, unavailable at the subsurface environments being complexed to lignins, phenolics, or adsorbed to clay minerals, become biodegradable through sampling by interaction at the surfaces. Since a volume effect has been reported, the major portion of the microbial activity should be linked to the attached bacteria. To date, little experimental evidence has been presented to demonstrate an attachment of bacteria with the inner surfaces of the bottles of mineral water. Bischofberger et al. (1990) reported no visible colonization with PVC mineral water bottles, using scanning electron microscopy. Low levels of adhesion have been shown by Jones et al. (1999). Viable counts on the surface (polyethylene-terephthalate (PET) bottles and high-density polyethylene caps) ranged from 11 to 632 cfu/cm^2 representing only 0.03%–1.79% of the total viable counts in the 1.5 l bottles, depending on the brand examined. In contrast, within the studies of Jayasekara et al. (1999), who reported considerable variation between bottles for a given producer of water, up to 83% of the total microbial population within a bottle was found to be adhered to the bottle surface, representing a population of about 10^6–10^7 cfu scattered over the surface. However, the counts of the attached bacteria were insufficient to constitute a real biofilm as representing an interdependent community-based existence (Davey & O'Toole, 2000). The studies cited above are not all directly comparable because there are differences in sampling and methods used for viable cell numbers. In the mineral water bottle systems studied by Jones et al. (1999), surface roughness appeared to be most significant in determining adhesion, while surface hydrophobicity and electrostatic charge had no significant role.

12.3.2 Other factors influencing the plate count

There has been some debate on higher bacterial counts that are generally found in PVC compared with those in glass bottles (Baldini et al., 1973; Masson & Chavin, 1974; De Felip et al., 1976; Yurdusev & Ducluzeau, 1985; Oger et al., 1987; Bischofberger et al., 1990). According to these authors, the major cause of the lower colony counts of the same mineral water bottled in mechanically cleaned glass than in plastic bottles is due to the bacteriostatic effect of residual cleaning agents. Other possible causes for the different colony counts of the same water bottled in glass and in PVC may be due to the migration of organic nutrients from PVC, higher rates of diffusion of oxygen through PVC, or the rougher surface of plastic bottles promoting adhesion and colonization.

The storage temperature of the bottles has never been thoroughly studied. Usually the maximum bacterial density was observed in samples stored at room temperature (about 20°C). Nevertheless, storage at low temperatures, such as that of refrigeration (4–6°C), does not stop bacterial multiplication.

Photodegradation of dissolved organic matter is a common phenomenon. Thus, the exposure time of recalcitrant organic substances in water samples to daylight, and moreover to sunlight, may again stimulate the growth of micro-organisms since complexed substances, such as carbohydrates, fatty acids and amino acids, may become bioavailable. Photochemical processes generate low molecular mass, readily biodegradable molecules from high molecular mass humic complexes (Morita, 1997).

The count of cultivable bacteria that can be recovered from mineral waters also depends to a large degree on the culture methods used. Counting colonies of bacteria in rich nutrient media, as done in medical microbiology, has dominated past research for a long time. Fortunately from the 1970s to the 1980s, the concept of substrate shock (too many nutrients) has been successfully addressed (Morita, 1997). Thus, when low-nutrient medium such as R2A is used for environmental samples, a significant increase in viable counts can be generally observed compared to the use of regular-strength medium such as standard plate count agar (Reasoner & Geldreich, 1985). R2A agar incubated at 20°C has proven to be especially suitable. There is dramatically less or even no bacterial growth at 37°C compared to 20°C. It is possible that high plate counts at 37°C, such as staphylococci, coryneforms or Gram-positive bacilli, indicate alloch-thonous populations in the mineral water. These thermotrophic bacteria which can grow in mineral water at 42°C were below $10^3/1$ (Leclerc *et al.*, 1985). The choice of incubation time is probably the most important factor for isolating bacteria from bottled mineral water because many of these organisms are slow-growing. Thus, for species distribution studies it is important to incubate the cultures for longer periods (up to 14 days at 20°C has frequently been used) than for monitoring purposes (three days).

12.3.3 Growth or resuscitation

Bacteria living in groundwater systems are subject to constantly changing, and frequently stressful, conditions such as reduced temperature, fluctuation in pH, change of osmotic pressure, oxidative shock and nutrient limitation. In nutrient-poor groundwater systems, autochthonous bacteria survive for prolonged periods under conditions designated starvation survival by Morita (1982, 1987, 1993). They may lose the ability to grow on media in which they are routinely cultured, while remaining viable (VBNC). As mentioned earlier, it is now well established that plate counts of still mineral waters at the source, or immediately after bottling, range between about 1/ml increasing to 10^4–10^5/ml within 3–7 days after bottling. This population remains high and is generally stable for months. It remains unclear whether the ultimate large population of culturable bacteria in mineral water is due to resuscitation of a large number of nonculturable dormant (VBNC) cells present in the water source or in the bottling system, or is the result of cell division and growth of a few culturable cells initially present.

Resuscitation is defined here as a reversal of the metabolic and physiological processes that result in nonculturability, i.e. the restoration of the ability of the cells to be culturable in media normally supporting growth of the organisms. Resuscitation would appear to be essential to the VBNC state if this is truly a survival strategy. Whereas the nonculturable state may in some way protect the cell against one or more environmental stresses, resuscitation of the cell would allow it to compete actively in the environment. However, according to Bogosian et al. (1998), recovery of culturable cells from a population of nonculturable cells, via a process of resuscitation, can be confounded by the presence of a low level of culturable cells, which can grow in response to the addition of nutrients and give the illusion of resuscitation. The data of Oger et al. (1987) seem to demonstrate that the culturable population arising within one week of storage is derived from a large number of bacteria in a 'minicell' state but stainable by AO (approximately 10^3 cells/ml) and, therefore, from resuscitation processes. Ferreira et al. (1993) assume that such large populations of culturable bacteria are the result of growth from a very low number of organisms enumerated with ethidium bromide, initially present at the source and/or the bottling system.

Compared to cultivation-based methods, nucleic acid probes currently allow the taxonomically most precise and quantitative description of microbial community structures. Over the last decade, rRNA-targeted probes have become a handy tool for microbial ecologists (Amann & Ludwig, 2000). The fluorescence in situ hybridization (FISH) with rRNA-targeted probes helps to detect and identify bacteria even at a single cell level without prior cultivation and purification.This method had been optimized (Wallner et al., 1993; Amann et al., 1995) and applied to mineral water by combining it with membrane filtration (Fig. 12.4). Probes specific for the bacteria, and the alpha, beta and gamma subclass of proteobacteria were used. The applied fluorescent probes were targeted to rRNAs and the number of cells detected with fluorescently labeled rRNA-targeted probes was called 'probe active count'. The development of the bacterial community in an uncarbonated water sample in a PET bottle was monitored during nine days after bottling, using the FISH method and DNA staining with DAPI (W. Beisker, personal communication):

(1) As measured by AODC, the number of bacterial cells increased from 1000 ml/1 to 8×10^4 within seven days after PET bottling (Fig. 12.4), which is similar to the others studies (Oger et al., 1987; Bischofberger et al., 1990; Morais & da Costa, 1990; Ferreira et al., 1993; Vachée et al., 1997).

(2) As only 5% of total counts (DAPI) were detected on the first day by the bacteria-specific probe, the number of physiologically active bacteria

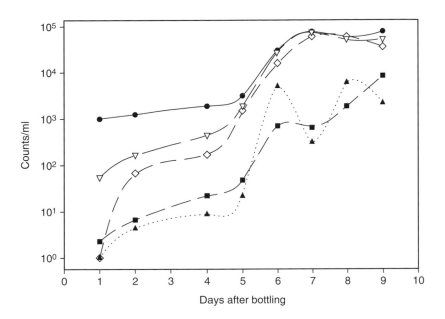

Fig. 12.4 Total counts and counts for active bacteria and bacterial subgroups in French uncarbonated NMW during nine days after bottling (PET bottles). Total counts (number of bacteria stained by DAPI) (●). Active bacteria (▽); active alpha proteobacteria (■); active beta proteobacteria (◇); active gamma proteobacteria (▲) (number of bacteria detected by respective Cy3-labeled rRNA-targeted probe (From W. Beisker, personal communication).

(viable and culturable) can be assumed to be significant while the plate count of still mineral water is generally in the order of a few cfu/ml in the R2A medium (about 1–5 cfu/ml). This portion increases slowly up to day 5, then rapidly between days 5 and 7. It appears that the increase of total count might essentially be due to physiological growing of active bacteria that have been detected by the eubacterial probe. These results suggest that the apparent resuscitation was merely due to the growth of the culturable cells from day 1.

(3) The appearance of biphasic growth or a double growth cycle (diauxie) is typical of media that contain mixtures of substances. The first substrate will induce the synthesis of those enzymes required for its utilization and at the same time will repress the synthesis of enzymes required for second substrate. The latter enzymes are only produced when all of the first substrate has been metabolized. However, recent results obtained for the kinetics of growth during 'mixed substrate growth' suggest that simultaneous utilization of carbon sources will result at the low substrate concentrations present in these systems, which allows growth at very low concentrations of individual carbon substrates (Kovavora-Kovar & Egli, 1998).

(4) The bacterial population in bottled mineral water is dominated by proteobacteria and beta proteobacteria were found to be the most abundant
group of detected bacteria (see Section 12.4).

12.3.4   Genetic diversity before and after bottling

Mineral water ecosystems, including those in aquifers, exhibit a high degree of
phenotypic and genetic microbial diversity that cannot always be supported by
species identification (Chapelle, 1993). Phenotypic characteristics that rely on
physiological activities have been shown to be less important for estimating
bacterial diversity than genetic characteristics, because many metabolic traits
may be induced or repressed by different environmental conditions. Restriction
fragment length polymorphism (RFLP) patterns of rDNA regions (ribotyping)
therefore constitute a more reliable method for assessing genetic diversity within
autochthonous bacterial associations of mineral water.

In an investigation on the fate of the bacterial flora at source before bottling
and upon bottling, Vachée et al. (1997) isolated 890 strains from five springs and
observed 378 distinct ribotypes. RFLP analysis detected a large number of
polymorphisms combined with unequivocal band resolution in all groups, but
particularly high in a set of isolates producing a fluorescent pigment (72 patterns
for 174 strains). Mineral water samples from the five springs were analyzed
during one year in order to assess whether the isolates specific for each spring
could be separated repeatedly throughout the period of storage. In the case of
spring A, 155 isolates provided 75 RFLP patterns. Indistinguishable or closely
related isolates were found 114 times in the samples examined and among these
103 (90%) isolates had been obtained from samples before and after bottling
(Table 12.3). The results from four springs are presented in Table 12.4. These
data suggest that, within the ground mineral water bacterial community, a high
percentage of indistinguishable or closely related isolates (identical ribopatterns)
is retained through the bottling system and storage.

12.4   Microbial community

Community structure is generally considered to be related to the types of
organisms present in an environment and to their relative proportions. As
discussed earlier, many approaches are used to determine community structure
for subsurface environments. For Natural Mineral Waters all the data have been
obtained, thus far, by culture methods. It must also be stressed that the study of
the microbial populations of mineral water will undergo important developments as new molecular biological and other highly technical approaches are
likely to be used to study this environment.

Noncarbonated mineral water contains very low numbers of culturable
bacteria at the source. However, after bottling, colony counts of more than

Table 12.3 Isolates from samples before and after bottling

Isolates	Samples[a]			Ribopattern
	I	II	III	
	Number of indistinguishable isolates[b]			
5	—	2	3	b2
3	1	1	1	c3
3	1	1	1	c4
4	2	—	2	c6
2	1	—	1	c7
17	9	5	3	m1
10	5	3	2	m2
3	1	1	1	m3
2	1	—	1	m4
5	1	3	1	m5
2	1	—	1	m6
2	1	—	1	m7
2	1	—	1	m8
2	1	—	1	m9
3	2	—	1	m10
3	2	—	1	m11
3	1	—	1	m12
2	1	—	1	m13
2	—	1	1	m14
2	—	1	1	m15
3	1	1	1	m16
4	3	—	1	m17
	Number of closely related isolates[b]			
4	2	1	1	m'1
2	—	1	1	m'2
3	1	1	1	m'3
3	1	1	1	m'4
2	1	—	1	m'5
3	1	1	1	m'6

[a] I, source; II, before bottling; III, after bottling.
[b] Isolates are designated genetically indistinguishable when their restriction patterns have the same numbers of bands and the corresponding bands are of the same apparent size. The ecological interpretation of these results is that the isolates are all considered to represent the same strain. Two isolates are considered to be closely related when their RFLP pattern differs by one band difference which is consistent with a single genetic event, i.e. a point mutation or an insertion or deletion of DNA.
(From Vachée et al., 1997.)

10^4 cfu/ml can be enumerated after one week of storage at 20°C. Most of the results on the diversity of microorganisms has been obtained from bottled mineral water, taken from the bottling plant or purchased at retail outlets, and stored for at least one week.

Table 12.4 Indistinguishable isolates (identical ribopatterns) recovered during storage (%)

	A	B	C	D
	155	133	141	173
Indistinguishable isolates recovered more than once	114 (74%)	96 (72%)	93 (70%)	94 (55%)
Indistinguishable isolates recovered before and after bottling	103 (90%)	50 (52%)	62 (65%)	61 (66%)

Source: From Vachée *et al.*, 1997.

12.4.1 Algae, fungi and protozoa

The microbiological investigations of groundwater systems indicate that pro-karyotes are, as discussed previously, the dominant microorganisms present and that eukaryotes might be absent altogether or present in low numbers. In subsurface environments, where photosynthesis does not occur, cyanobacteria and algae will not be found, unless the geological stratum is hydrologically connected to surface water. These organisms may be present in the form of cysts or other resistant states of development, and their presence in bottled water is most likely to be the result of contamination during the bottling process.

The occurrence of pathogenic protozoans or slightly thermophilic amebae has never been demonstrated in mineral waters (Dive *et al.*, 1979; Rivera *et al.*, 1981). The populations of fungi in groundwater habitats appear to be low, but not necessarily absent (Sinclair & Ghiorse, 1987, 1989; Sinclair *et al.*, 1990; Madsen, 1991). Thus, it is possible to isolate fungi from Natural Mineral Water before or after bottling. Because of the possible presence of fungi in groundwater, it is difficult to have accurate knowledge of the origin of these organisms in bottled mineral water (Fujikawa *et al.*, 1996).

12.4.2 Heterotrophic bacteria

Because photosynthesis is not possible in groundwater environments, the food chain is primarily heterotrophic, depending either on DOC from the hydro-logical flow path, or on organic compounds of sedimentary origin. The compos-ition of the bacterial flora that can be recovered depends largely on the culture techniques used and on the physicochemical parameters of the aquifers.

The heterotrophic plate count (HPC) was introduced by the Standard Methods of the USA (American Public Health Association, 1985). This method enumerates aerobic and facultative aerobic bacteria found in potable water that are capable of growth on simple organic compounds found in the culture medium, for a specific incubation period and at specific temperatures. This method has been applied successfully to bottled, noncarbonated mineral water, with some modifications (Bischofberger *et al.*, 1990). It is important, for example, to use the surface culture procedure (spread plate) that yields higher counts than the pour plate method (Schmidt-Lorenz, 1974; Schwaller & Schmidt-Lorenz, 1982). R2A, a low nutrient medium, devised by Reasoner and Geldreich (1985) has proved to be especially suitable, but the 1.10 diluted plate count agar can yield even higher colony counts after an adequate incubation period (Bischofberger *et al.*, 1990). An incubation temperature of 20°C has generally prevailed in the monitoring of drinking water. Most laboratories examining bottled mineral water find this temperature to be better suited than higher incubation temperatures for obtaining higher numbers of isolates from mineral water. However, the choice of incubation time is probably the most important factor for isolating bacteria from bottled mineral water because many of these organisms are slow-growing. Thus, for species distribution studies it is important to incubate the cultures for longer periods than for monitoring purposes (three days), and periods of up to 14 days at 20°C have frequently been used.

The vast majority of the heterotrophic bacteria isolated from Natural Mineral Waters can be classified in a restricted number of phylogenetic divisions. Prosthecate bacteria belonging to the alpha subclass of the proteobacteria, pseudomonads of the alpha, beta and gamma subclasses of the proteobacteria, members of the *Cytophaga–Flavobacterium–Bacteroides* (CFB) phylum, and gram-positive bacteria of the actinomycetes subclass are the most common bacteria isolated from bottled mineral water.

12.4.3 Prosthecate bacteria

Prosthecate bacteria *sensu stricto* are characterized by a cellular extension (appendage), designated a prostheca, i.e. continuous with the main body of the cell (Staley, 1968). Prosthecate bacteria are dimorphic, resulting, upon cell division, in the formation of two cells that are morphologically and behaviourally different from each other. One sibling is nonmotile and prosthecate, possessing at least one appendage. In natural habitats, this prosthecate cell is sessile because of an adhesive material associated with a cell pole or with the prostheca. The other sibling is flagellated, bearing (typically) one polar or subpolar flagellum, and is actively motile. This mode of reproduction has an important ecological function as a means of dispersing the population at each generation and thereby minimizing competition between siblings for nutrients (Poindexter, 1992). Thus, each normal reproductive event in these bacteria produces two siblings: one to grow, and one to go. This strategy is also consistent with the oligotrophic nature of these organisms, adapted to prolonged nutrient scarcity (Poindexter, 1981a). It is experimentally clear that these organisms are highly successful scavengers of very low concentrations of nutrients (Morgan & Dow, 1985), and it has been suggested that competitive advantages for an oligotrophic mode of existence are due to efficient uptake of nutrients and the possession of a high surface area-to-volume ratio (Hirsch, 1979; Jannash, 1979; Shilo, 1980; Poindexter, 1981b). As typical freshwater nonphototrophic bacteria, prosthecate bacteria such as

Caulobacter and *Hyphomicrobium* can be regarded as second only to pseudomonads in distribution and numbers (Lapteva, 1987).

The occurrence of prosthecate bacteria has rarely been reported in Natural Mineral Waters, but these bacteria have not usually been sought because of their special medium requirements. It is, nevertheless, interesting to note the predominence of appendaged and/or budding bacteria in all the springs examined in the survey by Gonzalez *et al.* (1987). *Caulobacter* was the most frequently isolated organism in both the bottled water and the water collected at source. A large appendaged and budding bacterium similar to *Hyphomicrobium* or *Hyphomonas* was also recovered from some samples. However, it may be that a low tolerance of decreased oxygen concentration could limit the occurrence of prosthecate bacteria in groundwater habitats.

Sheathed iron-related bacteria, isolated on a specific medium, were observed in all samples collected at one of the springs and in 76% of commercial samples (Gonzalez *et al.*, 1987).

12.4.4 *Pseudomonads, Acinetobacter, Alcaligenes*

The classification of gram-negative aerobic rods is becoming more and more complex owing to the creation of many new genera and the description of large numbers of species. The species of the genus *Pseudomonas* comprise a substantial proportion of the microflora of free-living saprophytes in soils, fresh water, groundwater, marine environments and many other natural habitats, especially plants. The general chracteristics of the genus *Pseudomonas* described by Palleroni (1984) are:

- straight or slightly curved, rod-shaped gram-negative cells;
- many species accumulate poly-b-hydroxybutyrate;
- motile by one or several polar flagella, rarely nonmotile;
- aerobic, having a strictly respiratory type of metabolism with oxygen as terminal electron acceptor;
- in some cases, nitrate can be used as an alternate electron acceptor;
- some species are facultatively chemolithotrophic and capable of using H_2 as energy source.

The genus *Pseudomonas* was subdivided by Palleroni *et al.* (1973) into five distantly related rRNA groups (Table 12.5). During the last decade there has been a considerable revision of the phylogenetic relationships of these organisms leading to the description of new groups within the proteobacteria (Fig. 12.5). The major changes in the nomenclature of the pseudomonads can be summarized as shown in Table 12.5. The most frequently isolated organisms from Natural Mineral Waters belong to the alpha, beta and gamma subclasses of the proteobacteria and especially to the genus *Pseudomonas* belonging to rRNA group I (Kersters *et al.*, 1996). The species of *Acinetobacter*, on the other hand,

Table 12.5 Major changes in the nomenclature of the pseudomonads isolated from Natural Mineral Waters

Previous name in the rRNA groups of Palleroni et al. (1973)	New name within the different subclasses of the proteobacteria	Reference
Group I (fluorescent)	**Gamma subclass**	
P. aeruginosa[a]	Same	Palleroni et al. (1973)
P. fluorescens	Same	Palleroni et al. (1973)
P. putida	Same	Palleroni et al. (1973)
P. syringae[b]	Same	Palleroni et al. (1973)
P. rhodesiae[b]	Same	Coroler et al. (1996)
P. veronii[b]	Same	Elomari et al. (1996)
P. monteilii[c]	Same	Elomari et al. (1996)
P. jessenii[b]	Same	Verhille et al. (1999a)
P. mandelii[b]	Same	Verhille et al. (1999a)
P. gessardii[n]	Same	Verhille et al. (1999b)
P. migulae[b]	Same	Verhille et al. (1999b)
P. libanensis[d]	Same	Dabboussi et al. (1999a)
P. cedrella[d]	Same	Dabboussi et al. (1999b)
P. orientalis[d]	Same	Dabboussi et al. (1999b)
P. brennerii[b]	Same	Baida et al. (2001)
P. grimontii[b]	Same	Baida et al. (2002)
Group I (nonfluorescent)		
P. stutzeri	Same	Palleroni et al. (1973)
P. alcaligenes	Same	Palleroni et al. (1973)
P. pseudoalcaligenes	Same	Palleroni et al. (1973)
Group II	**Beta subclass**	
P. cepacia	Burkholderia cepacia	Yabuuchi et al. (1992)
P. pickettii	Ralstonia pickettii	Yabuuchi et al. (1992, 1995)
P. solanacearum	R. solanacearum	Yabuuchi et al. (1992, 1995)
P. lemoignei	Burkholderia-Ralstonia	De Vos and De Ley (1983)
Group III	**Beta subclass**	
P. acidovorans	Comamonas acidovorans	Tamaoka et al. (1987)
P. terrigena	C. terrigena	De Vos et al. (1985)
P. testosteroni	C. testosteroni	Tamaoka et al. (1987)
P. delafieldii	Acidovorax delafieldii	Williams et al. (1990)
Group IV	**Alpha subclass**	
P. diminuta	Brevundimonas diminuta	Segers et al. (1994)
P. vesicularis	Brevundimonas vesicularis	Segers et al. (1994)
Group V	**Gamma-beta subclass**	
P. maltophiia (Xanthomonas)	Stenotrophomonas maltophilia	Palleroni and Bradbury (1993)
No group	**Alpha subclass**	
P. paucimobilis	Sphingomonas paucimobilis	De Vos et al. (1989) Yabuuchi et al. (1990)

[a] *Pseudomonas aeruginosa* is not a normal component of the microbial flora of Natural Mineral Waters.
[b] From Natural Mineral Waters.
[c] From clinical samples.
[d] From Lebanese springs.

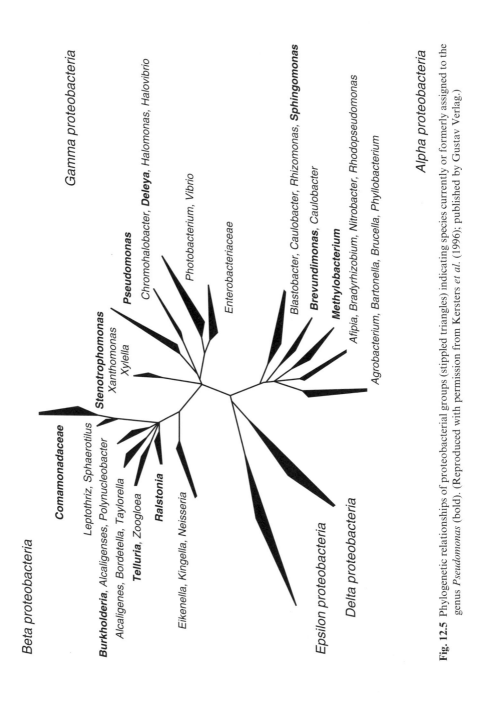

Fig. 12.5 Phylogenetic relationships of proteobacterial groups (stippled triangles) indicating species currently or formerly assigned to the genus *Pseudomonas* (bold). (Reproduced with permission from Kersters *et al.* (1996); published by Gustav Verlag.)

are nonmotile and oxidase negative, while *Alcaligenes* spp. are motile by peritrichous flagella. All species are aerobic, having a strictly respiratory type of metabolism with oxygen as the terminal electron acceptor.

The organisms most widely isolated from mineral water and representing major groups are shown in Table 12.6. These results were obtained in extensive studies by Schwaller and Schmidt-Lorenz (1981a, 1981b), Bischofberger *et al.* (1990), Manaia *et al.* (1990), Guillot and Leclerc (1993) and Vachée *et al.* (1997). It is difficult to compare the studies because of the different water types and sampling sites, techniques of isolation and identification and methods of colony selection. However, some features are common to all the studies. By far the most important members of the mineral water flora are fluorescent and nonfluorescent pseudomonad species. The genus *Pseudomonas*, now restricted to rRNA group I, encompasses some genuine *Pseudomonas* species that display genomic and phenotypic relationship to *Pseudomonas aeruginosa* (Moore *et al.*, 1996). In some instances, the strains producing fluorescent pigment constituted up to 50%

Table 12.6 Major groups of bacteria isolated from Natural Mineral Waters

Species, genus or group	Schwaller & Schmitz-Lorenz (1980)	Bischofberger *et al.* (1990)	Manaia *et al.* (1990)	Guillot & Leclerc (1993), Vachée *et al.* (1997)
Pseudomonas (fluorescent)[a]	++	++	++	++
Pseudomonas (nonfluorescent)[b]	++	+	++	+
Burkholderia cepacia	—	+	+	+
Ralstonia pickettii	+	—	+	+
Burkholderia-Ralstonia (*P. lemoigne*)	—	++	—	—
Comamonas acidovorans	+	—	++	+
Comamonas testosteroni	+	+	—	—
Acidovorax delafieldii	+	+	—	—
Stenotrophomonas maltophilia	—	+	+	+
Brevundimonas diminuta	—	—	—	+
Brevundimonas vesicularis	—	—	—	+
Sphingomonas paucimobilis	—	—	+	+
Acinetobacter	++	+	+	+
Alcaligenes	+	+	++	+
Cytophaga, Flexibacter, Flavobacterium	++	++	++	+
Arthrobacter, Corynebacterium	+	++	—	+

+, less than 10% of isolates; ++, between 10% and 50%; +++ more than 50%.
[a] *P. chlororaphis, P. fluorescens, P. putida, P. rhodesiae, P. veronii.*
[b] *P. stutzeri, P. alcaligenes.*

of all the isolates (Guillot & Leclerc, 1993). The strains of the genera *Acineto-bacter* and *Alcaligenes* were isolated in all studies in numbers that sometimes rivalled those of the genus *Pseudomonas*. In decreasing order of importance, species of *Comamonas*, *Burkholderia*, *Ralstonia* and *Stenotrophomonas* were also isolated, followed by species of *Sphingomonas*, *Acidovorax* and *Brevundimonas*.

In some studies, such as those of Guillot and Leclerc (1993) and Vachée *et al.* (1997), the unidentified isolates reached about 80%. These results are not surprising because of the large phenotypic and genotypic diversity of bacteria from groundwater and the large number of unclassified species in this environment. Furthermore, some common species are remarkably heterogeneous; this is the case with *P. fluorescens*, which can be subdivided by various criteria into subspecies or biovars. The adaptability of *Pseudomonas* and related bacteria makes them ideal candidates for colonizing groundwater systems where organic carbon compounds are largely limited to dissolved organic carbon leaching out of the soil zone above (local flow system) or present in the sediments as a primary carbon source (intermediate flow system).

12.4.5   *Cytophaga, Flavobacterium, Flexibacter*

It is not uncommon to observe yellow, orange or brick-red colonies on agar plated with mineral water samples. Sometimes these form films that may cover the whole plate within a few days. In other cases, the colonies expand slowly or remain more or less compact. In few instances, rhizoid growth is also observed. Many of the strains produce flexirubin-type pigments in addition to carotenoids (Reichenbach, 1992). These bacteria generally belong to the genera *Cytophaga*, *Flavobacterium* and *Flexibacter*, which are regularly isolated from most Natural Mineral Waters, sometimes even as dominant populations (Schwaller & Schmitz-Lorenz, 1981a, 1981b; Quevedo-Sarmiento *et al.*, 1986; Bischofberger *et al.*, 1990).

The overall characteristics of these groups of bacteria suggest that they can adapt to the groundwater environment, but perhaps more readily to shallow aquifers (local flow systems) where dissolved oxygen concentrations are relatively high and open to sources of nutrients from the surface or from the unsaturated zone.

12.4.6   *Gram-positive bacteria*

Gram-positive bacteria ocurring in Natural Mineral Waters have been frequently reported to belong to 'arthrobacter-like' or 'coryneform-like' bacteria (Schwaller & Schmidt-Lorenz, 1981a; Gonzalez *et al.*, 1987; Bischofberger *et al.*, 1990), and more rarely to *Bacillus*, *Staphylococcus* and *Micrococcus* (Leclerc *et al.*, 1985; Quevedo-Sarmiento *et al.*, 1986; Hunter & Burge, 1987; Mavridou, 1992). However, it is advisable to be cautious in the identification of gram-positive bacteria. When these are isolated from bottled mineral water, it is possible that they are derived from the bottling plant, since gram-positive bacteria such as *Micrococcus* and *Staphylococcus* are common inhabitants on the skin and mucous membranes of mammals. All these bacteria may, indeed, be part of the ambient microflora.

The distribution of gram-positive bacteria is a critical issue in groundwater systems. Transmission electron microscopy showed, in fact, that about two-thirds of the bacterial cells from subsurface environments had gram-positive cell walls, whereas isolation of microorganisms in culture medium revealed a preponderance of gram-negative cells (Ghiorse & Balkwill, 1983). This observation was also corroborated by Wilson et al. (1983). In addition to direct microscopic observation, biochemical techniques can also give an indication of the relative abundance of gram-positive and gram-negative microorganisms in samples. For example, the amount of ribitol, which is a part of teichoic acids of gram-positive bacteria, is a rough measure of their relative abundance. Likewise, the abundance of gram-negative bacteria can be estimated by the level of hydroxy fatty acids in the lipopolysaccharides.

The ability to form endospores when growing cells are subjected to nutritional deficiency or excessive heat or dryness is characteristic of some gram-positive bacteria such as *Bacillus* and *Clostridium*. Endospores are particularly well adapted to environments subjected to wide variations in water and low-nutrient conditions such as subsurface environments but, with some exceptions, species of *Bacillus* or *Clostridium* have not been reported widely from aquifer systems (Hirsch & Rades-Rohkohl, 1983; Chapelle et al., 1988). These observations indicate that spore formation *per se* might not be a major feature for bacteria in groundwater habitats.

12.5   Inhibitory effect of autochthonous bacteria

Natural Mineral Water is not subjected to antibacterial treatments of any kind and, after bottling, it is often stocked for several months before it is distributed and sold. To assess public health risks it is therefore important to know the survival capacity of pathogens and indicator bacteria. Much of the early literature on bacterial survival in water suggested that die off or decay was the only functional response of bacteria of fecal origin exposed to marine or freshwater. This was primarily attributed to the 'bactericidal' property of seawater (Ketchum et al., 1952) or the self-purifying power of freshwater (Leclerc & Mossel, 1989). Dilution, light and temperature, as well as biological parameters such as inhibition, antagonism and predation, could be important factors affecting the fate of pathogens or indicator bacteria. The use of the terms 'die off' or 'decay' to describe changes in contaminant bacteria densities over time is probably inappropriate. The apparent reduction of recoverable counts from marine or freshwaters is not only the result of 'true' cell death, i.e. cells that become nonviable, but is also a function of physiological adaptation to an adverse environment and complex interactions of physical, biological and chemical processes. Changes in bacterial density may be expressed as loss of viability, alteration in culturability, persistence or aftergrowth. Under certain conditions of metabolic stress such as starvation, bacterial cells may enter into a VBNC. In this state the cells are not culturable in standard media but retain certain features of living cells such as respiratory activity (Zimmermann et al., 1978) and substrate uptake (Kogure et al., 1979). The VBNC state has been suggested to represent an escape strategy for survival (Colwell et al., 1985; Roszak & Colwell, 1987a, 1987b) and a large number of bacterial species, including pathogens and indicators, have been recognized to enter VBNC states under laboratory or field conditions (Oliver, 1993).

The available data on the survival of bacteria in surface waters cannot be extrapolated to bottled mineral waters. It is important, for example, to take into account some specific factors such as the impact of drilling, the bottling stress, the selective attachment of some populations to solid surfaces, the fate of autochthonous populations which can reach very high numbers a few days after bottling, the effect of an enclosed environment (bottle effect), used for influence of PVC, PET or glass used for bottles.

Ducluzeau (Delabroise & Ducluzeau, 1974; Ducluzeau et al., 1976a, 1976c; Lucas & Ducluzeau, 1990a, 1990b) was the first to study the survival of enterobacteria in mineral water to assess the influence of autochthonous bacteria on indicator bacteria. In the most significant experiment, Escherichia coli was inoculated into sterile water at a concentration of 1.2×10^5 cfu/ml. The plate counts of E. coli were reduced by less than one log over a three-month period, and more than 10^2 cfu/ml were still detected five months later. On the other hand, when this experiment was repeated with mineral water, i.e. in the presence of the autochthonous mineral flora (between 5×10^4 and 5×10^5 cfu/ml), the complete loss of viability of E. coli took place between 35 and 55 days depending on the experiment. The same loss of viability of the test organism exerted by the whole autochthonous flora was exhibited by some strains from the dominant flora. Therefore, it appears that these authors observed an antagonistic activity by autochthonous flora of the mineral water, and not an effect due to the physical or chemical properties of the water itself. The antagonistic activity could possibly be related to an inhibitory substance accumulating in the water during the successive cycles of growth and death of the autochthonous bacterial population (Fig. 12.6).

More recent studies by Moreira et al. (1994) with E. coli and other coliform indicators such as Enterobacter cloacae and Klebsiella pneumoniae showed that the viable counts of the three test enterobacteria decreased under all experimental conditions, but the decrease depended on the organism and the conditions in which they were examined (Fig. 12.7). The population of E. coli decreased rapidly in mineral water, especially when bottled in PVC, irrespective of the presence or absence of autochthonous bacterial flora. In sterile tap water, after an initial decrease, the viable counts remained almost constant during the experimental period. Of the three enterobacteria tested, Enterobacter cloacae had the slowest decrease in viable counts in any of the conditions examined, although the decrease was slightly more pronounced in sterile mineral water bottled in PVC than in the other test conditions. A small constant decrease in the viable

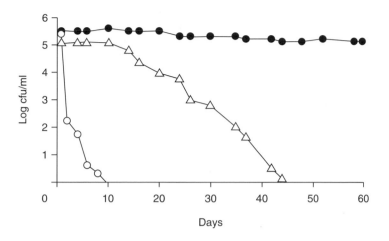

Fig. 12.6 Antagonistic effect of the microbial flora of a mineral water on *Escherichia coli*. Filtered water that had contained the autochthonous flora for one week (●); non-filtered water containing the autochthonous flora (△); filtered water that had contained the autochthonous flora for 50 days (○). These observations indicate that it takes several weeks before antagonistic substances accumulate in the water in toxic levels sufficient to inhibit the recovery of the target organism. Redrawn from Ducluzeau *et al.* (1976c).

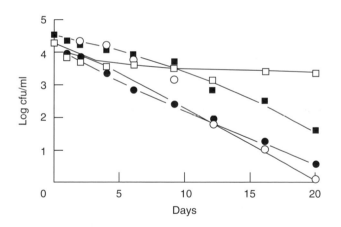

Fig. 12.7 Survival determined by viable counts on triptic soy agar of *Escherichia coli* inoculated in mineral water with the autochthonous flora in PVC bottles (●); in sterile mineral water in PVC bottles (○); in sterile mineral water in glass containers (■); and in sterile tap water in glass containers (□). Redrawn from Moreira *et al.* (1994).

counts of *Klebsiella pneumoniae* was observed in mineral water bottled in PVC with indigenous flora and in sterile tap water. On the other hand, this strain was rapidly inactivated in sterile mineral water bottled in PVC and glass, resulting in very low viable counts after the 20-day experimental period. In this study, however, the autochthonous flora of the mineral water, which reached 1.03×10^6 cfu/ml during the experimental period, did not appear to have an effect on the survival of *E. coli* and *Enterobacter cloacae*, but it did appear to have an effect on the survival of *Klebsiella pneumoniae* compared to the other conditions tested. The contradiction between the results of Ducluzeau and his colleagues and those of Moreira *et al.* (1994) can only be extended to the effect of the autochthonous flora on *E. coli*, since the former authors did not examine the effect of the normal flora of mineral waters on the other coliform bacteria.

Several other studies on the fate of *E. coli* and pathogenic bacteria in mineral water have been performed irrespective of the influence of autochthonous flora. Burge and Hunter (1990) showed that *E. coli* was able to survive in bottled mineral water for about 42 days and that *Salmonella typhimurium*, *Aeromonas hydrophila* and *P. aeruginosa* persisted for at least 70 days; on the other hand, *Campylobacter jejuni* was recovered for only 2–4 days. In carbonated waters, the survival of the same bacteria was reduced by 25–50%. Another study reported that *E. coli* declined by one log every two weeks in sterile spring water stored at 4°C; the behaviour of *Yersinia enterocolitica* was different from that of *E. coli*, increasing for the first eight weeks and still being recoverable 64 weeks later (Karapinar & Gönül, 1991). Several *E. coli* O157:H7 outbreaks associated with both drinking and recreational water raised concerns about waterborne illness caused by these bacteria (Chalmers *et al.*, 2000). The survival characteristics of this pathogen have been described in inoculated drinking and recreational water (Wang & Doyle, 1998), bottled water (Warburton *et al.*, 1998) and natural mineral water (Kerr *et al.*, 1998). Overall, these studies indicate that *E. coli* O157:H7 is hardly a pathogen that can survive for long periods of time in water. Some contradictory results concerning the survival of enteric pathogens have been found that may be due to the different techniques used between laboratories, different strains used and different waters (Afonso *et al.*, 1998a, 1998b).

P. aeruginosa is frequently isolated from surface water and is also a major concern in mineral water bottling plants because it is an opportunistic pathogen and can contaminate boreholes and bottling plants. The ability of *P. aeruginosa* to grow in water even with low concentrations of organic substrates has been studied in relation to its presence in tap water and distilled water (Botzenhart & Röpke, 1971; Favero *et al.*, 1971; Botzenhart & Kufferath, 1976; Dickgiesser & Rittweger, 1979; Van der Kooij *et al.*, 1982a), and to define the biological stability of mineral waters (Van der Kooij, 1990). In contrast to most enterobacteria, *P. aeruginosa* is able to grow in water. The behaviour of *P. aeruginosa* based on the culturability was assessed in still mineral water with or without autochthonous flora (Moreira *et al.*, 1994). Immediately after inoculation of this organism in mineral water, there was a sharp decrease in the viable counts. Later, there was a very slow decrease in viable counts in PVC-bottled mineral water with autochthonous flora, whereas the viable counts in sterile mineral water remained constant for the duration of the experiment. Gonzalez *et al.* (1987) showed a significant inhibitory effect of the autochthonous flora of mineral water on *P. aeruginosa*. The bacterial generation time at 30°C was about 19 h in water with intact flora, while it was about 6 h in mineral water sterilized by filtration. The *P. aeruginosa* population became unculturable more quickly as the storage temperature was lowered to 6°C, the initial population of about 100 cfu/ml becoming undetectable after 210 days. On the other hand, at 37°C *P. aeruginosa* density increased from 10^2–10^4 cfu/ml at the end of the one-year experiment.

The effect of the utilization of laboratory-adapted allochthonous pathogens or indicators, the effect of the size of the inoculum, the biological state of the inoculum and the physicochemical composition of water are among the concerns of the validity of these studies. Therefore, the antagonistic effect of the autochthonous flora on *P. aeruginosa* was examined in three types of Natural Mineral Water (very low mineral content, low mineral content, rich in mineral content) with an inoculum that gave a final concentration of approximately one organism per ml in the bottled water (Vachée & Leclerc, 1995). Four test strains were used: one obtained from a culture collection, one from a patient with septicemia and two from surface water. The test bacteria were inoculated immediatly after sampling from the source of the mineral waters. Overall experimental conditions mimicked natural contamination before bottling. In the filter-sterilized waters, *P. aeruginosa* attained more than 10^4 cfu/ml a few days after inoculation, and remained almost constant during the nine months of the experiment. In mineral water with the autochthonous flora, the initial inoculum did not increase at all during the experiment (Fig. 12.8).

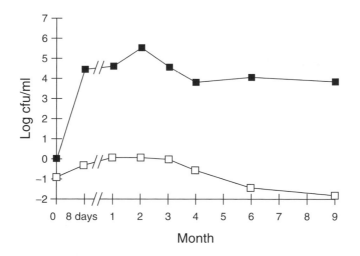

Fig. 12.8 Survival or growth determined by viable counts (cfu) of *P. aeruginosa* (wild-type strain) on a selective medium after inoculation into mineral water maintained at room temperature containing the autochthonous flora (■) and without the autochthonous flora (□). The results show that the normal flora exerts a strong antagonistic effect on a low inoculum of *P. aeruginosa*. (Redrawn from Vachée & Leclerc (1995).)

The inhibitory ability of the autochthonous flora observed in some experiments cannot be extrapolated to all waters and test pathogens or indicators of fecal pollution owing to the limited number of studies using different experimental conditions, waters with different chemical compositions and different types of bottling material. In addition to these considerations, the results do not take into account the ability of many bacteria to enter into a VBNC state. It is also important to remember that the predominant bacterial flora of mineral water belongs to genus *Pseudomonas* or related genera, and that these bacteria produce secondary metabolites with toxic or inhibitory activity for competitors (Leisinger & Margraff, 1979; O'Sullivan and O'Gara, 1992; Budzikiewicz, 1993). In contrast to carbohydrates, lipids, proteins or nucleic acids, secondary metabolites are not present during all stages of the growth cycle. They are not important as sources of energy or reserve substances, and they are only slowly metabolized. The two functional classes typical of secondary metabolites of *Pseudomonas* spp. are siderophores and antibiotically active substances.

With rare exceptions, iron is available to microorganisms only in its trivalent form. Owing to the low dissociation constants of different oxide hydrates, the concentration of free Fe^{3+} at pH 7.0 is at best 10^{-17} mol/l, while about 10^{-6} mol/l would be needed to maintain the necessary supply for the living cells. Soil and water bacteria, as well as those infecting animals or humans (where the iron supply is limited because it is bound to peptidic complexes), produce a variety of complexing agents (usually called siderophores) capable of making inorganic Fe^{3+} available or securing it by transcomplexation. Most known siderophores can be grouped into hydroxamate- and phenolate/catecholate-type structures and have different affinities for ferric iron. Water and soil pseudomonads generally produce fluorescent, yellow-green, water-soluble siderophores named pyoverdins or pseudobactins with both a hydroxamate and a phenolate group.

Antibiotic production by some fluorescent *Pseudomonas* spp. is now recognized as an important factor in microbial competition. The diversity of the antibiotics produced by different species is now being fully recognized (Leisinger & Margraff, 1979; Budzikiewicz, 1993). Compounds such as phenazines (Thomashow & Weller, 1988), pyoluteorin (Howell & Stipanovic, 1980), pyrrolnitrin (Howell & Stipanovic, 1979), tropolone (Lindberg, 1981), pyocyanin (Vandenbergh *et al.*, 1983) and 2,4-diacetylphloroglucinol (Keel *et al.*, 1990; Shanahan *et al.*, 1992) fall into the class of N-containing heterocycles and have been shown to originate from intermediates or end products of the aromatic amino acid biosynthesis. Another important class of secondary products of *Pseudomonas* comprises unusual amino acids and peptides. In addition to these two major groups of secondary metabolites, some glycolipids, lipids and aliphatic compounds with a broad spectrum of activity against bacteria and fungi have been isolated from *Pseudomonas* cultures (Lindberg, 1981). These observations were substantiated by genetic studies, whereby mutants defective in the production of

Table 12.7 Inhibitory activity of strains isolated from mineral water against nine target organisms on King B medium with or without the addition of iron

Number of strains tested[a]	Target organisms[b]	Activity (%)[c]	
		King B	King B with iron
148	*Escherichia coli*, ATCC 10536	49.2	3.9
148	*Pseudomonas aeruginosa*, ATCC 10490	19.1	5.4
80	*Staphylococcus aureus*, ATCC 65388	67.5	0.0
80	*Enterococcus faecalis* (clinical isolate)	1.25	0.0
148	*Salmonella typhimurium*, ATCC 13311	14.0	2.0
102	*Aeromonas hydrophila*, ATCC 9071	32.3	7.0
102	*Shigella sonnei*, ATCC 29930	26.0	0.0
54	*Yersinia enterocolitica*, CUETM 82–52	16.6	0.0
54	*Bacillus cereus*, CIP 6624	5.5	0.0

[a] 382 strains were examined of which 50 belonged to *P. fluorescens*, 20 to *Sphingomonas paucimobilis*, 18 to *Brevundimonas vesicularis*, 10 to *Comamonas testosteroni*, and 154 were unidentified gram-negative rods.
[b] ATCC, American Type Culture Collection, Rockville, MD, USA; CIP, Collection de l'Institut Pasteur, Paris, France; CUETM, Collection de l'Unité d'Ecotoxicologie Microbienne, Villeneuve d'Ascq, France.
[c] Mineral water strains were grown in nutrient broth (0.5% peptone, 0.3% meat extract) at 30°C. Agar diffusion assays to screen for growth inhibition of the test bacteria were performed in King B medium with or without iron (0.3% ferric citrate).

A total of 916 assays were peformed of which 25% inhibited the growth of one or more target organism in medium without added iron, and only 3.3% inhibited growth in the medium with additional iron.

certain antibiotics were directly compared with their otherwise wild-type parental strains. These reports showed that antibiotic-negative mutants had lost their specific inhibitory activity against pathogenic fungi (Gutterson *et al.*, 1986; Gill & Warren, 1988; Thomashow & Weller, 1988; Shanahan *et al.*, 1992).

However, fluorescent *Pseudomonas* spp. have emerged as the largest and possibly most promising group of bacteria because of their potential for rapid and aggressive colonization and for preventing the invasion of detrimental or pathogenic microorganisms. These organisms have been used in the biocontrol of plant diseases due to bacteria or fungi (Kloepper *et al.*, 1980, 1988; Suslow, 1982; Keel *et al.*, 1990; O'Sullivan & O'Gara, 1992). The relative importance of the production of antibiotics, siderophores, hydrogen cyanide and direct competition for nutrients may differ considerably among strains, and conflicting results have been reported with regard to the role of pyoverdins or antibiotics produced by fluorescent pseudomonads in control of soilborne pathogens (Hamdan *et al.*, 1991; Henry *et al.*, 1991; Laine *et al.*, 1996; Raaijmakers *et al.*, 1997).

12.6   Assessing health risk from autochthonous microflora

The qualities of autochthonous bacteria in mineral water, referred to as the 'heterotrophic plate count', and can be investigated by the HPC test, leads us to believe that they do not cause disease. Their metabolism (oligotrophic) and their nutrition (prototrophic) mean that they are not adapted to living in humans or animals. Their psychrotrophy (maximum growth temperature of 25–30°C) renders them particularly vulnerable to invasion of human tissue. The digestive tract with its natural barriers (gastric trap, mucus and gastric mucous membrane, intestinal motricity, intestinal cytoprotection), its complex multiple microbial flora exerting powerful antagonisms, its differentiated lymphoid tissue, and general and local immune response precludes colonization by these stressed bacteria.

Owing to various circumstances, one or more host resistance mechanisms may be lost, thus increasing the probability of infection. The term immunosuppressed host is used to refer to hosts in which (one or more) resistance mechanisms are malfunctioning. Interest has turned to infections that arise with increasing frequency in such 'compromised hosts'; such infections have been called 'opportunistic infections' by Von Graevenitz (1977). Almost all people at risk to environmental bacteria are hospitalized patients with profound specific defects in the immune system. There are principally two categories of high-risk patients for whom water ingestion is restricted. These include bone marrow transplant patients and those with acute leukemia. It can be recommended that HIV-infected persons who wish to take independent action to reduce the risk of infections such as cryptosporidiosis, microsporidiosis or disseminated infection with *Mycobacterium avium* complex, possibly occurring in tap water, boil the water for 1 min or use bottled water such as Natural Mineral Water.

There are several approaches to detecting bacterial populations such as those autochthonous to mineral waters that could have public health importance but are not known to be pathogenic. The methods available include animal model systems and epidemiological studies. Another approach is to search for virulence factors from bacterial isolates.

12.6.1   Inoculation of the digestive tract of axenic mice
The purpose of the work carried out by Ducluzeau *et al.* (1976b) was to determine whether the microflora of mineral water was able to establish itself in the digestive tract of axenic mice, to multiply in great numbers and to entail pathological disorders in mice receiving mineral water or bacterial species from mineral water. In the axenic animal, the 'barrier effect' of the digestive microflora does not exist and any bacterial strain potentially capable of growing in the digestive tract reaches, within 12–24 h, between 10^9 and 10^{10} cells/g of fresh feces (Ducluzeau & Raibaud, 1974, 1976, 1979). Axenic animals are a first choice for determining whether autochthonous bacteria of mineral water are able to adhere, penetrate and multiply in epithelial cells or produce toxins

or other substances causing tissue damage. The most stringent experiments were devised to compare the transit of an inoculum of several autochthonous strains and that of the spores used as markers (Fig. 12.9). In spite of the presence of an equivalent number of *Pseudomonas* (strain P1) cells and of the inert marker in the inoculum, the maximum number of *Pseudomonas* in the feces was lower than that of the spores, and the former disappeared from the feces more rapidly than the latter. Thus, a partial destruction of *Pseudomonas* P1 was shown during its transit through the digestive tract. Other assays, performed with other strains of *Pseudomonas* or *Acinetobacter* provided similar results.

12.6.2   Randomized trials in infants

The quality of water for the preparation of babies' feeding bottles is universally recognized as an essential choice. In the past, mineral water conditioned in glass bottles was used. Since 1970, PVC conditioning has been used, and some people have wondered about the modifications in the microbial populations that may have resulted from using water bottled in PVC, as well as effects on the health of babies. To answer this question, a study was carried out from January 1984 to January 1986 by Leclerc (1990) with two groups of 30 babies each, using the double-blind method. One group was fed milk made up from mineral water in PVC bottles; the other group was fed milk made up from the same pasteurized mineral water in glass bottles. The babies chosen corresponded to very precise inclusion criteria, such as the absence of any disease (including icterus or diarrhea), absence of any antibiotic treatment, and babies being exclusively fed on milk. Diarrhea was defined according to the usual criteria, namely the number

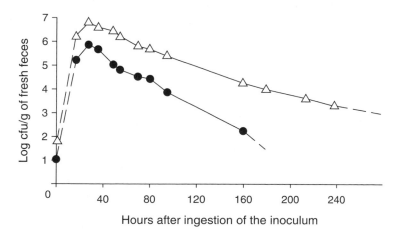

Fig. 12.9 Transit of *Bacillus subtilis* spore markers ($\triangledown$) as compared to *Pseudomonas* strain P1 (●) through the digestive tract of axenic mice. (Redrawn from Ducluzeau *et al.* (1976b).)

of stools, consistency, odor, effect on weight of the baby and hyperthermia (as a criterion of an infectious origin). The bacteriological criteria included the search for pathogenic bacteria from coprocultures, and in particular *Salmonella*, *Shigella*, *Yersinia*, enteropathogenic *E. coli* and the enumeration of all gram-negative rods to allow the evaluation of a possible modification of the intestinal microbial ecosystem.

In no case was it possible to isolate mineral water–derived bacteria from rhinopharyngeal samples analyzed 1 or 2 h after drinking bottled milk. Nor was there evidence for digestive tract colonization from the analysis of the stool samples. On rare occasions, bacteria from mineral water were isolated, but at levels that were 10–100 times smaller than in the original water. The intestinal flora of the neonate acts as a 'permissive barrier'. The waterborne bacteria travel through the intestine in low numbers, then disappear little by little when feeding bottles prepared with mineral water are substituted with those prepared using sterile water. It has been demonstrated experimentally that body temperature, transit time, gastric acidity and the oxidation–reduction potential of the intestine, working in conjunction, result in decrease in numbers of bacteria ingested and their total elimination.

From an epidemiological point of view, no difference could be found in the two groups. Of the 60 babies, nine showed signs of diarrhea as defined in the protocol, of which five received milk prepared with sterile mineral water and four, milk conditioned with Natural Mineral Water. In no case did the risks appear sufficiently severe to justify the suspension of milk feeding. It may be concluded that there was no difference in the pathology observed in the two groups.

12.6.3 Virulence characteristics of bacteria

The development of an infection is related to three basic parameters (Edberg *et al.*, 1997): the number of microbes and the target organ of the host; the virulence characteristics of the microbe; and the immune status of the host and target organ (Duncan & Edberg, 1995). Several studies have been conducted to test the invasive or cytotoxic activity of drinking water on cultured cell lines (Lye & Dufour, 1991; Payment *et al.*, 1994; Edberg, 1996; Edberg *et al.*, 1996). In all cases, a small percentage (1–2%) of bacteria examined were cytotoxic. Payment *et al.* (1994) examined the cytotoxicity from HPC bacteria isolated on blood-containing medium incubated at 35°C, to mimic what was considered nearest to the human physiological environment. A high percentage of the cytotoxic isolates belonged to the genus *Bacillus*, which might be related to low-level gastroenteritis. Only Edberg *et al.* (1997) studied HPC bacteria isolated from bottled water. Health risks were estimated by the determination of cytotoxicity and invasiveness in a human enterocyte cell line. More than 95% of naturally occurring HPC bacteria showed low invasiveness and cytotoxicity. When either invasiveness or cytotoxicity was demonstrated, only a small number of cells from the culture were positive.

A study was conducted in our laboratory to determine the virulence charac-teristics of natural mineral water bacteria. The tests selected determine the ability of bacteria to attach, invade and injure Hep-2 cells. The method used was the one described by Edberg *et al.* (1996). A total of 240 representative strains isolated from five French springs were selected, including *P. fluorescens* and all new species cited in Table 12.5. Results showed that none of the bacteria studied is capable of growing and attaching to Hep-2 cells or producing cytotoxin at a temperature of 37°C. The detection of bacterial activity in one or several of the tests for putative virulence factors may be useful in showing potential health hazards posed by bacteria isolated from potable water. Nevertheless, the exact relationship between putative virulence factors and their potential health effects remains to be investigated.

Overall experimental and epidemiological data show that autochthonous bacteria in Natural Mineral Waters have never brought about detectable patho-logical disorders in humans or animals and, *in vitro*, are incapable of directly damaging human cells in tissue culture. Since the existence of European regula-tions (EC, 1980), no outbreak or single case of disease due to the consumption of natural mineral water has been recorded in the literature or by the health authorities of the countries within the European Community.

12.7 Assessment and management of microbial health risks

All microbial hazards occurring in drinking water from distribution systems must be taken into account in the case of Natural Mineral Water. In the past decade, many outbreaks attributed to protozoan or viral agents have been reported in conventionally treated water supplies, all of which met coliform standards. Viruses have been shown to persist longer in these waters than fecal coliforms and many are more resistant to water and wastewater treatment processes. A similar situation exists for protozoan cysts. These findings repeatedly suggest the inadequacy of treatment processes for safe water and the inadequacy of coliforms as indicators. Natural Mineral Waters are recognized as not being vulnerable to fecal contamination after a strict procedure that requires a few years of evidence to confirm stability of physical and chemical characteristics. The water must also be shown to be microbiologically wholesome, i.e. requiring no treatment. No outbreak or single case of disease due to the consumption of natural mineral water, in line with European microbiological standards, has been recorded. Other epidemiological data including cohort study in infants, animal tests and cell tests have never showed adverse effects (see Section 12.6).

The most distinctive factor of mineral waters might be the very low amount of DOC with an available fraction and its identifiable compounds such as labile amino acids, carbohydrates and carboxylic acids (Lacoste, 1992). Organic substances in the distribution network water originate from the raw water (generally surface water) used for its production and from materials (pipes, lubricants, joints, sealants, etc.) that may release biodegradable compounds (Geldreich, 1996). These nutrients are a major factor for the colonization of pipes by heterotrophic bacteria. The so-formed biofilms are capable of retaining pathogens including environmental pathogens (*Legionella* spp., *M. avium*), viruses and protozoa entering a distribution network.

For mineral water sources claiming to be protected, an inherent feature is that the physical and chemical nature of water is constant over time. Therefore, simple measurements such as temperature, ionic strength, anions, cations and trace elements have great meaning in sampling source water, whereas they would have little meaning when sampling tap water.

Mineral water bacterial communities, identified by culture or with specific probes, are primarily aerobic gram-negative rods. These bacteria belong to alpha, beta and gamma proteobacterial groups as well as to the phylum of *Flavobacterium–Cytophaga*. In contrast, the general population in water supplies includes many gram-negative and gram-positive bacteria, sporeformers, acid-fast bacilli, free-living amebae and nematodes, opportunistic fungi and yeasts (Geldreich, 1996). Many authors have observed antibacterial activity by autochthonous flora of mineral waters; however, the issue is widely debated (see Section 12.5).

12.7.1   Identifying microbial hazards in drinking water

A large variety of bacterial, viral and protozoan pathogens are capable of initiating waterborne infections.

(1) There are primarily the enteric bacterial pathogens including classic agents such as *Vibrio cholerae*, *Salmonella* spp., *Shigella* spp. and newly recognized pathogens from fecal sources like *Campylobacter jejuni* and enterohemorrhagic *E. coli*. The survival potential of these bacteria increases in biofilms due to their ability to form a VBNC state.

(2) Several new bacterial pathogens such as *Legionella* spp., *Aeromonas* spp., *P. aeruginosa* and *M. avium* have a natural reservoir in the aquatic environment and soil. These organisms are introduced from the surface water into the drinking water system usually in low numbers. They may survive and grow within the distribution system biofilm.

(3) More than 15 different groups of viruses, encompassing more than 140 distinct types, can be found in the human gut. These enteric viruses are excreted by patients and find their way into sewage. Hepatitis A and E viruses cause illnesses unrelated with gut epithelium. Another specific group of viruses has been incriminated as a cause of acute gastroenteritis in humans that includes rotavirus, calicivirus, the most notorious being norovirus, astrovirus and some enteric adenoviruses. These viruses

cannot grow in receiving water and may only remain static in number or die off.

(4) The most prevalent enteric protozoa, associated with waterborne disease, include *Giardia lamblia* and *Cryptosporidium parvum*. In addition, protozoa like *Cyclospora*, *Isospora* and many microsporidian species are emerging as opportunist pathogens and may have waterborne routes of transmission. Like viruses, protozoa cannot multiply in receiving waters. With the exception of *Salmonella*, *Shigella* and hepatitis A virus, all the other organisms can be called 'new or emerging pathogens'.

There are a number of reasons for the emergence of these new pathogens, analyzed in every detail by Szewzyk *et al.* (2000), including high resistance of viruses and protozoan cysts, a lack of identification methods for viruses, changes in water use habits (*Legionella*) and human populations at risk. Another striking epidemiological feature is the low number of bacteria that may trigger disease. The infectious dose of *Salmonella* is in the range of 10^7–10^8 cells while only around 100 cells are required to cause clinical illness with *E. coli* O157:H7 and *Campylobacter* (Leclerc *et al.*, 2002). The infective dose of enteric viruses is low, typically in the range of 1–10 infectious units; it is about 10–100 oocysts for *Cryptosporidium* (Meinhardt *et al.*, 1996).

12.7.2 Assessment of microbial risks

The view on the microbiological safety of drinking water is changing. The demand for the total absence of any pathogenic organism is no longer significant in light of the new pathogens, some of which are capable of growing in drinking water systems. According to the new European Union Council directive 98/83/EC (EC, 1998), water for human consumption must be free from any microorganisms and parasites and from any substances which, in numbers or concentrations, constitute a potential danger to human health. To deal with this issue, the US Environmental Protection Agency for the first time used a microbial risk assessment approach. It has been defined that an annual risk of 10^{-4} (one infection per 10 000 consumers per year) should be acceptable for diseases acquired through potable water, this value being close to the annual risk of infection from waterborne disease outbreaks in the USA (4×10^{-3}) (Gerba, 2000).

Microbiological risk assessment is a major tool for decision making in the regulatory area. The problem is, however, that the key data to perform this assessment are mostly missing. Few epidemiological studies correlating the incidence of disease with pathogen densities have been reported. Several outcomes, from asymptomatic infection to death, are possible through exposure to microbes. The issue of dose–response relationships is particularly striking: these relationships are only available for a few pathogens. When infectious doses are low as is the case for some viruses and protozoan cysts, the calculated tolerable concentrations are also low and monitoring of these pathogens in drinking water becomes impracticable.

Natural Mineral Waters are subject to the general rules laid down by Council Directive 80/777/EEC (EC, 1980). At source and when sold, an natural mineral water must be free from parasites and pathogenic microorganisms. These requirements are therefore distinct from tap water.

12.7.3   Management of microbial risks

12.7.3.1   Heterotrophic plate counts

HPC measurements have been used to gain better information on the effects of water treatment processes and distribution on the bacteriological quality of drinking water. Various methods and the application of HPC monitoring have been analyzed in considerable depth by Reasoner (1990). In certain epidemiological studies reported by Calderon and Mood (1988, 1991) and Payment et al. (1991a, 1991b, 1997), there is a debate on the potential negative human health from the consumption of treated water containing high HPC levels of bacteria. The available body of evidence supports the conclusion that, in the absence of fecal contamination, there is no direct relationship between HPC values in ingested water and human health effects in the population at large (WHO, 2002).

Natural Mineral Waters cannot be subjected to any type of disinfection that modifies or eliminates their biological components; therefore, they always contain the bacteria that are primarily a natural component of these waters. It is also clearly stated that, after bottling, the recoverable bacterial counts should only result from the normal increase of bacteria present in the source. The studies described in Section 12.6 have not been capable of identifying any microbiological risk from examined bottled mineral waters. To date, there has been no association between human disease and the natural bacteria found in natural mineral water.

Measurement of HPC in bottled mineral water is useful for several reasons (Moreau, 2001). First, it proves that no disinfection has occurred. Second, it helps to ensure that, from the spring to the finished product, no major quantitative changes have occurred in the microbial status of the water. Indeed modification from counts normally found at a particular location may give an early warning of significant microbial alteration.

The bacterial species that make up HPC in natural mineral water are psychrotrophic. In the study by Reasoner and Geldreich (1979) on treated distribution water, incubation at 20°C yielded the highest counts in all media when incubation was extended to 12–14 days, whereas 28°C appeared to be the best temperature from day 2 through day 6 of incubation. The 37°C plate count was believed to give an indication of the presence of rapid-growing bacteria more likely to be related to pathogenic or fecal types that might be present from sewage pollution. It can be stated that the measurement at 37°C is unsuitable and unnecessary to determine inadequate processing for safety reasons because there are other appropriate indicators for this purpose, including the indicators of fecal contamination.

12.7.3.2 Marker organisms and enteric pathogens

In the microbiological monitoring of water and foods, Ingram (1977) introduced the distinction between 'index organisms' for markers whose presence indicates the possible occurrence of ecologically similar pathogens, and 'indicator organisms' for those whose presence points to inadequate processing for safety. In short, index markers indicate a potential health risk, whereas indicators reveal process failure. The terms 'indicator of fecal contamination' (index) and 'indicator of quality' (indicator) are also commonly used. *E. coli* is now the sole recognized indicator of fecal contamination, being a direct public health threat (Edberg *et al.*, 2000). The other indicators of quality include coliforms other than *E. coli*, commonly fecal streptococci, *P. aeruginosa* and sulfite-reducing anaerobes: in the case of treated drinking water, they demonstrate treatment effectiveness and water quality in the distribution system (biofilm development); in the case of untreated groundwater (mineral water), they indicate possible deficiency in natural hydrogeological protection mechanisms (indicator of vulnerability). No indicators can indicate the occurrence of environmental pathogens such as *Legionella* or *P. aeruginosa*.

Innovative taxonomic approaches in the bacteriology of the coliform group and comprehensive studies of their habitats allowed the ecological positioning of coliforms in one of the three following groups: (1) the thermotrophic and true fecal *E. coli*; (2) the thermotrophic and ubiquitous coliforms (e.g. *Klebsiella pneumoniae*, *Enterobacter cloacae*), which may form part of the intestinal flora of humans and warm-blooded animals, but also occur in the natural environment; and (3) psychrotrophic, purely environmental coliforms (e.g. *Serratia fonticola*, *Rahnella aquatilis*), which proliferate in polluted or pristine waters and mostly originate from vegetable or small animal sources. From a public health point of view, both the ubiquitous and environmental groups are quality indicators. The controversy over the value of fecal coliform or thermotolerant coliform as fecal indicators, associated with the heterogeneity of the group, has led to the suggestion that the term 'fecal coliform' or 'thermotolerant coliform' should be redefined to be synonymous with *E. coli* (Leclerc *et al.*, 2001).

Basically, an acceptable indicator of pathogens such as *E. coli* must only have two attributes: it must be present when the pathogens are present, and it must be easy to detect and to quantify. It is sometimes important, but not imperative for the protection of the public health, that the indicator is absent when the pathogen is absent. The most significant change over the last two decades is the general recognition that the coliform test including *E. coli* in treated water supply is not so much a measure of sanitary significance but more an indication of treatment effectiveness. Coliform bacteria as well as other bacterial indicators are easily captured and inactivated in conventional treatment processes but the more resistant enteric viruses and protozoan pathogens are not. It is now considered that the use of *E. coli* may be the most appropriate to indicate the presence of enteric bacterial pathogens and that viruses and protozoan pathogens must be analyzed separately (Gleeson & Gray, 1997).

With natual mineral waters that are untreated, the problem of a relationship existing between marker organisms and pathogens must be discussed specifically, as this relationship is governed by key factors and processes that control the mobility and fate of suspended microbes in soil and groundwater environments. The first category of factors focuses primarily on characteristics of the microbes such as size, adhesion and inactivation or die-off rate. The second category pertains to abiotic factors such as porous medium characteristics, filtration effects and water flow. The implication of microbial transport relative to the safety of groundwater has been closely analyzed by Robertson and Edberg (1997) and Newby *et al.* (2000). In general, the larger the suspended microorganism, the more readily it will be physically filtered by the subsurface material. Thus, parasite cysts or oocysts, such as *Giardia* and *Cryptosporidium*, are relatively large and so much more readily filtered than viruses and bacteria. However, the die-off rate of *E. coli* through transport in subsurface environments is certainly higher than that of *Giardia* or *Cryptosporidium* cysts. Taking into account the half-life of *E. coli* as conservatively estimated to be at least eight days under groundwater conditions, Edberg *et al.* (1997) recommended the use of *E. coli* as indicator of fecal protozoan. (Deuret *et al.*, 2000, Gassilloud *et al.*, 2003)

The relatively high mobility of viruses in subsurface material is primarily due to their smaller size, lower inactivation or die-off rates, and physical properties compared with bacteria. Enteric viruses have been detected in many groundwater supplies, usually those in close proximity to surface water or septic tanks (vulnerable groundwater). Thus viral contamination of groundwater is of special concern. There is no absolute correlation between bacterial indicators and enteric viruses due to the essentially unpredictable behavior of viruses. Coliphages do not yet fulfill enough of the criteria to be reliably employed (Leclerc *et al.*, 2000). Viral pathogens including hepatitis A virus, enterovirus and calicivirus should be detectable by a combination of cell culture and molecular methods (awaiting validation by international groups of experts). At present there is a large discrepancy in the results of studies that compare infectivity, molecular and combined methods (Yates *et al.*, 1999; Abbaszadegan *et al.*, 1999). The report of Norwalk-like virus sequences in bottled mineral waters, in absence of disease outbreaks, shows the difficulty in choosing appropriate methods and the very high risk of methodological contamination of samples (Beuret *et al.*, 2000; Gassilioud *et al.*, 2003).

Common bacteria from soil and vegetation, unrelated to fecal contamination, may be the best indicators of the quality of natural mineral waters. In the absence of *E. coli*, their presence in a water sample does not indicate an imminent health threat. However, they are very sensitive indicators of surface contamination and can appear as the first agents of water quality change. Their occurrence in mineral water at source and after bottling should be limited to a low frequency of events and should be followed by a study to determine their origin. A single sampling procedure allows no flexibility in the interpretation of positive findings. There are objections against such a procedure in the sense that it may become technologically impracticable. A three class–sampling plan that incorporates so-called tolerances as used for microbiological safety requirements is much more rational (Codex Alimentarius, 1997). The recognition process for a new source of natural mineral water requires a few years' evidence of stability that must be demonstrated by continuous monitoring at the source of physicochemical and microbiological parameters listed in European directives (EC, 1980). Monitoring should include periodic sampling of water (at least four times a year) at the source point, with analysis for new pathogens such as *Cryptosporidium* and enteric viruses.

In the light of epidemiological and ecological data, it appears that the combination of two categories of markers, i.e. indicators of fecal contamination and indicators of vulnerability, may be the most appropriate for characterization of a microbiologically safe natural mineral water. However, taking into account the numerous outbreaks occurring in the world, especially in the US, it is recommended that microbiological monitoring be intensified to detect regularly (for instance, once a year), but not routinely, viral and protozoan pathogens.

12.7.3.3   Pathogens growing in water

There is a variety of environmental opportunistic human pathogens that can pass through water treatment barriers at very low densities and take advantage of selected sites in the water supply systems to colonize. They are typical biofilm organims that grow at the periphery of distribution systems (long pipes leading to dead ends) and throughout the pipe network where the water can become stagnant. The most important organisms to consider are *Legionella, P. aeruginosa, Aeromonas* and *M. avium* complex. Their significance in treated drinking water has been discussed in detail by Leclerc (2003).

It is now well established that legionellae are ubiquitous in engineered water supplies plumbing systems of hospitals and other large buildings, being an important cause (5–15%) of community-acquired and hospital-acquired pneumonia (Atlas, 1999). Epidemiological and genetic studies demonstrated that environmental amebae have acted as an evolutionary incubator for the emergence of *Legionella pneumophila* as an opportunistic pathogen for humans (Swanson & Hammer, 2000). The occurrence of *Legionella* within biofilms and its ability to enter a VBNC state contribute to its survival. *Legionella* spp. are being looked at as bacteria able to contaminate mineral water. However, the occurrence of *Legionella* has never been reported from mineral water either at source or in bottles. But the problem is highly relevant for the use of mineral water in hydrothermal areas where warm spa water can promote the growth of legionellae (Bornstein *et al.*, 1989; Rocha *et al.*, 1995; Verissimo *et al.*, 1991).

Here, various care categories for patients, including nebulizers, hot whirlpool spas, baths or other aerosols generating mechanical devices, can increase the risk of acquiring legionnaires' disease. Risk assessment has important implications fort the maintenance of adequate standards of hygiene, bacteriological monitoring and clinical surveillance in the establishments.

P. aeruginosa is a ubiquitous environmental bacterium. It can be isolated, often in high numbers, in common food, especially vegetables. Its presence is constant in surface waters and sometimes at low levels in drinking water. Other than certain specific hosts at risk, the general population is resistant to infection with *P. aeruginosa* (Hardalo & Edberg, 1997). Since *P. aeruginosa* is capable of growing abundantly in the purest of freshwaters and since it has major opportunistic pathogen capability, its occurrence in natural mineral water should be limited as far as possible. So there are two reasons to monitor *P. aeruginosa* in mineral water: on the one hand, as an indicator of vulnerability and/or poor control of the bottling environment, on the other hand, as an opportunistic pathogen.

Aeromonas are widespread in surface waters. Their presence in sediment accumulated in pipelines in the water supply is an indication of biofilm development. Their significance in drinking water relative to the occurrence of gastrointestinal infections is a much debated question (Leclerc, 2003). *Aeromonas* spp. are sometimes able to contaminate mineral water in low numbers for the same reason as coliforms or *P. aeruginosa*. Their significance is the same as that of quality indicators.

It has been shown recently that members of the genus *Mycobacterium* are present in drinking waters (Falkinham *et al.*, 2001; LeChevallier *et al.*, 2001; Le Dantec *et al.*, 2002); however, the numbers and frequencies of recovery of *M. avium* and *M. intracellulare* are usually low. The occurrence of nontuberculous mycobacteria such as *M. avium*, *M. kansasii* or *M. intracellulare* has never been reported for mineral water samples (Covert *et al.*, 1999).

12.8   Conclusion

The microbiology of Natural Mineral Waters is almost completely dependent on the hydrogeology and the microbial ecology of groundwater. Our knowledge of the natural flora of mineral water also depends on our ability to assess microorganisms in the environment, and is further compounded by the heterogeneity and the inaccessibility of these subterranean environments. Moreover, the bacteriological methods used to study mineral waters have often been chosen according to public health demands in detriment to our need to understand the ecological aspects of this type of environment.

The study of groundwater microbiology has progressed markedly since the 1970s. It is not, therefore, surprising to find viable microbial communities in the groundwater habitat, even at enormous depths. Within the limits of water availability, temperature and other factors such as the levels of organic components that affect living cells, groundwater habitats are expected to contain primarily bacteria and to a lesser extent fungi and protozoa. The abundance and diversity of the microbial communities is expected to vary depending on the geochemical and hydrogeological properties of the aquifer. The bacterial populations, which are by far the most abundant members of the groundwater community, have a large variety of metabolic capabilities that confer selective advantages in this environment. On the other hand, we know almost nothing about the microbiological succession and development of community structure in the groundwater habitat.

Most of the studies on the microbiology of Natural Mineral Waters have been performed on heterotrophic bacteria isolated in standardized growth media for HPC. The majority of these organisms belongs to the genus *Pseudomonas* but cannot be identified at the species level, or to other closely related genera, and may represent the predominant flora at the source. However, other bacterial groups, namely the prosthecate bacteria or species belonging to the genera *Cytophaga*, *Flavobacterium* or *Flexibacter* may also constitute major populations of heterotrophic bacteria in groundwater. The occurrence and, above all, the importance of gram-positive bacteria remain the object of conjecture and further study.

In shallow or deep aquifers, the supply of available carbon and energy sources may be extremely small. Nutrient limitation or starvation is common for most bacteria in groundwater. Starved bacteria may form minicells that are able to escape starvation by more efficient scavenging of nutrients; they may also become more resistant by synthesizing stress proteins. Bacterial cells of natural mineral water aquifers respond to nutrient deprivation by entering the VBNC state, with the probability that the VBNC state represents an additional response to starvation displayed by bacteria for survival.

After bottling, the number of viable and culturable bacteria increases appreciably within 3–7 days attaining 10^4–10^5 cfu/ml. Many factors such as use of glass or plastic containers, storage temperature, the level of organic carbon, incubation temperature of media and others can influence the fate of the bacterial flora in the bottle. Whether the culturable cells are a result of true resuscitation or of regrowth of a few initial cells remains questionable. It is now fairly certain that the genetic diversity of groundwater bacteria is maintained after bottling, as has been demonstrated by molecular-genetic studies.

The predominant populations isolated from mineral waters, and specifically those of the genus *Pseudomonas* exhibit *sensu stricto* an antagonistic activity on test pathogens or indicators of fecal contamination under some experimental conditions. This antagonistic activity may be due to the synthesis of siderophores or other antibiotic substances.

The bacteria isolated from Natural Mineral Waters do not belong to species known to be pathogenic or to have public health importance. Overall experimental data from animal model systems and epidemiological studies show that these bacteria have never been responsible for detectable pathological disorders in man, and *in vitro*, they are incapable of damaging human cells in tissue culture. After the implementation of European regulations in 1980, no outbreak or disease case due to the consumption of Natural Mineral Water has been recorded in the literature.

Ecological data, especially the diversity and physiological properties of bacterial communities, are essential together with epidemiological studies in order to perform a risk analysis for natural mineral waters. On a continuing basis, the management of microbial risks has to rely on assessment of the HPC and, more specifically, on detection of marker organisms, i.e. the classic fecal contamination indicators that have to be absent and vulnerability indicators for which the occurrence should be as low as possible. It is also recommended to search regularly, but not routinely, for viral and protozoan pathogens.

References

Abbaszadegan, M., Stewart, P. & LeChevallier, M. (1999) A strategy for detection of viruses in groundwater by PCR. *Applied and Environmental Microbiology*, **65**, 444–449.

Afonso, A., Teixeira, G., Cunha, J., Lobo, M., Barbosa, M., Teixeira, P., Ramalho, R. & Gibbs, P. (1998a) Factors affecting survival of pathogens in a Portuguese bottled mineral water. Poster presented at the 67th Annual Meeting of the Society for Applied Microbiology. *Supplement to Journal of Applied Microbiology*, **85**(1), **P15**.

Afonso, A., Teixeira, G., Cunha, J., Lobo, M., Barbosa, M., Teixeira, P., Ramalho, R. & Gibbs, P. (1998b) Survival of pathogens in a Portuguese bottled mineral water. Poster presented at the 67th Annual Meeting of the Society for Applied Microbiology. *Supplement to Journal of Applied Microbiology*, **85**(1), P14.

Amann, R. I. (1995) Fluorescently labelled, rRNA-targeted oligonucleotide probes in the study of microbial ecology. *Molecular Ecology*, **4**, 543–553.

Amann, R. I., Binder, B. J., Olson, R. J., Chisholm, S. W., Devereux, R. & Stahl, D. A. (1990a) Combination of 16S rRNA-targeted oligonucleotide probes with flow cytometry for analyzing mixed microbial populations. *Applied and Environmental Microbiology*, **56**, 1919–1925.

Amann, R. I., Krumholz, L. & Stahl, D. A. (1990b) Fluorescent-oligonucleotide probing of whole cells for determinative, phylogenetic, and environmental studies in microbiology. *Journal of Bacteriology*, **172**, 762–770.

Amann, R. I. & Ludwig, W. (2000) Ribosomal RNA-targeted nucleic acid probes for studies in microbial ecology. *FEMS Microbiological Review*, **24**, 555–565.

Amann, R. I., Ludwig, W. & Schleifer, K. H. (1995) Phylogenetic identification and in situ detection of individual microbial cells without cultivation. *Microbiological Reviews*, **59**, 143–169.

American Public Health Association (1985) *Standard Methods for the Examination of Water and Wastewater*, 16th edn. American Public Health Association, Washington, DC, pp. 853–917.

Atlas, R. M. (1999) *Legionella*: from environmental habitats to disease pathology, detection and control. *Environmental Microbiology*, **1**, 283–293.

Baida, N., Yazourh, A., Singer, E. & Izard, D. (2001) *Pseudomonas brenneri* sp. nov., a new species isolated from natural mineral waters. *Research in Microbiology*, **152**, 493–502.

Baida, N., Yazourh, A., Singer, E. & Izard, D. (2002) *Pseudomonas grimontii* sp. nov., a new species isolated from natural mineral waters. *International Journal of Systematic Bacteriology*, **52**, 1497–1503.

Baker, R. M., Singleton, F. L. & Hood, M. A. (1983) Effects of nutrient deprivation on *Vibrio cholerae*. *Applied and Environmental Microbiology*, **46**, 930–940.

Baldini, I., Current, E. & Scandelari, E. (1973) Rilievi e osservazioni sulle modificazioni; nel tempo, della flora microbica naturale di un' acqua minerale imbottigliata in relazione al giudizio igienico. *Annali di sanita publica*, **33**, 1009–1021.

Balkwill, D. L. (1989) Numbers, diversity, and morphological characteristics of aerobic, chemoheterotrophic bacteria in deep subsurface sediments from a site in South Carolina. *Geomicrobiology Journal*, **7**, 33–52.

Balkwill, D. L. & Ghiorse, W. C. (1985) Characterization of subsurface bacteria associated with two shallow aquifers in Oklahoma. *Applied and Environmental Microbiology*, **50**, 580–588.

Balkwill, D. L., Leach, F. R., Wilson, J. T., McNabb, J. F. & White, D. C. (1988) Equivalence of microbial biomass measures based on membrane lipid and cell wall components, adenosine triphosphate, and direct counts in subsurface aquifer sediments. *Microbial Ecology*, **16**, 73–84.

Bar-Or, Y. (1990) Hydrophobicity in the aquatic environment. In *Microbial Cell Surface Hydrophobicity*, R. J. Doyle & M. Rosenberg (eds.). American Society for Microbiology, Washington, DC, pp. 211–228.

Berg, J., Aieta, E. M. & Roberts, P. V. (1979) Effectiveness of chlorine dioxide as a wastewater disinfectant. In *Progress in Wastewater Technology, Proceedings of the National Symposium, Cincinnati, Ohio, September 1978*, A. D. Venosa (ed.). EPA-600/9-018, US Environmental Protection Agency, Cincinnati, pp. 61–71.

Berg, J., Hoff, J. C., Roberts, P. V. & Matin, A. (1984) Growth of *Legionella pneumophila* in continuous culture and its sensitivity to inactivation by chloride dioxide. *Applied and Environmental Microbiology*, **49**, 2465–2467.

Beuret, C., Kohler, D. & Luthi, T. (2000) Norwalk-like virus sequences detected by reverse transcription-polymerase chain reaction in mineral waters imported into or bottled in Switzerland. *Journal of Food Protection*, **63**, 1576–1582.

Bischofberger, T., Cha, S. K., Schmitt, R., König, B. & Schmidt-Lorenz, W. (1990) The bacterial flora of non-carbonated, Natural Mineral Water from the springs to reservoir and glass and plastic bottles. *International Journal of Food Microbiology*, **11**, 51–72.

Blom, A., Harder, W. & Matin, A. (1992) Unique and overlapping pollutant stress proteins of *Escherichia coli*. *Applied and Environmental Microbiology*, **58**, 331–334.

Bogosian, G., Morris, P. J. L. & O'Neil, J. (1998) A mixed culture recovery method indicates that enteric bacteria do not enter the viable but nonculturable state. *Applied and Environmental Microbiology*, **64**, 1736–1742.

Bornstein, N., Marmet, D., Surgot, M., Nowicki, M., Arslan, A., Esteve, J. & Fleurette, J. (1989) Exposure to Legionellaceae at a hot spring spa: a prospective clinical and serological study. *Epidemiology and Infection*, **102**, 31–36.

Botzenhart, K. & Kufferath, R. (1976) On the growth of various *Enterobacteriaceae*, *Pseudomonas aeruginosa* and *Alcaligenes* spp. in distilled water, deionized water, tap water and mineral salt solution. *Zentralblatt fur Bakteriologie Parasitenkunde Infektionskrankheiten und Hygiene 1 Abt. Omg.* **B163**, 470–485.

Botzenhart, K. & Röpke, S. (1971) Lebensfähigkeit und Vermehrung von *Pseudomonas aeruginosa* in anorganischen Salzlösungen. *Archives of Hygiene*, **154**, 509–516.

Budzikiewicz, H. (1993) Secondary metabolites from fluorescent pseudomonads. *FEMS Microbiology Reviews*, **104**, 209–228.

Burge, S. H. & Hunter, P. R. (1990) The survival of enteropathogenic bacteria in bottled mineral water. *Rivista Italiana d'Igiene*, **50**, 401–406.

Buttiaux, R. & Boudier, A. (1960) Comportement des bactéries autotrophes dans les eaux minérales conservées en récipients hermétiquement clos. *Annales de l'Institut Pasteur de Lille*, **11**, 43–52.

Calderon, R. L. & Mood, E. W. (1988) Bacterial colonizing point-of-use, granular activated carbon filters and their relationship to human health. *US Environmental Protection Agency*, CR-811904-01-0.

Calderon, R. L. & Mood, E. W. (1991) Bacterial colonizing point-of-entry, granular activated carbon filters and their relationship to human health. *US Environmental Protection Agency*, CR-813978-01-0.

Chalmers, R. M., Aird, H. & Bolton, F. J. (2000) Waterborne *Escherichia coli* O157. *Journal of Applied Microbiology*, **88**, 124S–132S.

Chapelle, F. H. (1993) *Groundwater Microbiology & Geochemistry*, Wiley, New York.

Chapelle, F. H., Morris, J. T., McMahon, P. B. & Zelibor, J. L. Jr (1988) Bacterial metabolism and the del-13C composition of groundwater, Floridan aquifer, South Carolina. *Geology*, **16**, 117–121.

Codex Alimentarius (1997) Draft revised standard for Natural Mineral Waters. Alinorm 97/20, appendix II.

Colwell, R. R., Brayton, P. R., Grimes, D. J., Roszak, D. B., Huq, S. A. & Palmer, L. M. (1985) Viable but non-culturable *Vibrio cholerae* and related pathogens in the environment: implications for release of genetically engineered microorganisms. *Biotechnology*, **3**, 817–820.

Coroler, L., Elomari, M., Hoste, B., Gillis, M., Izard, D. & Leclerc, H. (1996) *Pseudomonas rhodesiae* sp. nov., a new species isolated from Natural Mineral Waters. *Systematic and Applied Microbiology*, **19**, 600–607.

Costerton, J. W. & Colwell, R. R. (1979) *Native Aquatic Bacteria: Enumeration, Activity, and Ecology*. American Society For Testing and Materials, STP 695, Philadelphia.

Covert, T. C., Rodgers, M. R., Reyes, A. L. & Stelma, G. N. (1999) Occurrence of nontuberculous mycobacteria in environmental samples. *Applied and Environmental Microbiology*, **65**, 2492–2496.

Dabboussi, F., Hamze, M., Elomari, M., Verhille, S., Baida, N., Izard, D. & Leclerc, H. (1999a) *Pseudomonas libanensis* sp. nov., a new species isolated from Lebanese spring waters. *International Journal of Systematic Bacteriology*, **49**, 1091–1101.

Dabboussi, F., Hamze, M., Elomari, M., Verhille, S., Baida, N., Izard, D. & Leclerc, H. (1999b) Taxonomy study of bacteria isolated from Lebanese spring waters: proposal for *Pseudomonas cedrella* sp. nov. and *P. orientalis* sp. nov. *Research in Microbiology*, **150**, 303–316.

Daley, R. J. & Hobbie, J. E. (1975) Direct counts of aquatic bacteria by a modified epifluorescence technique. *Limnology and Oceanography*, **20**, 875–882.

Davey, M. E. & O'Toole, G. A. (2000) Microbial biofilms: from ecology to molecular genetics. *Microbiology and Molecular Biological Review*, **64**, 847–867.

Davis, S. N. & De Wiest, R. J. M. (1966) *Hydrogeology*. Wiley, New York.

De Felip, G., Toti, L. & Iannicelli, P. (1976) Osservazzioni comparative sull' andamento della flora microbica delle acque minerali naturali confezionate in vetro 'PVC' e 'tetrabrik'. *Annali dell Istituto Superiore di Sanita*, **12**, 203–209.

De Vos, P. & De Ley, J. (1983) Intra- and intergeneric similarities of *Pseudomonas* and *Xanthomonas* ribosomal ribonucleic acid cistrons. *International Journal of Systematic Bacteriology*, **33**, 487–509.

De Vos, P., Kersters, K., Falsen, E., Pot, B., Gillis, M., Segers, P. & De Ley, J. (1985) *Comamonas* Davis and Park 1962 gen. nov., nom. rev. emend., and *Comamonas terrigena* Hugh 1962 sp. nov. nom. rev. *International Journal of Systematic Bacteriology*, **35**, 443–453.

De Vos, P., Van Landschoot, A., Segers, P., Tytgat, R., Gillis, M., Bauwens, M., Rossau, R., Goor, M., Pot, B., Kersters, K., Lizzaraga, P. & De Ley, J. (1989) Genotypic relationships and taxonomic localization of unclassified *Pseudomonas* and *Pseudomonas*-like strains by deoxyribonucleic acid: ribosomal ribonucleic acid hybridizations. *International Journal of Systematic Bacteriology*, **39**, 35–49.

Delabroise, A. M. & Ducluzeau, R. (1974) Le microbisme naturel de l'eau minérale, son développement, son innocuité sur l'organisme. *Annales d'Hygiène, Médecine et Nutrition*, **10**, 189–192.

Dickgiesser, N. & Rittweger, F. (1979) Examinations on the behaviour of Gram-positive and Gram-negative bacteria in aqua bidest and tap water at different initial colony counts and different temperatures. *Zentralblatt fur Bakteriologie Parasitenkunde Infektionskrank heiter und Hygiene Abt. 1 Originale*, **169**, 308–319.

Dive, D., Picard, J. P. & Leclerc, H. (1979) Les amibes dans les eaux d'alimentation: évaluation des risques. *Annales de Microbiologie (Institut Pasteur)*, **130A**, 487–498.

Ducluzeau, R., Bochand, J. M. & Dufresne, S. (1976a) La microflore autochtone de l'eau minérale: nature, caractères physiologiques, signification hygiénique. *Médecine et Nutrition*, **12**(2), 115–119.

Ducluzeau, R., Dufresne, S. & Bochand, J. M. (1976b) Inoculation of the digestive tract of axenic mice with the autochthonous bacteria of mineral water. *European Journal of Applied Microbiology*, **2**, 127–134.

Ducluzeau, R., Hudault, S. & Galpin, J. V. (1976c) Longevity of various bacterial strains of intestinal origin in gas-free mineral water. *European Journal of Applied Microbiology*, **3**, 227–236.

Ducluzeau, R. & Raibaud, P. (1974) Interaction between *Escherichia coli* and *Shigella flexneri* in the digestive tract of 'gnotobiotic' mice. *Infection and Immunology*, **9**, 730–733.

Ducluzeau, R. & Raibaud, P. (1976) La compétition bactérienne dans le tube digestif. *La Recherche*, **65**, 270–272.

Ducluzeau, R. & Raibaud, P. (1979) *Ecologie microbienne du tube digestif: ces microbes qui nous protègent*. Masson, Paris.

Duncan, H. E. & Edberg, S. C. (1995) Host–microbe interaction in the gastrointestinal tract. *Critical Reviews of Microbiology*, **21**, 85–100.

Edberg, S. C. (1996) Assessing health risk in drinking water from naturally occurring microbes. *Journal of Environmental Health*, **58**, 18–24.

Edberg, S. C., Gallo, P. & Kontnick, C. (1996) Analysis of the virulence characteristics of bacteria isolated from bottled, water cooler, and tap water. *Microbial Ecology in Health and Disease*, **9**, 67–77.

Edberg, S. C., Kops, S., Kontnick, C. & Escarzaga, M. (1997) Analysis of cytotoxicity and invasiveness of heterotrophic plate count bacteria (HPC) isolated from drinking water on blood media. *Journal of Applied Microbiology*, **82**, 455–461.

Edberg, S. C., Rice, E. W., Karlin, R. J. & Alen, M. J. (2000) *Escherichia coli*: the best biological drinking water indicator for public health protection. *Journal of Applied Microbiology*, **88**, 106S–116S.

EC (1980) Council Directive relating to the quality of water intended for human consumption (80/778/EEC). *Official Journal of European Community*, L229, 11–29.

EC (1998) Council Directive relating to the quality of water intended for human consumption (98/83/EEC). *Official Journal of European Community*, L330, 32–54.

Elomari, M., Coroler, L., Hoste, B., Gillis, M., Izard, D. & Leclerc, H. (1996) DNA relatedness among *Pseudomonas* strains isolated from Natural Mineral Waters and proposal of *Pseudomonas veronii* sp. nov. *International Journal of Systematic Bacteriology*, **46**, 1138–1144.

Falkinham, III J. O., Norton, C. D. & LeChevallier, M. W. (2001) Factors influencing numbers of *Mycobacterium avium*, *Mycobacterium intracellulare*, and other mycobacteria in drinking water distribution systems. *Applied and Environmental Microbiology*, **67**(3), 1225–1231.

Favero, M. S., Carson, L. A., Bond, W. W. & Petersen, N. J. (1971) *Pseudomonas aeruginosa*: growth in distilled water from hospitals. *Science*, **177**, 836–838.

Ferreira, A. C., Morais, P. V. & da Costa, M. S. (1993) Alterations in total bacteria, iodonitrophenyltetrazolium (INT)-positive bacteria, and heterotrophic plate counts of bottled mineral water. *Canadian Journal of Microbiology*, **40**, 72–77.

Fletcher, M. (1979) The attachment of bacteria to surfaces in aquatic environments. In *Adhesion of Microorganisms to Surfaces*, D. G. Ellwood, J. Melling & P. Rutter (eds.). Academic Press, London, pp. 87–108.

Fred, E. B., Wilson, F. C. & Davenport, A. (1924) The distribution and significance of bacteria in Lake Mendota. *Ecology*, **5**, 322–339.

Frederikson, J. K., Garland, T. R., Hicks, R. J., Thomas, J. M., Li, S. W. & McFadden, S. M. (1989) Lithotrophic and heterotrophic bacteria in deep subsurface sediments and their relation to sediment properties. *Geomicrobiology Journal*, **7**, 53–66.

Frederikson, J. K. & Phelps, T. J. (1997) Subsurface drilling and sampling. In *Manual of Environmental Microbiology*, C. J. Hurst & G. R. Knudsen (eds.). American Society for Microbiology, Washington, DC, pp. 527–540.

Freeze, R. A. & Cherry, J. A. (1979) *Groundwater*, Prentice-Hall, Englewood Cliffs, NJ.

Fujikawa, H., Wauke, T., Kusunoki, J., Noguchi, Y., Hashimoto, Y., Ohta, K. & Itoh T. (1996) Surveys on microbial foreign matter contamination in bottled mineral water in Tokyo. *Journal of Food Microbiology*, **13**, 41–44.

Gannon, J. T., Manilal, V. B. & Alexander, M. (1991a) Relationship between cell surface properties and transport of bacteria through soil. *Applied and Environmental Microbiology*, **57**, 190–193.

Gannon, J. T., Tan, Y., Baveye, P. & Alexander, M. (1991b) Effect of sodium chloride on transport of bacteria in a saturated aquifer material. *Applied and Environmental Microbiology*, **27**, 2497–2501.

Gassilloud, B., Schwarzbrod, L. & Gantzer, C. (2003) Presence of viral genomes in Mineral Waters: a sufficient condition to assume infectious risk? *Applied and Environmental Microbiology*, **69**, 3965–69.

Geldreich, E. E. (1996) Microbial quality of water supply in distribution systems. Lewis Publishers, CRD Press, Boca Raton, FL.

Gerba, C. P. (2000) Risk assessment. In *Environmental Microbiology*, C. P. Gerba, R. M. Maier & I. L. Pepper (eds.). Academic Press, London, pp. 557–571.

Ghiorse, W. C. & Balkwill, D. L. (1983) Enumeration and characterization of bacteria indigenous to subsurface environments. *Developments in Industrial Microbiology*, **24**, 213–224.

Ghiorse, W. C. & Wilson, J. L. (1988) Microbial ecology of the terrestrial subsurface. *Advances in Applied Microbiology*, **33**, 107–172.

Gill, P. R. & Warren, G. J. (1988) A iron-antagonized fungistatic agent that is not required for iron assimilation from a fluorescent rhizosphere pseudomonad. *Journal of Bacteriology*, **170**, 163–170.

Giovannoni, S. J., DeLong, E. F., Olsen, G. J. & Pace, N. R. (1988) Phylogenetic group-specific oligonucleotide probes for identification of single microbial cells. *Journal of Bacteriology*, **170**, 720–726.

Gleeson, C. & Gray, N. (1997) The coliform index and waterborne disease, problems of microbial drinking water assessment. E&FN Spon, London.

Gonzalez, C., Gutierrez, C. & Grande, T. (1987) Bacterial flora in bottled uncarbonated mineral drinking water. *Canadian Journal of Microbiology*, **33**, 1120–1125.

Groat, R., Schultz, J., Zychlinsky, E., Bockman, A. & Matin, A. (1986) Starvation proteins in *Escherichia coli*: kinetics of synthesis and role in starvation survival. *Journal of Bacteriology*, **168**, 486–493.

Guillot, E. & Leclerc, H. (1993) Bacterial flora in Natural Mineral Waters: characterization by ribosomal ribonucleic acid gene restriction patterns. *Systematic and Applied Microbiology*, **16**, 483–493.

Gutterson, N. I., Layton, T. J., Ziegle, J. S. & Warren, G. J. (1986) Molecular cloning of genetic determinants for inhibition of fungal growth by a fluorescent pseudomonad. *Journal of Bacteriology*, **165**, 696–703.

Hamdan, H., Weller, D. M. & Thomashow, L. S. (1991) Relative importance of fluorescent siderophores and other factors in biological control of *Gaeumannomyces graminis* var. *tritici* by *Pseudomonas fluorescens* 2-79 and M4-80R. *Applied and Environmental Microbiology*, **57**, 3270–3277.

Hardalo, C. & Edberg, S. C. (1997) *Pseudomonas aeruginosa*: assessment of risk from drinking water. *Critical Reviews of Microbiology*, **23**, 47–75.

Henry, M. B., Lynck, J. M. & Fermor, T. R. (1991) Role of siderophores in the biocontrol of *Pseudomonas tolaasii* by fluorescent pseudomonad antagonists. *Journal of Applied Bacteriology*, **70**, 104–108.

Heukelekian, H. & Heller, A. (1940) Relation between food concentration and surface for bacterial growth. *Journal of Bacteriology*, **40**, 547–558.

Hightower, L. E. (1991) Heat shock, stress proteins, chaperones, and proteotoxicity. *Cell*, **66**, 191–197.

Hirsch, P. (1979) Life under conditions of low nutrient concentration. In *Strategies of Microbial Life in Extreme Environments*, M. Shilo (ed.). Verlag Chemie, Weinheim, pp. 357–372.

Hirsch, P. & Rades-Rohkohl, E. (1983) Microbial diversity in a groundwater aquifer in northern Germany. *Developments in Industrial Microbiology*, **24**, 183–200.

Hirsch, P. & Rades-Rohkohl, E. (1988) Some special problems in the determination of viable counts of groundwater microorganisms. *Microbial Ecology*, **16**, 99–113.

Hirsch, P. & Rades-Rohkohl, E. (1990) Microbial colonization of aquifer sediment exposed in a groundwater well in northern Germany. *Applied and Environmental Microbiology*, **56**, 2963–2966.

Hobbie, J. E., Daley, R. J. & Jasper, S. (1977) Use of nucleopore filters for counting bacteria by fluorescence microscopy. *Applied and Environmental Microbiology*, **33**, 1225–1228.

Hoff, K. A. (1988) Rapid and simple method for double staining of bacteria with 4'6-diamidino-2-phenylindole and fluorescein isothiocyanate-labeled antibodies. *Applied and Environmental Microbiology*, **54**, 2949–2952.

Howell, C. R. & Stipanovic, R. D. (1979) Control of *Rhizoctonia solani* on cotton seedlings with *Pseudomonas fluorescens* and with an antibiotic produced by the bacterium. *Phytopathology*, **69**, 480–482.

Howell, C. R. & Stipanovic, R. D. (1980) Suppression of *Pythium ultimum*-induced damping off of cotton seedlings by *Pseudomonas fluorescens* and its antibiotic pyoluteorin. *Phytopathology*, **70**, 712–715.

Hunter, P. R. & Burge, S. H. (1987) The bacteriological quality of bottled Natural Mineral Waters. *Epidemiology and Infection*, **99**, 439–443.

Hunter, P. R., Burge, S. H. & Hornby, H. (1990) An assessment of the microbiological safety of bottled mineral waters. *Rivista Italiana d'Igiene*, **50**, 394–400.

Ingram, M. (1977) The significance of index and indicator organisms in foods. Presented at International Symposium IUMS. Comm. Food Microbiol. Hygiène, 10th, Szczecin, Poland. *Lancet* 2, 1425.

Jannash, H. W. (1979) Microbial ecology of aquatic low nutrient habitats. In *Strategies of Microbial Life in Extreme Environments*, M. Shilo (ed.). Verlag Chemie, Weinheim, pp. 243–260.

Jayasekara, N. Y., Heard, G. M., Cox, J. M. & Fleet, G. H. (1999) Association of microorganisms with the inner surfaces of bottles of non-carbonated mineral waters. *Food Microbiology*, **16**, 115–128.

Jones, C. R., Adams, M. R., Zhdan, P. A. & Chamberlain, A. H. L. (1999) The role of surface physicochemical properties in determining the distribution of the autochthonous microflora in mineral water bottles. *Journal of Applied Microbiology*, **86**, 917–927.

Jones, R. E., Beeman, R. E. & Suflita, J. M. (1989) Anaerobic metabolic processes in deep terrestrial subsurface. *Geomicrobiology Journal*, **7**, 117–130.

Karapinar, M. & Gönül, S. A. (1991) Survival of *Yersinia enterocolitica* and *Escherichia coli* in spring water. *International Journal of Food Microbiology*, **13**, 315–320.

Keel, C., Wirthner, P. H., Oberhansli, T. H., Voisard, C., Burger, D., Haas, D. & Defago, G. (1990) Pseudomonads as antagonists of plant pathogens in the rhizosphere: role of the antibiotic 2,4-diacetylphloroglucinol in the suppression of black root rot of tobacco. *Symbiosis*, **9**, 327–341.

Kerr, M., Fitzgerald, M. & Sheridan, J. J. (1998) A study on the survival of *E. coli* O157:H7 in natural mineral water. Poster presented at the 67th Annual Meeting of the Society for Applied Microbiology. *Supplement to Journal of Applied Microbiology*, **85**(1), 17.

Kersters, K., Ludwig, W., Vancanneyt, M., De Vos, P., Gillis, M. & Schleifer, K. H. (1996) Recent changes in the classification of the Pseudomonads: an overview. *Systematic and Applied Microbiology*, **19**, 465–477.

Ketchum, B. H., Ayers, J. C. & Vaccaro, R. F. (1952) Processes contributing to the decrease of coliform bacteria in a tidal estuary. *Ecology*, **33**, 247–258.

Kjelleberg, S., Flärdh, K. B. G., Nyström, T. & Moriarty, D. J. W. (1993) Growth limitation and starvation of bacteria. In *Aquatic Microbiology: an Ecological Approach*, T. E. Ford (ed.). Blackwell Scientific Publications, Boston, pp. 289–320.

Kloepper, J. W., Hume, D. J., Scher, F. M., Singleton, C., Tipping, B., Laliberte, M., Frauley, K., Kutchaw, T., Simonson, C., Lifshitz, R., Zaleska, I. & Lee, L. (1988) Plant growth-promoting rhizobacteria on canola (rapeseed). *Plant Diseases*, **72**, 42–46.

Kloepper, J. W., Schroth, M. N. & Miller, T. D. (1980) Effects of rhizosphere colonization by plant growth-promoting rhizobacteria on potato plant development and yield. *Phytopathology*, **70**, 1078–1082.

Kogure, K., Simidu, U., Taga, N. & Colwell, R. R. (1979) A tentative direct microscopic method for counting living marine bacteria. *Canadian Journal Microbiology*, **25**, 415–420.

Kölbel-Boelke, J., Anders, E. & Nehrkorn, A. (1988) Microbial communities in the saturated groundwater environment. II. Diversity of bacterial communities in a pleistocene sand aquifer and their in vitro activities. *Microbial Ecology*, **16**, 31–48.

Kovavora-Kovar, K. & Egli, T. (1998) Growth kinetics of suspended microbial cells from single-substrate-controlled growth to mixed-substrate kinetics. *Microbial Molecular Biology Review*, **62**, 646–667.

Lacoste, T. (1992) La matière organique des eaux thermales: caractérisation par des méthodes globales et spécifiques et évolution lors des traitements de désinfection. *Thèse de Doctorat d'Université*, Poitiers.

Laine, M. H., Karwoski, M. T., Raaska, L. B. & Mattila-Sandholm, T. M. (1996) Antimicrobial activity of *Pseudomonas* spp. against food poisoning bacteria and moulds. *Letters in Applied Microbiology*, **22**, 214–218.

Lapteva, N. A. (1987) Ecological characteristics of *Caulobacter* incidence in fresh-water basins. *Mikrobiologiya*, **56**, 677–684.

LeChevallier, M. W., Norton, C. D., Falkinham III, J. O., Williams, M. D., Taylor, R. H. & Cowan, H. E. (2001) Occurrence and control of *Mycobacterium avium* complex. AWWA Research Foundation and American Water Works Association, Denver, CO, pp. 1–115.

Leclerc, H. (1990) Les qualités bactériologiques de l'eau minérale d'Evian. In *De l'eau minérale d'Evian*. S.A. des Eaux Minérales d'Evian, Evian, France, pp. 24–27.

Leclerc, H. (1994) Les eaux minérales naturelles: flore bactérienne native, nature et signification. In *Colloque International: les eaux minérales conditionnées, maîtrise qualité*, 26–27 May, Tunis.

Leclerc, H. (2003) Y a-t-il des infections bactériennes opportunistes transmises par les eaux d'alimentation? *European Journal of Water Quality*, **34**(1), 11–45.

Leclerc, H., Edberg, S. C., Pierzo, V. & Delattre, J. M. (2000) Bacteriophages as indicators of enteric viruses and public health risk in groundwaters. *Journal of Applied Microbiology*, **55**, 5–21.

Leclerc, H. & Mossel, D. A. A. (1989) *Microbiologie: le tube digestif, l'eau et les aliments*. Doin ed, Paris.

Leclerc, H., Mossel, D. A. A. & Savage, C. (1985) Monitoring non-carbonated (still) mineral waters for aerobic colonization. *International Journal of Food Microbiology*, **2**, 341–347.

Leclerc, H., Mossel, D. A. A., Edberg, S. C. & Struijk, C. B. (2001) Advances in the bacteriology of the coliform group: their suitability as markers of microbial water safety. *Annual Review of Microbiology*, **55**, 134–201.

Leclerc, H., Schwartzbrod, L. & Dei Cas, E. (2002) Microbial agents associated with waterborne diseases. *Critical Reviews in Microbiology*, **28**(4), 371–409.

Le Dantec, C., Duguet, J. P., Montiel, A., Dumoutier, N., Dubrou, S. & Vincent, V. (2002) Occurrence of mycobacteria in water treatment lines and in water distribution systems. *Applied and Environmental Microbiology*, **68**(11), 5318–5325.

Leisinger, T. & Margraff, R. (1979) Secondary metabolites of the fluorescent pseudomonads. *Microbiological Reviews*, **43**, 422–442.

Lindberg, G. D. (1981) An antibiotic lethal to fungi. *Plant Disease*, **65**, 680–683.

Lindqvist, R. & Bengtsson, G. (1991) Dispersal dynamics of groundwater bacteria. *Microbial Ecology*, **21**, 49–72.

Lucas, F. & Ducluzeau, R. (1990a) Antagonistic role of various bacterial strains from the autochthonous flora of gas-free mineral water against *Escherichia coli*. *Sciences des Aliments*, **10**, 62–73.

Lucas, F. & Ducluzeau, R. (1990b) Behaviour of the autochthonous flora of mineral water from Vittel in the digestive tract of axenic mice, its antagonistic effect on *Escherichia coli in vitro*. *Rivista Italiana d'Igiene*, **50**, 383–393.

Lye, D. J. & Dufour, A. P. (1991) A membrane filter procedure for assaying cytotoxic activity in heterotrophic bacteria isolated from drinking water. *Journal of Applied Bacteriology*, **70**, 89–94.

Madsen, E. D. (1991) Determining *in situ* biodegradation: facts and challenges. *Environmental Sciences and Technology*, **25**, 1662–1673.

Madsen, E. L. & Ghiorse, W. C. (1993) Groundwater microbiology: subsurface ecosystem processes. In *Aquatic Microbiology: an Ecological Approach*, T. E. Ford (ed.). Blackwell Scientific Publications, Boston, pp. 167–213.

Manaia, C. M., Nunes, O. C., Morais, P. V. & da Costa, M. S. (1990) Heterotrophic plate counts and the isolation of bacteria from mineral waters on selective and enrichment media. *Journal of Applied Bacteriology*, **69**, 871–876.

Masson, A. & Chauvin, D. (1974) Etude de quelques phénomènes pouvant influencer la charge microbienne des eaux minérales en bouteilles plastique et verre. *Industries alimentaires et agricoles*, **5**, 531–540.

Matin, A. (1990) Molecular analysis of the starvation stress in *E. coli*. *Journal of Bacteriology*, **170**, 3910–3914.

Matin, A. (1991) The molecular basis of carbon starvation–induced general resistance in *Escherichia coli*. *Molecular Microbiology*, **5**, 3–11.

Matin, A., Auger, E., Blum, P. & Schultz, J. (1989) Genetic basis of starvation survival in non-differentiating bacteria. *Annual Reviews Microbiology*, **43**, 293–316.

Matin, A. & Harakeh, S. (1990) Effect of starvation on bacterial resistance to disinfectants. In *Drinking Water Microbiology*, G. A. McFeters (ed.). Springer Verlag, New York, pp. 88–103.

Mavridou, A. (1992) Study of the bacterial flora of a non-carbonated Natural Mineral Water. *Journal of Applied Bacteriology*, **73**, 355–361.

McCann, M. P., Kidwell, J. & Matin, A. (1992) Microbial starvation survival, genetics. In *Encyclopedia of Microbiology*, vol. 3. Academic Press, London, p. 159.

Meinhardt, P. L., Casemore, D. P. & Miller, K. B. (1996) Epidemiologic aspects of human cryptosporidiosis and the role of waterborne transmission. *Epidemiological Reviews*, **18**, 118.

Mergaert, J., Schirmer, A., Hauben, L., Mau, M., Hoste, B., Kersters, K., Jendrossek, D. & Swings, J. (1996) Isolation and identification of poly(3-hydroxyvalerate) degrading strains of *Pseudomonas legmoignei*. *International Journal of Systematic Bacteriology*, **46**, 769–773.

Moore, E. R. B., Mau, M., Arnscheidt, A., Van de Peer, Y., De Wachter, R., Collins, M. D., Boettger, E. C. & Timmis, K. N. (1996) Determination and comparison of the 16S rRNA gene sequences of species of *Pseudomonas* (*sensu stricto*) and estimation of intrageneric relationships. *Systematic and Applied Microbiology*, **19**, 478–492.

Morais, P. V. & da Costa, M. S. (1990) Alterations in the major heterotrophic bacterial populations isolated from a still bottled mineral water. *Journal of Applied Bacteriology*, **69**, 750–757.

Moreau, A. (2001) Surveillance microbiologique des eaux minérales naturelles. *Bulletin de la Société Francaise de Microbiologie*, **16**, 13–19.

Moreira, L., Agostinho, P., Morais, P. V. & da Costa, M. S. (1994) Survival of allochthonous bacteria in still mineral water bottled in polyvinyl chloride (PVC) and glass. *Journal of Applied Bacteriology*, **77**, 334–339.

Morgan, B., Christman, M., Jacobson, F., Storz, G. & Ames, B. (1986) Hydrogen peroxide-inducible proteins in *Salmonella typhimurium* overlap with heat shock and other stress proteins. *Proceeding of the National Academy of Sciences of the United States of America*, **83**, 8059–8063.

Morgan, P. & Dow, C. S. (1985) Environmental control of cell-type expression in prosthecate bacteria. In *Bacteria in their Natural Environments*, M. Fletcher & G. D. Floddgate (eds.). Academic Press, London, pp. 131–169.

Morita, R. Y. (1982) Starvation-survival of heterotrophs in the marine environment. *Advances in Microbiology and Ecology*, **6**, 117–198.

Morita, R. Y. (1987) Starvation-survival: the normal mode of most bacteria in the ocean. In *Current Perspectives in Microbial Ecology*, F. Megusar & M. Gantar (eds.). Slovene Society of Microbiology, Ljubljana, Yugoslavia, pp. 243–248.

Morita, R. Y. (1993) Bioavailability of energy and the starvation state. In *Starvation in Bacteria*, S. Kjelleberg (ed.). Plenum Press, New York and London, pp. 1–23.

Morita, R. Y. (1997) Starved bacteria in oligotrophic environments. In *Bacteria in Oligotrophic Environments*, Chapman and Hall Microbiology Series. International Thomson Publishing, London, pp.193–246.

Mossel, D. A. A., Corry, J. E. L., Struijk, C. B. & Baird, R. M. (1995) *Essentials of the Microbiology of Foods: a Textbook for Advanced Studies*. Wiley, Chichester.

Moyer, C. L. & Morita, R. Y. (1989) Effect of growth rate and starvation-survival on the viability and stability of a psychrophilic marine bacterium. *Applied and Environmental Microbiology*, **55**, 1122–1127.

Neidhard, F., VanBogelen, R. & Vaughn, V. (1984) The genetics and regulation of heat-shock proteins. *Annual Reviews of Genetics*, **18**, 295–329.

Newby, D. T., Pepper, I. L. & Maier, R. M. (2000) Microbial transport. In *Environmental Microbiology*, C. L. Gerda, R. M. Maier & I. L. Pepper (eds.). Academic Press, London, pp. 147–176.

Norvatov, A. M. & Popov, O. V. (1961) Laws of the formation of minimum stream flow. *Bulletin of the International Association of Science and Hydrology*, **6**(1), 20–27.

O'Sullivan, D. J. & O'Gara, F. (1992) Traits of fluorescent *Pseudomonas* spp. involved in suppression of plant root pathogens. *Microbiological Reviews*, **56**, 662–676.

Oger, C., Hernandez, J. F., Delattre, J. M., Delabroise, A. H. & Krupsky, S. (1987) Etude par épifluorescence de l'évolution de la microflore totale dans une eau minérale embouteillée. *Water Research*, **21**(4), 469–474.

Oliver, J. D. (1993) Formation of viable but nonculturable cells. In *Starvation in Bacteria*, S. Kjelleberg (ed.). Plenum Press, New York and London, pp. 239–272.

Oliver, J. D., Hite, F., McDougald, D., Andon, N. L. & Simpson, L. M. (1995) Entry into and resuscitation from the viable but nonculturable state by *Vibrio vulnificus* in an estuarine environment. *Applied and Environmental Microbiology*, 61, 2624–2630.

Palleroni, N. J. (1984) Genus I. *Pseudomonas* Migula 1894. In *Bergey's Manual of Systematic Bacteriology*, vol. 1, N. R. Krieg & J. G. Holt (eds.). Williams and Wilkins, Baltimore, pp. 141–199.

Palleroni, N. J. & Bradbury, J. F. (1993) *Stenotrophomonas*, a new bacterial genus for *Xanthomonas maltophilia* (Hugh, 1980; Swings *et al.* 1983). *International Journal of Systematic Bacteriology*, 43, 606–609.

Palleroni, N. J., Kunisawa, R., Contopoulou, R. & Doudoroff, M. (1973) Nucleic acid homologies in the genus *Pseudomonas*. *International Journal of Systematic Bacteriology*, 23, 333–339.

Payment, P., Coffin, E. & Paquette, G. (1994) Blood agar to detect virulence factors in tap water heterotrophic bacteria. *Applied and Environmental Microbiology*, 60, 1179–1183.

Payment, P., Fanco, E., Richardson, L., Siemiatycki, J. (1991a) Gastrointestinal health effects associated with the consumption of drinking water produced by point-of-use domestic reverse-osmosis filtration units. *Applied and Environmental Microbiology*, 57, 945–948.

Payment, P., Richardson, L., Siematycki, J., Dewar, R., Edwards, M. & Franco, E. (1991b) A randomized trial to evaluate the risk of gastrointestinal disease due to the consumption of drinking water meeting currently accepted microbiological standards. *American Journal of Public Health*, 81, 703–708.

Payment P., Siematycki, J., Richardson, L., Renaud, G., Franco, E. & Prévost, M.(1997) A prospective epidemiological study of gastrointestinal health effects due to the consumption of drinking water. *International Journal of Environmental Health Research*, 7, 5–31.

Poindexter, J. S. (1981a) Oligotrophy: fast and famine existence. In *Microbial Ecology*, vol. 5, M. Alexander (ed.). Plenum Press, New York, pp. 63–89.

Poindexter, J. S. (1981b) The caulobacters: ubiquitous, unusual bacteria. *Microbiological Reviews*, 45, 123–179.

Poindexter, J. S. (1992) Dimorphic prosthecate bacteria: the genera *Caulobacter*, *Asticcacaulis*, *Hyphomicrobium*, *Pedomicrobium*, *Hyphomonas*, and *Thiodendron*. In *The Prokaryotes*, 2nd edn, A. Balows & H. G. Trüper (eds.). Springer Verlag, New York, pp. 2176–2196.

Porter, K. G. & Feig, Y. S. (1980) The use of DAPI for identifying and counting aquatic microflora. *Limnology and Oceanography*, 25, 943–948.

Quevedo-Sarmiento, J., Ramos-Cormenzana, A. & Gonzalez-Lopez, J. (1986) Isolation and characterization of aerobic heterotrophic bacteria from natural spring waters in the Lanjaron area (Spain). *Journal of Applied Bacteriology*, 61, 365–372.

Raaijmakers, J. M., Weller, D. M. & Thomashow, L. S. (1997) Frequency of antibiotic-producing *Pseudomonas* spp. in natural environments. *Applied and Environmental Microbiology*, 63, 881–887.

Reasoner, D. J. (1990) Monitoring heterotrophic bacteria in potable water. In *Drinking Water Microbiology*, G. A. McFeters (ed.). Springer Verlag, New York, pp. 452–477.

Reasoner, D. J. & Geldreich, E. E. (1979) A new medium for the enumeration and subculture of bacteria from potable water. In *Abstract of the Annual Meeting of the American Society for Microbiology*. American Society for Microbiology, Washington, DC, p. 180.

Reasoner, D. J. & Geldreich, E. E. (1985) A new medium for the enumeration and subculture of bacteria from potable water. *Applied and Environmental Microbiology*, 49, 1–7.

Reeve, C. A., Amy, P. S. & Matin, A. (1984a) Role of protein synthesis in the survival of carbon-starved *Escherichia coli* K-12. *Journal of Bacteriology*, 160, 1041–1046.

Reeve, C. A., Bockman, A. T. & Matin, A. (1984b) Role of protein degradation in the survival of carbon-starved *Escherichia coli* and *Salmonella typhimurium*. *Journal of Bacteriology*, **157**, 758–763.

Reichenbach, H. (1992) The order Cytophagales. In *The Prokaryotes*, 2nd edn, A. Balows, H. G. Trüper, M. Dworkin, W. Harder & K. H. Schleifer (eds.). Springer Verlag, New York, pp. 3631–3676.

Reynolds, P. J., Sharma, P., Jenneman, G. E. & Mc Inerney, M. J. (1989) Mechanisms of microbial movement in subsurface materials. *Applied and Environmental Microbiology*, **55**, 2280–2286.

Rivera, F., Galvan, M., Robles, E., Leal, P., Gonzalez, L. & Lacy, A. M. (1981) Bottles mineral waters polluted by protozoa in Mexico. *Journal of Protozoology*, **28**, 54–56.

Robertson, J. B. & Edberg, S. C. (1997) Natural protection of spring and well drinking water against surface microbial contamination. I. Hydrogeological parameters. *Critical Reviews of Microbiology*, **23**, 143–178.

Rocha, G., Veríssimo, A., Bowker, R. G., Bornstein, N. & da Costa, M. S. (1995) Relationship between *Legionella* spp. and antibody titres at a therapeutic spa in Portugal. *Epidemiology and Infection*, **115**, 79–88.

Rodriguez, G. G., Phipps, D., Ishiguro, K. & Ridgway, H. F. (1992) Use of a fluorescent redox probe for direct visualization of actively respiring bacteria. *Applied and Environmental Microbiology*, **58**, 1801–1808.

Roszak, D. B. & Colwell, R. R. (1987a) Metabolic activity of bacterial cells enumerated by direct viable count. *Applied and Environmental Microbiology*, **53**, 2889–2983.

Roszak, D. B. & Colwell, R. R. (1987b) Survival strategies of bacteria in the natural environment. *Microbiological Reviews*, **51**, 365–379.

Savage, D. C. & Fletcher, M. (1993) *Bacterial Adhesion: Mechanisms and Physiological Significance*. Plenum Press, New York.

Schmidt-Lorenz, W. (1974) Untersuchungen über den Keimgehalt von unkarbonisiertem, natürlichem mineralwasser und überlegungen zum bakteriologisch-hygienischem beurteilen von unkarbonisiertem mineralwasser. Teil. I. *Chemical Technology Lebensmittel*, **4**, 97–107.

Scholl, M. A., Mills, A. L., Herman, J. S. & Hornberger, G. M. (1990) The influence of mineralogy and solution chemistry on the attachment of bacteria to representative aquifer materials. *Journal Contamination Hydrology*, **6**, 321–326.

Schwaller, P. & Schmidt-Lorenz, W. (1980) Flore microbienne de quatre eaux minérales non gazéifiées et mises en bouteilles. I. Dénombrement de colonies, compositions grossière de la flore, et caractères du groupe des bactéries Gram négatif pigmentées en jaune. *Zentralblatt fur Bakteriologie und Hygiene I. Abt. Originale*, **C1**, 330–347.

Schwaller, P. & Schmidt-Lorenz, W. (1981a) Flore microbienne de quatre eaux minérales non gazéifiées et mises en bouteilles. II. Les *Pseudomonas* et autre bactéries à Gram négatif—Composition fine de la flore. *Zentralblatt für Bakteriologie und Hygiene I. Abt. Originale*, **C2**, 179–196.

Schwaller, P. & Schmidt-Lorenz, W. (1981b) Dénombrements de colonies bactériennes dans les eaux minérales naturelles non gazéifiées et mises en bouteilles. 1re communication: Etude de méthodes spécifiques pour le dénombrement de colonies bactériennes. *Brauerei Rundschau*, **92**, 281–285.

Schwaller, P. & Schmidt-Lorenz, W. (1982) Dénombrements de germes dans les eaux minérales naturelles non gazéifiées et mises en bouteilles. 2ème communication: Nombres de colonies dans 4 eaux minérales françaises. Comparaisons de méthodes officielles avec d'autres méthodes. *Brauerei Rundschau*, **93**, 25–33.

Segers, P., Vancanneyt, M., Pot, B., Torck, U., Hoste, B., Dewettinck, D., Falsen, E., Kersters, K. & De Vos, P. (1994) Classification of *Pseudomonas diminuta* (Leifson & Hugh, 1954) and *Pseudomonas vesicularis* (Büsing et al., 1953) in *Brevundimonas* gen. nov. as *Brevundimonas diminuta* comb. nov. and *Brevundimonas vesicularis* comb. nov., respectively. *International Journal of Systematic Bacteriology*, **44**, 499–510.

Shanahan, P., O'Sullivan, D. J., Simpson, P., Glennon, J. & O'Gara, F. (1992) Isolation of 2,4-diacetylphloroglucinol from a fluorescent pseudomonad and investigation of physiological parameters influencing its production. *Applied and Environmental Microbiology*, **58**, 353–358.

Shilo, M. (1980) Strategies of adaptation to extreme conditions in aquatic microorganisms. *Naturwissenschaften*, **67**, 384–389.

Sinclair, J. L. & Ghiorse, W. C. (1987) Distribution of protozoa in subsurface sediments of a pristine groundwater study site in Oklahoma. *Applied and Environmental Microbiology*, **53**, 1157–1163.

Sinclair, J. L. & Ghiorse, W. C. (1989) Distribution of aerobic bacteria, protozoa, algae, and fungi in deep subsurface sediments. *Geomicrobiology Journal*, **7**, 15–32.

Sinclair, J. L., Randtke, S. J., Denne, J. E., Hathaway, L. R. & Ghiorse, W. C. (1990) Survey of microbial populations in buried-valley aquifer sediments from northeastern Kansas. *Ground-Water*, **28**, 369–377.

Smith, R. L. (1997) Determining the terminal electron-accepting reaction in the saturated subsurface. In *Manual of Environmental Microbiology*, C. J. Hurst & G. R. Knudsen (eds.). American Society for Microbiology, Washington, DC, pp. 577–585.

Spector, M., Alibada, Z., Gonzales, T. & Foster, J. (1986) Global control in *Salmonella typhimurium*: two-dimensional electrophoretic analysis of starvation-, anaerobiosis-, and heat shock-inducible proteins. *Journal of Bacteriology*, **168**, 420–424.

Stackebrandt, E., Murray, R. G. E. & Trüper, G. H. (1988) *Proteobacteria* classis nov., a name for the phylogenetic taxon that includes the 'purple bacteria and their relatives'. *International Journal of Systematic Bacteriology*, **38**, 321–325.

Staley, J. T. (1968) *Prosthecomicrobium* and *Ancalomicrobium*: new prosthecate freshwater bacteria. *Journal of Bacteriology*, **95**, 1921–1942.

Suslow, T. V. (1982) Role of root colonizing bacteria in plant growth. In *Phytopathogenic Prokaryotes*, M. S. Mount & G. H. Lacy (eds.), vol. 1. Academic Press, London, pp. 187–223.

Swanson, M. S. & Hammer, B. K. (2000) *Legionella pneumophila* pathogenesis: a fateful journey from *Amoebae* to macrophages. *Annual Review of Microbiology*, **54**, 567–613.

Szewzyk, U., Szewzyk, R., Manz, W. & Schleifer, K. H. (2000) Microbiological safety of drinking water. *Annual Review of Microbiology*, **54**, 81–127.

Tamaoka, J., Ha, D. M. & Komagata, K. (1987) Reclassification of *Pseudomonas acidovorans* (den Dooren de Jong, 1926) and *Pseudomonas testosteroni* (Marcus & Talalay, 1956) as *Comamonas acidovorans* comb. nov. and *Comamonas testosteroni* comb. nov. with and emended description of the genus *Comamonas*. *International Journal of Systematic Bacteriology*, **37**, 52–59.

Thomashow, L. S. & Weller, D. M. (1988) Role of a phenazine antibiotic from *Pseudomonas fluorescens* in biological control of *Gaeumannomyces graminis* var. *tritici*. *Journal of Bacteriology*, **170**, 3499–3508.

Toth, J. (1963) A theoretical analysis of groundwater flow in small drainage basins. *Journal of Geophysical Research*, **68**(16), 4795–4812.

Vachée, A. & Leclerc, H. (1995) Propriétés antagonistes de la flore autochtone des eaux minérales naturelles vis-à-vis de *Pseudomonas aeruginosa*. *Journal Européen d'Hydrologie*, **26**, 327–338.

Vachée, A., Vincent, P., Struijk, C. B., Mossel, D. A. A. & Leclerc, H. (1997) A study of the fate of the autochthonous bacterial flora of still mineral waters by analysis of restriction fragment length polymorphism of genes coding for rRNA. *Systematic and Applied Microbiology*, **20**, 492–503.

Vandenbergh, P. A., Gonzalez, C. F., Wright, A. M. & Kunka, B. S. (1983) Iron-chelating compounds produced by soil pseudomonads: correlation with fungal growth inhibition. *Applied and Environmental Microbiology*, **46**, 128–132.

Van der Kooij, D. (1990) Growth measurements with *Pseudomonas aeruginosa*, *Aeromonas hydrophila* and autochthonous bacteria to determine the biological stability of drinking water. *Rivista Italiana d'Igiene*, **5**(6), 375–383.

Van der Kooij, D., Oranje, J. P. & Hijnen, W. A. M. (1982a) Growth of *Pseudomonas aeruginosa* in tap water in relation to utilization of substrates at concentrations of a few micrograms per liter. *Applied and Environmental Microbiology*, **44**, 1086–1095.

Van der Kooij, D., Visser, A. & Hijnen, W. A. M. (1982b) Determining the concentration of easily assimilable organic carbon in drinking water. *Journal of the American Water Works Association*, **74**, 540–545.

Verhille, S., Baida, N., Dabboussi, F., Izard, D. & Leclerc, H. (1999a) Taxonomic study of bacteria isolated from natural mineral waters: proposal of *Pseudomonas jessenii* sp. nov. and *Pseudomonas mandelii* sp. nov. *Systematic and Applied Microbiology*, **22**, 45–58.

Verhille, S., Baida, N., Dabboussi, F., Hamze, M., Izard, D. & Leclerc, H. (1999b) *Pseudomonas gessardii* sp. nov. and *Pseudomonas migulae* sp. nov., two new species isolated from natural mineral waters. *International Journal of Systematic Bacteriology*, **49**, 1559–1572.

Veríssimo, A., Marrão, G., Gomes da Silva, F. & da Costa, M. S. (1991) Distribution of *Legionella* spp. in hydrothermal areas in Continental Portugal and on the Island of S. Miguel, Azores. *Applied and Environmental Microbiology*, **57**, 2921–2927.

Von Graevenitz, A. (1977) The role of opportunistic bacteria in human disease. *Annual Review of Microbiology*, **31**, 447–471.

Wagner, M., Amann, R., Lemmer, H. & Schleifer, K. H. (1993) Probing activated sludge with oligonucleotide specific for proteobacteria: inadequacy of culture-dependent methods for describing microbial community structure. *Applied and Environmental Microbiology*, **59**, 1520–1525.

Walch, M. & Colwell, R. R. (1994) Detection of nonculturable indicators and pathogens. In *Environmental Indicators and Shellfish Safety*, C. R. Hackney & M. D. Pierson (eds). Chapman and Hall, New York and London, pp. 258–273.

Walker, G. (1984) Mutagenesis and inductible responses to deoxyribonucleic acid damage in *Escherichia coli*. *Microbiological Reviews*, **48**, 60–93.

Wallner, G., Amann, R. & Beisker, W. (1993) Optimizing fluorescent in situ hybridization of suspended cells with rRNA-targeted oligonucleotide probes for flow cytometric identification of microorganisms. *Cytometry*, **14**, 136–143.

Wallner, G., Erhart, R. & Amann, R. (1995) Flow cytometric analysis of activated sludge with rRNA-targeted probes. *Applied and Environmental Microbiology*, **61**, 1859–1866.

Wallner, G., Steinmetz, I., Bitter-Suermann, D. & Amann, R. (1996) Combination of rRNA-targeted hybridization probes and immuno-probes for the identification of bacteria by flow cytometry. *Systematic and Applied Microbiology*, **19**, 569–576.

Wang, G. & Doyle, M. P. (1998) Survival of enterohemorrhagic *Escherichia coli* O157:H7 in water. *Journal of Food Protection*, **61**, 662–667.

Warburton, D. W. (1993) A review of the microbiological quality of bottled water sold in Canada. Part 2. The need for more stringent standards and regulations. *Canadian Journal of Microbiology*, **39**, 158–168.

Warburton, D. W., Austin, J. W., Harrisson, B. W. & Sanders, G. (1998) Survival and recovery of *Escherichia coli* O157:H7 in inoculated bottled water. *Journal of Food Protection*, **61**, 948–952.

Warburton, D. W., Dodds, K. L., Burke, R., Johnston, M. A. & Laffey, P. J. (1992) A review of the microbiological quality of bottled water sold in Canada between 1981 and 1989. *Canadian Journal of Microbiology*, **38**, 12–19.

Warburton, D. W., Peterkin, P. I., Weiss, K. F. & Johnston, M. A. (1986) Microbiological quality of bottled water sold in Canada. *Canadian Journal of Microbiology*, **32**, 891–893.

Weichart, D. & Kjelleberg, S. (1996) Stress resistance and recovery potential of culturable and viable but nonculturable cells of *Vibrio vulnificus*. *Microbiology*, **42**, 845–853.

White, D. C., Fredrickson, J. F., Gehron, M. H., Smith, G. A. & Martz, R. F. (1983) The groundwater aquifer microbiota: biomass, community structure, and nutritional status. *Developments in Industrial Microbiology*, **24**, 189–199.

WHO (2002) Heterotrophic plate count measurement in drinking water safety management. WHO/SDE/WHS/02.10. Report of an expert meeting. Geneva, 24–25 April.

Willems, A., Falsen, E., Pot, B., Jantzen, E., Hoste, B., Vandamme, P., Gillis, M., Kersters, K. & De Ley, J. (1990) *Acidovorax*, a new genus for *Pseudomonas facilis, Pseudomonas delafieldii*, E. Falsen (EF) Group 13, EF Group 16, and several clinical isolates, with the species *Acidovorax facilis* comb.nov., *Acidovorax delafieldii* comb. nov., and *Acidovorax temperans* sp. nov. *International Journal of Systematic Bacteriology*, **40**, 384–398.

Wilson, N. (1995) *Soil, Water and GroundWater Sampling*. CRC Press, Boca Raton, FL.

Wilson, J. T., McNabb, J. F., Cochran, J. W., Wang, T. H., Tomson, M. B. & Bedient, P. B. (1983) Enumeration and characterization of bacteria indigenous to a shallow water-table aquifer. *GroundWater*, **21**, 134–142.

Wimpenny, J. W. T., Kinniment, S. L. & Scourfield, M. A. (1993) The physiology and biochemistry of biofilm. In *Microbial Biofilms: Formation and Control*, S. P. Denyer, S. P. Gorman & M. Sussman (eds). Blackwell Scientific Publications, Oxford, pp. 51–94.

Xu, H. S., Roberts, N. C., Singleton, F. L., Atwell, R. W., Grimes, D. J. & Colwell, R. R. (1982) Survival and viability of nonculturable *Escherichia coli* and *Vibrio cholerae* in the estuarine and marine environment. *Microbial Ecology*, **8**, 313–323.

Yabuuchi, E., Kosako, Y., Oyaizu, H., Yano, I., Hotta, H., Hashimoto, Y., Ezaki, T. & Arakawa, M. (1992) Proposal of *Burkholderia* gen. nov. and transfer of seven species of the genus *Pseudomonas* homology group II to the new genus, with the type species *Burkholderia cepacia* (Palleroni and Holmes 1981) comb. nov. *Microbiology and Immunology*, **36**, 1251–1275.

Yabuuchi, E., Kosako, Y., Yano, I., Hotta, H. & Nishiuchi, Y. (1995) Transfer of two *Burkholderia* and an *Alcaligenes* species to *Ralstonia* gen. nov.: proposal of *Ralstonia pickettii* (Ralston *et al.*, 1973) comb. nov., *Ralstonia solanacearum* (Smith, 1896) comb. nov. and *Ralstonia eutropha* (Davis, 1969) comb. nov. *Microbiology and Immunology*, **39**, 897–904.

Yabuuchi, E., Yano, I., Oyaizu, H., Hashimoto, Y., Ezaki, T. & Yamamoto, H. (1990) Proposals of *Sphingomonas paucimobilis* gen. nov. and comb. nov., *Sphingomonas parapaucimobilis* sp. nov., *Sphingomonas yanoikuyae* sp. nov., *Sphingomonas adhaesiva* sp. nov., *Sphingomonas capsulata* comb. nov., and two genospecies of the genus *Sphingomonas*. *Microbiology and Immunology*, **34**, 99–119.

Yates, M. V., Citek, R. W., Kamper, M. F. & Salazar, A. M. (1999) Detecting enteroviruses in water: comparing infectivity, molecular, and combination methods. Presented at International Symposium AWWA on waterborne pathogens, 29 August–1 September.

Yurdusev, N. & Ducluzeau, R. (1985) Qualitative and quantitative development of the bacterial flora of Vittel mineral water in glass or plastic bottles. *Sciences des Aliments*, **5**, 231–238.

Zimmerman, R., Iturriaga, R. & Becker-Birck, J. (1978) Simultaneous determination of the total number of aquatic bacteria and the number thereof involved in respiration. *Applied and Environmental Microbiology*, **36**, 926–935.

ZoBell, C. E. & Anderson, D. Q. (1936) Observations on the multiplication of bacteria in different volumes of stored sea water and the influence of oxygen tension and solid surface. *Biological Bulletin*, **71**, 324–342.

13 Microbiology of treated bottled water

Stephen C. Edberg

13.1 Introduction

Although this chapter examines the microbiology of treated bottled water, it is useful first to ask: 'Is it necessary for microbiological reasons to treat bottled water at all?' Since the inherent reason for treating bottled water is to remove undesirable constituents, if none are present then treatment is not needed. In Europe it is a long-established principle, which has been substantiated by practice, that high-quality, well-defined, sequestered source waters need not be treated. It is illegal in Europe to treat Natural Mineral Water microbiologically. How can the Europeans be so certain that such waters do not require treatment?

The answer lies in the rigid application of the multiple barrier concept for public health protection. The multiple barrier concept states that safe drinking water can be bottled if sequential high-level barriers to the entry of pathogens into the bottle are employed. Multiple barriers can be divided into the following basic types: source water protection; source water monitoring; ozonation; reverse osmosis; filtration; and distillation. In Europe, Natural Mineral Water sources are highly selected and qualified.

13.2 Source water protection and monitoring

Two multiple barriers are employed to ensure safety. First, the hydrogeology of the sources is well studied and established (source water protection). It is well known that the journey of water from rainfall to the aquifer contains numerous elements that naturally bring about pathogen removal. For example, passage through the various soil and clay substrata has profound adsorbing and filtering activities (Hopkins *et al.*, 1985). Time of transit itself diminishes microbial viability so that after approximately one year there should be no significant viable pathogens present in the water making its journey into the aquifer. Second, within the category of source water protection, is the rigid application of sanitary surveys. The source is protected by precluding human and animal activities that would contaminate it. Thus, the source water protection multiple barrier is very powerful when rigidly applied (Robertson and Edberg, 1997).

The second multiple barrier employed in Europe for source water protection of Natural Mineral Water is one of intensive monitoring (source water monitoring). The source water monitoring protocols utilized in Europe are significantly different from the protocols utilized for the monitoring of municipal waters for regulatory purposes. In Europe, source water monitoring has a number of inherent components and goals. First, intensive quality control monitoring is conducted to ensure that the water that enters the bottle is the same as that which is present in the source. In Europe, in order for water to be certified as Natural Mineral Water, the chemical and microbiological profile of the water in the source must be equivalent to that in the bottle. Second, intensive monitoring is conducted to ensure the water is pathogen free (Leclerc *et al.*, 1985). The intensive monitoring conducted for European NMWs is several orders of magnitude larger than that conducted for municipal drinking water under regulatory requirements. For example, in the industrialized countries, the most frequent routine monitoring sampling protocol requires approximately only one sample per 1000 population per month. By comparison, a bottler of Natural Mineral Water in Europe may conduct more than 100 tests per day. Tests will be conducted notably for pathogen surrogates (such as total coliforms and *Escherichia coli*), but also for indicators of external intrusion into the system (such as sulfite-reducing anaerobes and *Pseudomonas aeruginosa*). Therefore, as applied in the European Natural Mineral Water industry, source water monitoring is a powerful natural barrier (Edberg *et al.*, 1997).

Accordingly, the combination of natural source water protection and intensive source water monitoring in Europe provides two powerful multiple barriers, which when used in sequence, has produced microbiologically safe drinking water for many years.

Another situation exists in which safe drinking water is produced without treatment. In the United States, municipal waters served by sequestered, subterranean sources are not required to disinfect if they demonstrate, through the use of a combination of hydrogeology, sanitary surveys and monitoring, that there is no intrusion of pathogens. This protocol exists under the Ground Water Disinfection Requirement Act, which is administered by the US Environmental Protection Agency. In effect, groundwaters can avoid disinfection if they demonstrate safety. This municipal water situation should not be confused in one important respect with the European Natural Mineral Water industry. There is no requirement in the United States that there be standards of identity or quality control parameters associated with the final municipal drinking water product: what is required is pathogen-free drinking water.

13.3 Water treatment

If a bottler wishes to disinfect, there are a number of treatment options available. For practical purposes, each of the treatment options can be considered a multiple barrier. Therefore, only those that have wide applicability and provide high levels of anti-pathogen activity will be discussed here. Moreover, unlike municipal tap water, bottled water does not employ multiple barriers that produce disinfection residues. Therefore, multiple barriers such as the use of chlorine-based disinfectants are not applicable to bottled water.

There are four basic multiple barrier treatment options employed in the bottled water industry: ozonation, reverse osmosis, filtration and distillation. As Table 13.1 demonstrates, each has its own strengths for particular groups of microbial pathogens.

Table 13.1 Relative effectiveness of treatment types on pathogen groups

Pathogen group	Ozonation	Reverse Osmois	Filtration	Distillation
		Effectiveness		
Viruses	Good	Good	Low	High
Bacteria	Good	Good	Low	High
Protozoa	Fair	High	High	High

It is important at this level of discussion to define microbiological treatment as the employment of more than one multiple barrier sequentially to produce pathogen-free water. There will always be some number of autochthonous, non-pathogenic naturally-occurring bacteria that enter the bottle. These autochthonous bacteria are not only natural but may actually inhibit other bacteria in the bottle (see Chapter 12). It is beyond the scope of this chapter to discuss each of the treatment multiple barriers in detail. However, there are some salient points about each that should be considered while evaluating their effect on microbiological activity.

Ozone is a powerful, short-acting, high-energy disinfectant that is generally created on-site. There are a number of ways to employ ozone; sufficient for this chapter is to understand the concept of CT: this is a value associated with the disinfecting power of energy-yielding disinfectants such as ozone, chlorine and monochloramines, in which the disinfecting power of the individual disinfectant is related to the activity concentration of the disinfectant (C) multiplied by the time the disinfectant is in contact with the pathogen (T). In the literature relating to treatment of municipal water, ozone is said not to have a disinfection residual. A disinfection residual is the effective amount of disinfectant that is available throughout the water distribution system after the mixing of the water with the disinfectant. Because ozone is high-energy, rapidly acting and rapidly dissipated, it is not found in appreciable concentrations throughout municipal distribution systems. Conversely, chlorine, which is of considerably less energy than ozone, can travel and produce an effective residual throughout the distribution system. High-energy ozone is rapidly dissipated as it enters the municipal distribution system, and there is no effective residual. Bottled water presents a different scenario. In effect, ozonated water enters a bottle and is therein sealed. Accordingly, there is an effective disinfection residual for a period of time until the ozone dissipates.

Therefore, it is to the value be expected that the overall CT value for ozone in the bottle will be higher than the value calculated in the ozonator itself. As Table 13.1 indicates, ozone is most effective against viruses and bacteria and less so against parasite cysts, in particular *Cryptosporidium parvum*. However, for water sealed in a bottle, ozone should have additional activity against parasites. A side-effect of ozonation is that ozone can react with normal organic constituents in the water and break them down to produce greater amounts of assimilable organic carbon (AOC). One of the consequences of ozonation is to produce water with a higher AOC concentration, which then serves as food for autochthonous bacteria, increasing the potential concentration of these organisms.

Reverse osmosis is most commonly employed to change the mineral content of drinking water. Theoretically, it should produce pathogen-free water. However, bacteria can grow on the membranes and break through any small tears therein. In addition, the seals and plumbing associated with reverse osmosis engineering can, unless carefully monitored, serve as points of intrusion into the system. Accordingly, some manufacturers of reverse osmosis equipment will not certify systems as producing pathogen-free water.

Filtration has been employed for thousands of years to produce safe drinking water. Various types of clay and diatomaceous earth have been part of the civil engineering landscape from the time of the Pharaohs and were extensively employed by Roman water engineers. In the bottled water industry, filtration is generally utilized as a multiple barrier directed against parasites, particularly *Cryptosporidium parvum* and *Giardia lamblia*. A wide selection of technically developed media is now available. Filters for bottled water are divided into two basic types: nominal and absolute. Nominal filters are those that generally exclude particles, but are not rated to do so with a specific rating of certainty. This type of filter is being replaced by absolute filters, which are rated to remove particles of a specific size at a particular efficiency. In the United States, many bottled water companies have begun installing absolute filters of at least 1 μm to exclude *Cryptosporidium parvum*. Some bottled water companies are using filters of sequential sizes down to 0.5 μm. The NSF (Ann Arbor, Michigan; formerly known as the National Sanitation Foundation) is currently in the final stages of developing standards for the certification of absolute filters for retention of *Cryptosporidium* (MMWR, 1997).

Distillation should produce sterile water when operated properly. However, it should be noted that the water produced is sterile only at the time it leaves the still. As it travels through the pipes, the water will acquire microflora.

13.4 Naturally occurring bacteria

After treatment and passage through two or more multiple barriers, bottled water should be pathogen free. However, it will still have an autochthonous flora component; this is a term of European origin, generally referring to naturally occurring bacteria that have evolved an aqueous lifestyle. In the United States, a number of other terms are used to refer to bacteria of this nature; those used include, plate-count bacteria, standard plate count bacteria, heterotrophic plate count (HPC) bacteria and others. One of the great difficulties in comparing bacterial isolation studies performed in different laboratories is that the numbers and types of the naturally occurring bacteria reported are strongly related to the type of culture media employed, the temperature of incubation and the culture technique used. Results can vary several thousandfold with even one of these changes (Reasoner, 1990; Hunter, 1993). For example, Table 13.2 shows the effect of three different methods on the number of HPC bacteria.

The HPC analyses are conducted in three formats: pour plate, spread plate and membrane filter. In utilizing the pour plate method, one adds a drinking water sample to liquid agar, pours the liquid agar into a Petri plate, and then counts the colonies throughout the agar after incubation. In the spread plate method, the water sample is added to the surface of already solidified agar and then spread across its surface using a glass rod. In the membrane filter method, a drinking water sample is first filtered through a bacterial-exclusion membrane

Table 13.2 Comparison of HPC results by pour plate, spread plate, and membrane filter methods on the same medium within each referenced study

Temperature °C	Bacterial count cfu/ml[a]			Ratio			Reference
	PP[a]	SP[a]	MF[a]	PP/MF	PP/SP	MF/SP	
16	3100	6300	—	—	0.49	—	Buck & Cleverdon (1960)
RT[a]	42	—	113	0.37	—	—	Stapert et al. (1962)
37	108	115	—	—	0.94	—	van Soestbergen & Lee (1969)
35	23	110	—	—	0.21	—	Klein & Wu (1974)
	62	200	—	—	0.31	—	
20	170	230	—	—	0.74	—	Means et al. (1981)
35	210	240	—	—	0.87	—	
20	740	3100	—	—	0.24	—	Taylor et al. (1983)
20	1020	—	986	1.03	—	—	Maul et al. (1985)
35	430	510	270	1.59	0.84	0.53	Reasoner & Geldreich (1985)
28	4000	7000	4600	0.86	0.57	0.65	
20	2700	6100	3900	0.69	0.44	0.64	
35	3.3	4.9	—	—	0.67	—	Lombardo et al. (1986)
28	4.0	5.2	—	—	0.76	—	
28	—	5.2	4.8	—	—	0.92	
22	100	710[b]	—	—	0.14	—	Gibbs & Hayes (1988)

[a] cfu, colony-forming units. PP, pour plate. SP, spread plate. MF, membrane filter. RT, room temperature.
[b] Based on ratio given in the reference and arbitrarily assigning the value of 100 cfu/ml as the pour plate mean.
Source: Reasoner (1990).

and the filter is placed on the surface of solidified agar. As Table 13.2 shows, the spread plate method is more sensitive than the pour plate method, and both are more sensitive than the membrane filter method. Table 13.3 demonstrates that the type of culture medium also exerts a profound effect on the number of colony-forming units (cfu) recovered. In the United States, the low-nutrient R2A agar is now most commonly utilized in performing HPC studies. Even the length of incubation time can significantly affect the number of colony-forming units per milliliter (cfu/ml) of HPC isolated (see Table 13.4). A number of non-culture methods are also utilized to generate HPC concentration. These include the use of epifluorescence microscopy, with or without vital or other dyes (Newell et al., 1986; Kepner and Pratt, 1994) and ATP measurements (Zweifel and Hagström, 1995). Generally, the HPC results produced by non-culture methods are hundreds to thousands of times higher than those from culture methods (Suzuki et al., 1993).

Table 13.3 Comparison of HPC results using different media

| Temperature °C | Bacterial count cfu/ml[a] | | | Reference |
	PP[a]	SP[a]	MF[a]	
35	3137 (SPC)	—	4273.0 (m-HPC)	Taylor & Geldreich (1979)
20	170 (SPC)	440 (R2A)	510 (m-HPC)	Means et al. (1981)
20	—	4000 (R2A)	12 (m-HPC)	Fiksdal et al. (1982)
	—	1000 (R2A)	110 (m-HPC)	
35	—	20 (R2A)	6 (m-HPC)	
	—	4 (RDA)	< 1 (m-HPC)	
35	277 (SPC)	—	283 (m-HPC)	Green et al. (1983)
20	1123 (NA)	—	1217 (m-HPC)	Maul et al. (1985)
	1192 (NA)	1192 (R2A)		
35	22 (SPC)	200 (R2A)	32 (m-HPC)	Reasoner & Geldreich (1985)
28	80 (SPC)	360 (R2A)	140 (m-HPC)	
20	22 (SPC)	90 (R2A)	47 (m-HPC)	
35	53 (SPC)	—	66.7 (m-HPC)	Fujioka et al. (1986)
	53 (SPC)	—	57.1 (R2A)	
26	590 (SPC)	1550 R2A)	—	Stetzenbach et al. (1986)
22	100 (YEA)	710 (YEA)	—	Gibbs & Hayes (1988)
	440 (R2A)	—	—	
	100 (YEA)	3900 (R2A)	—	

[a] cfu, colony-forming units. PP, pour plate. SP, spread plate. MF, membrane filter.
Media: SPC, standard plate count; NA, nutrient agar; YEA, yeast extract agar, R2A medium (Reasoner & Geldreich, 1985). m-HPC medium (Taylor & Geldreich, 1979) was published originally as m-SPC medium.
Source: Reasoner (1990).

Table 13.4 Effect of length of incubation on HPC results

Temperature °C	Medium and method[b]	Bacterial count cfu/ml[a] Incubation time, days							Reference
		1	2	3	4	5	6	7	
35	SPC-PP	—	2300	—	5000	—	—	6200	Klein & Wu (1974)
35	R2A-SP	—	10	—	—	—	—	20	Fiksdal et al. (1982)
20	R2A-SP	—	0	—	—	—	—	4000	
35	R2A-MF	—	5	—	—	—	—	10	
20	R2A-MF	—	0	—	—	—	—	40	
20	m-HPC-MF	—	190	945	1217	—	—	—	Maul et al. (1985)
	R2A-MF	—	287	904	1192	—	—	—	
	NA-PP	—	389	855	1123	—	—	—	
35	SPC-PP	—	22	—	100	—	110	115	Reasoner & Geldreich (1985)
28	SPC-PP	—	90	—	640	—	950	1000	
20	SPC-PP	—	22	—	130	—	570	900	
35	R2A-SP	—	200	—	340	—	500	510	
28	R2A-SP	—	360	—	2800	—	6700	7200	
20	R2A-SP	—	90	—	1100	—	4700	6100	
35	R2A-MF	—	41	—	200	—	270	280	
28	R2A-MF	—	160	—	2200	—	3500	4000	
20	R2A-MF	—	75	—	650	—	3000	4900	
35	m-HPC-MF	—	32	—	140	—	150	150	
28	m-HPC-MF	—	140	—	1000	—	1700	1900	
20	m-HPC-MF	—	48	—	400	—	1600	2000	
35	PCA-PP	1.3	16.8	—	30.8	34.2	34.8	34.8	Silley (1985)
20	PCA-PP	1.7	21.2	—	101	114.3	121.3	125.8	
22	YEA-PP	—	—	100[c]	—	—	—	1800	Gibbs & Hayes (1988)

[a] cfu, colony-forming units.
[b] See Table 13.3 for abbreviations used for media and methods.
[c] Based on plating ratio given in the reference and arbitrarily assigning the value of 100 cfu/ml as the pour plate mean.

Accordingly, it is somewhat difficult to make definitive statements concerning the microbial content of bottled water because the results of any study are highly confounded by many different variables. However, there are certain points of commonality that have emerged. First, the autochthonous flora of treated and non-treated bottled water have a lifestyle that favors an aqueous and not a human ecology. Second, regardless of the individual species, the number of autochthonous flora undergo increases and decreases in the bottle after it is sealed (Ducluzeau et al., 1976a; Gonzalez et al., 1987; Manaia et al., 1990; Mavridou, 1992). Each individual species will increase in number and then decrease once its particular food supply has been exhausted (Schmidt-Lorenz, 1976; Quevedo-Sarmiento et al., 1986; Hunter et al., 1990). As it dies and decays and disintegrates, its constituents are then used as foodstuffs for other species of autochthonous flora (Schwaller and Schmidt-Lorenz, 1980; Warburton et al., 1986; Hunter and Burge, 1987; Lucas and Ducluzeau, 1990). Consequently, a species of autochthonous bacteria will first be found in low numbers and over time will increase to much higher numbers and then decrease, to be replaced by other species. Accordingly, a sampling of the bottle at any point in time is only a snapshot of events occurring in the bottle (Bischofberger et al., 1990; Morais and da Costa, 1990). Apparently, if bottled water is stored long enough, it may eventually self-sterilize. For example, bottled water sequestered in the United States during the Korean War was found to be sterile when opened and analyzed 40 years later (Glenn Davis, Abscopure Water Company, Ann Arbor, M1, USA, personal communication). Professor H. Ledere has described these events in chapter 12 and his thorough, considered review will not be repeated here.

13.5 Product safety

The primary question for consideration now focuses on post-treatment micro-biological events that relate to the safety of the final product and, therefore, the next two questions to address are 'What is the post-treatment microbiology of bottled water?' and 'What is the pathogenic potential of this microbiological content?'

As Table 13.5 demonstrates, virtually all bottled waters obtained from retail outlets have a microbial content. The microbial content of bottled water is higher than that of municipal water because of the former's lack of disinfection residual (see Table 13.6). In examining the individual species found in the various bottled water surveys, one is struck by the absence of pathogens that are normally associated with gastroenteritis. Moreover, indicators of these pathogens, such as total coliforms and Escherichia coli, are routinely absent. Post-treated and non-treated drinking water have yielded species whose microbes are associated with infection. For example, various species of the genus Pseudomonas have been isolated from bottled water (Table 13.7). However, it is a classic error to assume that because a bacterial species of a certain name is found associated with a particular infection, the acquisition of that infection is via drinking water (Ducluzeau et al., 1976). For example, it is well known that Pseudomonas aeruginosa can be a pathogen. However, it has been well established over the last thirty years that only certain patient groups with very specific deficits in their immune capacity are susceptible to infection (Hardalo and Edberg, 1997). The reason for this lack of association with drinking water can be found when one examines the equation that describes infection potential:

$$\text{Microbiological health risk} \propto \frac{\text{number of microbes} \times \text{virulence}}{\text{immune status of the host}}$$

Table 13.5 Percentage frequency distribution of total colony counts from studies of bottled water sampled at retail outlets

Reference	Water type	Temperature of incubation °C	Number examined	Total viable count				
				$<10^2$	10^2–10^3	10^3–10^4	10^4–10^5	$>10^5$
Hunter & Burge (1987)	Still	22	29	31	17	52[a]	—	—
		37	29	86	3	10[a]	—	—
	Carbonated	22	29	90	7	3[a]	—	—
		37	29	100			—	—
Hunter et al. (1990)	Still	22	44	18	11	18	36	16
		37	44	68	11	11	5	5
Warburton et al. (1986)	Mineral	35	49[b]	67	16	16	—	—
	Purified[c]	35	41[b]	41	17	17	15	10

[a] Greater than 1000.
[b] Lots, see text.
[c] For definition of 'purified', see text.
Source: Hunter (1993).

This relationship can be well defined in terms of its individual components (Duncan and Edberg, 1995). When examining a microbe such as *Pseudomonas aeruginosa*, for example, one sees that the gastrointestinal tract is not a portal of entry. The changes in immune status necessary to produce infection lie in other, but specific, organ systems. For *Pseudomonas aeruginosa*, the changes in immune status necessary to acquire infection fall into some very well-defined groups. These include cystic fibrosis patients, full-thickness burn patients, patients with polymorphonuclear leukocyte counts less than 500 and those with intravenous or in-dwelling lines. In none of these cases is drinking water a risk factor. Moreover, clinical microbiology laboratories do not isolate *Pseudomonas aeruginosa* in patients with diarrhea. Accordingly, one must conclude that gastroenteritis with acquisition via drinking water is not a route of infection by *Pseudomonas aeruginosa*.

Another component of the infection equation that has been examined in relationship to the microbial content of bottled water is the virulence component of its microbes. Because species identifications are not as accurate from environmentally isolated bacteria as they are from those isolated clinically, a way of looking at health risk is to examine the virulence properties of the bacteria without regard to their identification. According to this strategic pathway, the name of the bacterium is immaterial; what is important is its armamentarium (i.e. virulence factors) against the human host. A number of virulence

Table 13.6 Comparison of the microbial content of drinking water types

Species	Bottled water $(n = 150)^a$ [Percentage samples with species (concentration range cfu/ml)]	Cooler water $(n = 81)^b$ [Percentage samples with species (concentration range cfu/ml)]	Tap water $(n = 150)^c$ [Percentage samples with species (concentration range cfu/ml)]
Acinetobacter spp.	5 (2–30)	10 (100–350)	5 (6–21)
Achromobacter spp.	5 (8–31)	<5 (>20)	4 (5–25)
Agrobacterium spp.	2 (7–12)	20 (2–110)	0
Bacillus brevis	0	0	35 (3–360)
Bacillus cereus	0	0	15 (8–90)
Bacillus licheniformis	0	0	25 (15–650)
Bacillus circulans	50 (1900–21 000)	80 (3–17 000)	0
Bacillus firmus	35 (600–39 000)	20 (900–35 000)	45 (8–860)
Bacillus megaterium	15 (21–4500)	20 (55–600)	0
Bacillus polymyxa	15 (18–1700)	20 (80–2100)	15 (5–630)
Bacillus pumilus	0	0	30 (7–260)
Bacillus sphericus	0	0	30 (2–35)
Bacillus macerans	25 (660–68 000)	30 (200–45 000)	0
CDC Group IV (GNR)^d	1 (45)	<5 (<20)	0
Corynebacterium spp.	10 (8–40)	30 (20–280)	0
Coryneform spp.	0	0	20 (6–440)
Methanococcus spp.	85 (30–76 000)	60 (120–61 000)	0
Moraxella spp.	10 (3–38)	80 (13–48)	10 (1–10)
Comamonas acidovorans	6 (1–28)	3 (4–12)	2 (2, 18)
Pseudomonas aeruginosa	3 (2–26)	2 (3, 13)	2 (2–16)
Burkholderia cepacia	3 (1–15)	20 (20–40)	5 (1–15)
Pseudomonas fluorescens	12 (1–65)	3 (4–46)	0
Comamonas testosteroni	2 (3–10)	4 (7–38)	0
Staphylococcus spp.	2 (25–600)	2 (6, 19)	10 (3–65)
Streptomyces spp.	0	10 (2)	0
Xanthomonas maltophilia	2 (2–22)	2 (2, 4)	0
Total coliforms	2 Samples 'Supermarket'^{e} (2, TNTC)^f	0	0

^{a} Bottled water samples obtained from supermarket shelves.
^{b} Nine water coolers sampled weekly for nine weeks.
^{c} Tap water samples collected in sodium thiosulfate, 50 from the northeast, 50 from the west coast and 50 from the southeast.
^{d} Unidentified group of gram-negative rods clustered by biochemical characteristics.
^{e} 'Supermarket' refers to a bottled water made and sold by a supermarket.
^{f} TNTC, too numerous to count.
Source: Edberg *et al.* (1996).

Table 13.7 Distribution of *Pseudomonas* species in surveys of bottled water

| *Pseudomonas* species | No. of isolates (percentage of all *Pseudomonas* strains isolated in study) | | |
	American waters[a]	German waters[b]	Mainly Portuguese[c]
P. aeruginosa	—	—	45 (29)
P. stutzeri	30 (24)	22 (27)	24 (15)
P. putida	14 (11)	20 (25)	11 (7)
P. fluorescens	18 (14)	13 (16)	25 (16)
P. diminuta	21 (19)	1 (1)	—
P. cepacia	12 (10)	8 (10)	4 (3)
P. acidovorans	7 (6)	6 (7)	—
P. maltophilia	7 (6)	3 (4)	7 (5)
P. pickettii	9 (7)	—	25 (16)
P. paucimobilis	7 (6)	2 (3)	3 (2)
P. alcaligenes	—	—	11 (7)
P. pseudoalcaligenes	—	6 (7)	—

[a] Hernandez-Duquino and Rosenberg (1987).
[b] Hernandez-Duquino and Rosenberg (1989).
[c] Manaia *et al.* (1990).
Source: Hunter (1993).

factors have been described that are associated with the development of infection (Lye and Dufour, 1991). Studies that have examined the virulence factors of bacteria isolated from bottled water have not demonstrated significant pathogenic potential (Edberg *et al.*, 1996, 1997a). In one study of bacteria isolated on R2A agar, insignificant virulence factors were found from the HPC content (see Table 13.8). Payment further refined the strategic analysis of virulence factors by hypothesizing that those autochthonous bacteria that were able to grow under conditions analogous to the human host would be most likely risk factors (Payment *et al.*, 1991a, b). Accordingly, bacteria from bottled water samples were isolated on a high-nutrient medium containing blood (blood agar medium). As Table 13.9 demonstrates, the microbial contents of both natural and post-treated bottled water did not contain significant virulence factors when isolated on a medium that mimicked the physiological conditions of the human body.

13.6 Summary

Bottled water is not a sterile product but must be pathogen free. The autochthonous flora is natural and varies in an individual bottle over time. In human terms, the

Table 13.8 Comparison of virulence characteristics of bacteria isolated from the three water sources

Characteristic	Percentage of all species having bacteria isolated from the three water sources		
	Cooler	Bottled	Tap
Hemolysin	2[a]	1	1
Proteinase	22	20	23
Gelatinase	4	3	3
Lipase	0	0	0
Elastase	0	0	0
Coagulase	0	0	0
DNAse	3	2	6
Fibrinogen	0	0	0
Acid restraint at pH 3–5	0	0	0

[a] Percentage of all bacteria isolated from this source demonstrating the characteristic.
Source: Edberg *et al.* (1997a).

Table 13.9 Invasiveness activity of naturally occurring HPC bacteria from bottled and tap water

Species	Number tested	Invasiveness (> 5% per field)	
		Stationary phase	Log phase
A. faecalis	4	0	0
A. haemolyticus	2	1	1
A. junii/johnson	5	0	1
Actinomyces spp.	3	0	0
Bacillus licheniformis	3	0	0
Bacillus sp.	5	0	0
Micrococcus spp.	16	0	0
O. anthropi	3	1	3
P. aeruginosa	3	0	0
P. alcaligenes	6	0	1
P. diminuta	5	0	1
P. fluorescens	8	0	0
P. vesicularis	2	0	1
Staphylococcus spp.	8	0	0
Streptococcus spp.	3	0	0
Unidentified GNR	7	0	0
X. maltophilia	2	0	0
Total	85	2	8
Z value of *Salmonella typhi*[a]	3.92	5.19	—

[a] The Z test represents the difference between two independent counts. A Z of 1.96 or more is needed for the differences between the two counts to be considered significant at the 5% level.

numbers they achieve in the bottle may be considered 'high'; however, in bacterial terms, they are merely eating the dissolved assimilatable organic carbon available to them and multiplying to numbers based on this food availability. Because all bottled waters are expected to contain various concentrations of autochthonous flora or HPC, health-risk assessment centers around the question 'Are there pathogens present and can these pathogens multiply?' There has been no evidence to date of the presence of pathogens or their multiplication in bottled water that has been subjected to appropriate multiple barriers and meets appropriate regulation.

References

Bischofberger, T., Cha, S. K., Schmitt, R., Konig, B. & Schmidt-Lorenz, W. (1990) The bacterial flora of non-carbonated, natural mineral water from the springs to reservoir and glass and plastic bottles. *International Journal of Food Microbiology*, **11**, 51–72.

Buck, J. D. & Cleverdon, R. C. (1960) The spread plate as a method for enumeration of marine bacteria. *Limnology and Oceanography*, **5**, 78–80.

Ducluzeau, R., Bochand, J. M. & Dufresne, S. (1976a) Longevity of various bacterial strains of intestinal origin in gas-free mineral water. *European Journal of Applied Microbiology*, **3**, 227–236.

Ducluzeau, R., Dufresne, S. and Bochand, J. M. (1976b) Inoculation of the digestive tract of axenic mice with the autochthonous bacteria of mineral water. *European Journal of Applied Microbiology*, **2**, 127–134.

Duncan, H. E. and Edberg, S. C. (1995) Host–microbe interaction in the gastrointestinal tract. *Critical Reviews in Microbiology*, **21**, 85–100.

Edberg, S. C., Gallo, P. & Kontnick, C. (1996) Analysis of the virulence characteristics of bacteria isolated from bottled water cooler, and tap water. *Microbial Ecology in Health and Disease*, **9**, 67–77.

Edberg, S. C., Kops, S., Kontnick, C. and Escarzaga, M. (1997a) Analysis of cytotoxicity and invasiveness of heterotrophic plate count bacteria (HPC) isolated from drinking water on blood media. *Journal of Applied Microbiology*, **82**, 455–461.

Edberg, S. C., Leclerc, H. and Robertson, J. (1997b) Natural protection of spring and well drinking water against surface microbial contamination. II. Indicators and monitoring parameters for parasites. *Critical Reviews in Microbiology*, **23**, 179–206.

Fiksdal, L., Vik, E. A., Mills, A. and Staley, J. T. (1982) Nonstandard methods of enumerating bacteria in drinking water. *Journal of American Water Works Association*, **74**, 313–318.

Fujioka, R., Kungskulniti, N. and Nakasone, S. (1986) Evaluation of the presence–absence test for coliforms and the membrane filtration method for heterotrophic bacteria, in *Advances in Water Analysis and Treatment*, Technology Conference Proceedings, WQTC-13, 1985, American Water Works Association, Denver, CO, pp. 271–283.

Gibbs, R. A. and Hayes, C. R. (1988) The use of R2A medium and the spread plate method for the enumeration of heterotrophic bacteria in drinking water. *Letters in Applied Microbiology*, **6**, 19–21.

Gonzalez, C., Gutierrez, C. and Grande, T. (1987) Bacterial flora in bottled uncarbonated mineral drinking water. *Canadian Journal of Microbiology*, **33**, 1120–1125.

Green, B. L., Taylor, R. H. and Geldreich, E. E. (1982) The SPC Sampler: a simple procedure for monitoring the bacteriologic quality of water, in *Advances in Laboratory Techniques for Quality Control*, Technology Conference Proceedings, WQTC-9, 1981, American Water Works Association, Denver, CO, pp. 125–133.

Hardalo, C. and Edberg, S. C. (1997) *Pseudomonas aeruginosa*: Assessment of Risk from Drinking Water. *Critical Reviews in Microbiology*, **23**, 47–75.

Hopkins, R. S., Shillam, P., Gaspard, B., Eisnach, L. and Karlin, R. J. (1985) Waterborne disease in Colorado: three years' surveillance and 18 outbreaks. *American Journal of Public Health*, **75**, 254–257.

Hunter, P. R. (1993) The microbiology of bottled natural mineral waters. *Journal of Applied Bacteriology*, **74**, 345–352.

Hunter, P. R. and Burge, S. H. (1987) The bacteriological quality of bottled natural mineral waters. *Epidemiology and Infection*, **99**, 439–443.

Hunter, P. R., Burge, S. H. and Hornby, H. (1990) An assessment of the microbiological safety of bottled mineral waters. *Rivista Italiana D'Igiene*, **50**, 394–400.

Kepner, R. L. Jr and Pratt, J. R. (1994) Use of fluorochromes for direct enumeration of total bacteria in environmental samples: past and present. *Microbiology Review*, **58**, 603–615.

Klein, D. A. and Wu, S. (1974) Stress: a factor to be considered in heterotrophic microorganism enumeration from aquatic environments. *Applied Microbiology*, **27**, 427–431.

Leclerc, H., Mossel, D. A. A. and Savage, C. (1985) Monitoring non-carbonated ('still') mineral waters for aerobic colonization. *International Journal of Food Microbiology*, **2**, 341–347.

Lombardo, L. R., West, P. R. and Holbrook, J. L. (1986) A comparison of various media and incubation temperatures used in the heterotrophic plate count analysis, in *Advances in Water Analysis and Treatment*, Technology Conference Proceedings, WQTC-13, 1985, American Water Works Association, Denver, CO, pp. 251–270.

Lucas, F. and Ducluzeau, R. (1990) Antagonistic role of various bacterial strains from the autochthonous flora of gas-free mineral water against *Escherichia coli*. *Sciences des Aliments*, **10**, 62–73.

Lye, D. J. and Dufour, A. P. (1991) A membrane filter procedure for assaying cytotoxic activity in heterotrophic bacteria isolated from drinking water. *Journal of Applied Bacteriology*, **70**, 89–94.

Manaia, C. M., Munes, O. C., Morais, P. V. and da Costa, M. S. (1990) Heterotrophic plate counts and the isolation of bacteria from mineral waters on selective and enrichment media. *Journal of Applied Bacteriology*, **69**, 871–876.

Maul, A., Block, J. C. and El-Shaarawi, A. H. (1985) Statistical approach for comparison between methods of bacterial enumeration in drinking water. *Journal of Microbiological Methods*, **4**, 67–77.

Mavridou, A. (1992) Study of the bacterial flora of a non-carbonated natural mineral water. *Journal of Applied Bacteriology*, **73**, 355–361.

Means, E. G., Hanami, L., Ridgway, H. F. and Olson, B. H. (1981) Evaluating mediums and plating techniques for enumerating bacteria in water distribution systems. *Journal of American Water Works Association*, **73**, 585–590.

MMWR (1997) Morbidity and Mortality Weekly Report 1997 *USPHS/IDSA Guidelines for the Prevention of Opportunistic Infections in Persons Infected with Human Immunodeficiency Virus*, vol. 46, No. RR-12. US Department of Health and Human Services, Public Health Service, Centers for Disease Control and Prevention, Atlanta, GA.

Morais, P. V. and da Costa, M. S. (1990) Alterations in the major heterotrophic bacterial populations isolated from a still bottled mineral water. *Journal of Applied Bacteriology*, **69**, 750–757.

Newell, S. Y., Fallon, R. D. and Tabor, P. S. (1986) Direct microscopy of natural assemblages, in *Bacteria in Nature*, J. S. Poindexter and E. R. Leadbetter (eds). Plenum Press, New York, pp 1–48.

Payment, P., Franco, E., Richardson, L. and Siemiatycki, J. (1991a) Gastrointestinal health effects associated with the consumption of drinking water produced by point-of-use domestic reverse osmosis filtration units. *Applied and Environmental Microbiology*, **57**, 945–948.

Payment, P., Richardson, L. and Siemiatycki, J. (1991b) A randomized trial to evaluate the risk of gastrointestinal disease due to consumption of drinking water meeting current biological standards. *American Journal of Public Health*, **81**, 703–708.

Quevedo-Sarmiento, J., Ramos-Cormenza, A. and Gonzales-Lopes, J. (1986) Isolation and characterisation of aerobic heterotrophic bacteria from natural spring waters in the Lanjaron area (Spain). *Journal of Applied Bacteriology*, **61**, 365–372.

Reasoner, D. J. (1990) Monitoring heterotrophic bacteria in potable water, in *Drinking Water Microbiology*, G. A. McFeters (ed). Springer-Verlag, New York.

Reasoner, D. J. and Geldreich, E. E. (1985) A new medium for the enumeration and subculture of bacteria from potable water. *Applied and Environmental Microbiology*, **49**, 1–7.

Robertson, J. B. and Edberg, S. C. (1997) Natural protection of spring and well drinking water against surface microbial contamination. I. Hydrogeological parameters. *Critical Reviews in Microbiology*, **23**, 143–178.

Schmidt-Lorenz, W. (1976) Microbiological characteristics of natural mineral waters. *Annali dell' Istituto Superiore Sanità*, **12**, 93–112.

Schwaller, P. and Schmidt-Lorenz, W. (1980) Bacterial flora of four different French non-carbonated mineral waters in bottles. I. Colony counts, rough differentiation of the bacterial flora, and characterisation of the group of Gram-negative yellow-pigmented bacteria. *Zentralblatt für Bakteriologie und Hygiene, I. Abt. Orig. C*, **1**, 330–347.

Silley, P. (1985) Evaluation of total-count samples against the traditional pour plate method for enumeration of total viable counts of bacteria in a process water system. *Letters in Applied Microbiology*, **1**, 41–43.

Stapert, E. M., Sokolski, W. T. and Northam, J. I. (1962) The factor of temperature in the better recovery of bacteria from water by filtration. *Canadian Journal of Microbiology*, **8**, 809–810.

Stetzenbach, L. D., Kelley, L. M. and Sinclair, N. A. (1986) Isolation, identification, and growth of well-water bacteria. *Groundwater*, **24**, 6–10.

Suzuki, M. T., Sherr, E. B. and Sherr, B. F. (1993) DAPI direct counting underestimates bacterial abundances and average cell size compared to AO direct counting. *Limnology and Oceanography*, **38**, 1566–1570.

Taylor, R. H., Allen, M. J. and Geldreich, E. E. (1983) Standard plate count: a comparison of pour plate and spread plate methods. *Journal of American Water Works Association*, **75**, 35–37.

Taylor, R. H. and Geldreich, E. E. (1979) A new membrane filter procedure for bacterial counts in potable water and swimming pool samples. *Journal of American Water Works Association*, **71**, 402–405.

van Soestbergen, A. A. and Lee, C. H. (1969) Pour plates or streak plates. *Applied Microbiology*, **18**, 1092–1093.

Warburton, D. W., Peterkin, P. I., Weiss, K. F. and Johnston, M. A. (1986) Microbiological quality of bottled water sold in Canada. *Canadian Journal of Microbiology*, **32**, 891–893.

Zweifel, U. L. and Hagström, A. (1995) Total counts of marine bacteria include a large fraction of non-nucleoid-containing bacteria (ghosts). *Applied Environmental Microbiology*, **61**, 2180–2185.

Index